Design of Reinforced Concrete Sections Under Bending and Axial Forces

Helena Barros · Joaquim Figueiras · Carla Ferreira ·
Mário Pimentel

Design of Reinforced Concrete Sections Under Bending and Axial Forces

Tables and Charts According to
EUROCODE 2

Springer

Helena Barros
INESC Coimbra
University of Coimbra
Department of Civil Engineering
Coimbra, Portugal

Carla Ferreira
INESC Coimbra
University of Coimbra
Department of Civil Engineering
Coimbra, Portugal

Joaquim Figueiras [iD]
CONSTRUCT
University of Porto
Faculty of Engineering
Porto, Portugal

Mário Pimentel [iD]
CONSTRUCT
University of Porto
Faculty of Engineering
Porto, Portugal

ISBN 978-3-030-80138-0   ISBN 978-3-030-80139-7   (eBook)
https://doi.org/10.1007/978-3-030-80139-7

This Springer imprint is published by the registered company Springer Nature Switzerland AG
The registered company address is: Gewerbestrasse 11, 6330 Cham, Switzerland

Preface

This book, addressed to structural designers and students, contains auxiliary calculation tools to facilitate the safety assessment of reinforced concrete sections. Essential parameters in the design to the ultimate limit state of resistance such as the percentage of reinforcement and the position of the neutral axis in concrete cross sections, as well as the control of the maximum stresses in service limit states, are provided by these tools. A set of tables, charts and diagrams used to design cross sections of reinforced and prestressed concrete structures are supplied. The most current beams and columns cross sections, namely rectangular, circular and T-sections are considered. These tools have been prepared in line with the provisions of the new European regulations, with particular reference to Eurocode 2—Design of Concrete Structures.

The auxiliary tables and charts consider the concrete compressive strength, f_{cd}, without reduction coefficient, $\alpha_{cc} = 1.0$, for the ultimate limit states of cross sections under bending, biaxial bending and axial forces. The different types of tables and charts correspond to those currently in use. They widen the scope of application without loss of efficiency. Some of the calculation aids are extended to high strength concrete, $f_{ck} > 50$ MPa (classes C55/67 to C90/105), which constitutive relations, $\sigma-\varepsilon$, are specific of each class, being the normalization of the calculation tool unique for the concrete class. The verification of the concrete and steel stresses in serviceability limit states can be easily performed by using the dedicated auxiliary calculation diagrams.

Coimbra, Portugal
Porto, Portugal
Coimbra, Portugal
Porto, Portugal

Helena Barros
Joaquim Figueiras
Carla Ferreira
Mário Pimentel

Contents

Notations

Uppercase

A, A_s, A'_s, A_p	Area of steel section
$C12, C50, \ldots$	Concrete strength classes
$C12/15, C50/60, \ldots$	Concrete strength classes obtained in cylinder/cube
E_c	Concrete modulus of elasticity
E_{cm}	Mean value of the concrete modulus of elasticity
E_s	Steel modulus of elasticity
G, G_s	Center of concrete and reinforcing steel, respectively
M_{Ed}, M_{Edx}, M_{Edy}	Design value of acting bending moment
M_{Eds}	Design value of acting bending moment in relation to gravity center of tensile reinforcement
M_{Rd}	Design value of resisting bending moment
N_{Ed}	Design value of acting normal force
N_{Rd}	Design value of resisting normal force
$S400, S500$	Ordinary steel reinforcements
$Y1860$	Prestressing steel
SLS	Serviceability limit state
ULS	Ultimate limit state

Lowercase

a, a_1, c_1	Distance from the surface of tensile and compressive reinforcements
b	Section width
b_t	Mean value of tensile zone of a section
b_w	Width of T-section web
d	Effective depth of cross section
f_{cd}	Design value of concrete compressive strength
f_{ck}	Characteristic compressive cylinder strength of concrete at 28 days

f_{cm}	Mean value of concrete cylinder compressive strength
f_{ctm}	Mean value of axial tensile strength of concrete
f_{puk}	Characteristic tensile strength of the prestressing steel
f_{yd}	Design yield strength of reinforcement
f_t	Ultimate strength of reinforcement
f_{yk}	Characteristic yield strength of reinforcement
h	Total depth of cross section
h_f	Flange depth of T-section
x	Neutral axis depth of the section
z_s	Distance to the reinforcing steel
z	Lever arm of internal forces under bending

Greek Alphabets

α	Ratio of neutral axis depth $\alpha = x/d$ or $\alpha = x/h$
α_e	Ratio of steel and concrete Young's modulus $\alpha_e = E_s/E_c$
$\varepsilon_c,\ \varepsilon_{c2},\ \varepsilon_{cu2}$	Concrete strain
$\varepsilon_s,\ \varepsilon_{s1}$	Strain in the tensile reinforcement
ε'_s	Strain in the compressed reinforcement
$\varepsilon_{ud},\ \varepsilon_{uk}$	Design and characteristic strain in the steel
μ	Non-dimensional design value of bending moment
μ_e	Non-dimensional value of bending moment for service limit states
μ_s	Non-dimensional value of design bending moment in relation to center of gravity of tensile reinforcement
$\mu_x,\ \mu_y$	Non-dimensional value of design bending moment in x and y directions, respectively
ν	Non-dimensional design value of normal force
ν_e	Non-dimensional value of normal force for service limit states
ρ	Reinforcement ratio: $\rho = \frac{A}{bd}$
$\sigma_c,\ \sigma_{c,max}$	Compressive stress in the concrete
σ_s	Stress in the tensile reinforcement
σ'_s	Stress in the compressed reinforcement
ϖ	Mechanical reinforcement percentage: $\varpi = \frac{A_s}{bh}\frac{\sigma_{sd}}{f_{cd}}$, or $\varpi = \frac{A_s}{bd}\frac{f_{yd}}{f_{cd}}$
$\varpi_{1,s},\ \varpi_2$	Mechanical percentages of tensile and compressive reinforcements
ζ	Normalized lever arm of internal forces under bending: $\zeta = z/d$

Chapter 1
Introduction

This book deals with the design and the safety assessment of reinforced concrete sections subjected to bending and axial forces. Given the internal forces in the section, the tables and charts provided in this book allow the design of different reinforcement solutions in an expeditious way and their comparison in terms of the economy and structural performance. In case both the concrete section and the steel reinforcement are known and a safety verification is to be made, the tools provided allow the immediate determination of the section resistance (in terms of bending moment and axial force) and the corresponding depth of the neutral axis, which is a measure of the ductility. Both the aforementioned calculation types, dimensioning and safety verification, can be done using tables or charts with sufficient precision for practical applications, although the tables allow for more accuracy.

Software for the analysis and design of reinforced concrete structures is now a common integrated tool for engineers, but it requires a great deal of knowledge and caution from the user on how to use it, how to enter the data, how to execute it and how to analyze the results. A small error in any of these phases can lead to an error in the design of the structure with unpredictable consequences. A simple and quick way to detect design errors is to use alternative means of calculation when checking critical points of the project. The tools available in this book represent an alternative and practical means for the validation of solutions obtained by automatic computation.

The book presents a set of tables and design charts to assist in the ultimate limit design of reinforced concrete sections submitted to bending moments and axial forces, along with auxiliary design diagrams for stress analysis of the steel and the concrete at section level to verify conformity with serviceability limit states. The tables and charts are auxiliary tools for the design of reinforced and prestressed concrete building elements according to the requirements of the European rules for the structural design, the Structural Eurocodes, namely the Eurocode 2—Design of Concrete Structures, EC2 [2]. The extension of the EC2, to higher strength concrete and steel classes means that the verification of the Servicibility Limit States, which was already essential for prestressed structures, is also currently taken into account in the design of reinforced concrete structures. This is the reason why this publication includes auxiliary interaction diagrams for the verification of service conditions, as well as the usual tables and charts for ultimate limit design.

Certain differences of this work in relation to previous tables and charts should be noted, especially regarding compressive strength and the stress–strain relations of the concrete for cross-sections design. The design compressive strength, $f_{cd} = \alpha_{cc} f_{ck}/\gamma_c$ is obtained with $\alpha_{cc} = 1.0$ instead of the reduction coefficient $\alpha_{cc} = 0.85$ adopted in some previous books. The widening of the scope of Eurocode 2, EC2 [2], to higher strength concrete classes (C55/67 to C90/105) led to the consideration of a stress–strain diagram specific of each class of high strength concrete that exhibit a more brittle failure mode. This means that the design tools can be normalized for the C12/15 to C50/60 classes, but they will have to be adjusted individually for higher strength concrete classes.

Following Eurocode 2, EC2 [2], the serviceability limit states comprise the verification of the maximum compressive stresses in concrete and the tensile stresses in the steel reinforcement, as well as the crack-width and the deflection control for both reinforced and prestressed concrete members. These evaluations, based on linear elastic material stress–strain laws, are quite easy to perform in the case of uncracked cross-sections, but become more laborious in the case of cracked sections. This publication presents a set of diagrams to obtain quickly the neutral axis depth, the maximum stresses in compression concrete and tensile steel, considering cracked section under compressive axial force and bending moment, useful for verifying reinforced and prestressed concrete cross-sections. It should be noted that the determination of the reinforcing steel stress, σ_s, for a given load combination is a relevant parameter for the crack width control and the deflection control in beams and slabs.

Prior to presenting the tables and charts, Chap. 2 briefly presents the main theoretical aspects and fundamentals of the design. It is supposed that the user has enough knowledge in the design of reinforced concrete sections that can be found in academic manuals. If this is not the case some books are indicated in the following.

Suggested Readings to Support Theory and Design Methodology

1. Appleton, J.: Estruturas de Betão, vol. 1 e 2. Edições Orion (2013)
2. Arroyo, J.C., Morán F., Meseguer A.G., et al.: Jimenez Montoya Esencial. Hormigón armado. 14th edn. Cinter (2018)
3. Eibl, J. (ed): Concrete Structures Euro-Design Handbook 1994/96 Karlsruhe. Ernst & Sohn, Berlin, (1995)
4. Favre, R., Jaccoud, J.P., Burdet O., Charif H.: Presse Polytechnique et Universitaires Romandes, Traité de Génie Civil, Vol. 8, 3rd edn. Lausanne, Switzerland (2004)
5. Ghali A., Favre R., Elbadry M.: Concrete Structures: Stresses and Deformations: Analysis and Design for Serviceability, 3rd edn. CRC Press (2014)
6. Mosley W.H., Hulse R., Bungey J.H.: Reinforced Concrete Design to Eurocode 2, 7th edn. Springer (2012)
7. Nilson, A.H.: Design of Concrete Structures, 12th edn. Mc-Graw Hill (1997)

8. Paillé, J.M.L Calcul des structures en béton (2009). ISBN Eyrolles: 978-2-212-12043-1
9. Roux, J.: Maîtrise de l'eurocode 2: Guide d´application. Editions Eyrolles (2009). ISBN Eyrolles: 978-2-212-12160-5 (2009)
10. Toniolo G., di Prisco M.: Reinforced Concrete Design to Eurocode 2. Springer Editor (2017)
11. Walther, R., Miehlbradt, M.: Dimensionnement des structures en béton, bases et technologie, vol. 7. Presses Polytechniques Et Universitaires Romandes, Traité de génie civil de l'Ecole Polytechnique Fédérale de Lausanne (1990)

Chapter 2
Calculation Methods and Assumptions

2.1 Introduction

This chapter presents a brief description of the methods and assumptions used in the design of reinforced concrete cross-sections subjected to axial forces and bending moments. The solution to this problem requires establishing the equilibrium equations, the compatibility conditions, the adoption of constitutive laws for the materials. The latter are based on Eurocode 2 (EC2) [2] recommendations regarding to the stress–strain relationships for the concrete and the steel. The compatibility conditions are established based on the assumption that plane sections remain plane after the deformation and that concrete and steel share the same strain field. The ultimate limit states are defined by the limits to the maximum strain in the materials, according to the EC2. In his work an enlarged set of strength concrete classes can be considered, including high strength concrete, with the properties described in EC2 [2].

2.2 Material Properties

The basic properties of concrete are defined in the EC2 [2], and Table 2.1 specifies the relevant parameters used in the calculation.

The stress–strain relationship of compressed concrete for the ultimate limite state design of cross-sections is defined by the parabola-rectangle diagram, given by the following parametric equation:

$$\sigma_c = f_{cd}\left[1 - \left(1 - \frac{\varepsilon_c}{\varepsilon_{c2}}\right)^n\right] \quad \text{for } 0 \le \varepsilon_c \le \varepsilon_{c2}$$

$$\sigma_c = f_{cd} \quad \text{for } \varepsilon_{c2} \le \varepsilon_c \le \varepsilon_{cu2}$$

© The Author(s), under exclusive license to Springer Nature Switzerland AG 2022
H. Barros et al., *Design of Reinforced Concrete Sections Under Bending and Axial Forces*,
https://doi.org/10.1007/978-3-030-80139-7_2

Table 2.1 Strength and deformation characteristics for concrete

Strength															Analytical relation/explanation
f_{ck} (MPa)	12	16	20	25	30	35	40	45	50	55	60	70	80	90	
$f_{ck,cube}$ (MPa)	15	20	25	30	37	45	50	55	60	67	75	85	95	105	
f_{cm} (MPa)	20	24	28	33	38	43	48	53	58	63	68	78	88	98	$f_{cm} = f_{ck} + 8$ (MPa)
f_{ctm} (MPa)	1.6	1.9	2.2	2.6	2.9	3.2	3.5	3.8	4.1	4.2	4.4	4.6	4.8	5.0	$f_{ctm} = 0.30 \times f_{ck}^{2/3} \leq C50/60$ $f_{ctm} = 2.12 \times ln(1 + (f_{cm}/10)) > C50/60$
$f_{ctk;0.05}$ (MPa)	1.1	1.3	1.5	1.8	2.0	2.2	2.5	2.7	2.9	3.0	3.1	3.2	3.4	3.5	$f_{ctk;0.05} = 0.7 \times f_{ctm}$ 5% fractile
$f_{ctk;0.95}$ (MPa)	2.0	2.5	2.9	3.3	3.8	4.2	4.6	4.9	5.3	5.5	5.7	6.0	6.3	6.6	$f_{ctk;0.95} = 1.3 \times f_{ctm}$ 95% fractile
E_{cm} (GPa)	27	29	30	31	33	34	35	36	37	38	39	41	42	44	$E_{cm} = 22(f_{cm}/10)^{0.3}$ f_{cm} in MPa
ε_{c2} (‰)	2.0									2.2	2.3	2.4	2.5	2.6	See Fig. 2.1. If $f_{ck} \geq 50$ MPa $\varepsilon_{c2}(‰) = 2.0 + 0.085(f_{ck} - 50)^{0.53}$
ε_{cu2} (‰)	3.5									3.1	2.9	2.7	2.6	2.6	See Fig. 2.1 $\varepsilon_{cu2}(‰) = 2.6 + 35\left[\frac{90-f_{ck}}{100}\right]^4$ If $f_{ck} \geq 50$ MPa
n	2.0									1.75	1.6	1.45	1.4	1.4	If $f_{ck} \geq 50$ MPa $n = 1.4 + 23.4\left[\frac{90-f_{ck}}{100}\right]^4$

Adapted from the EC2 [2]
* see Eurocode 2

Fig. 2.1 Parabola-rectangle diagram in characteristic (**A**), and design (**B**) values for compression concrete [2]

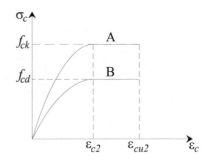

Fig. 2.2 Idealised (**A**) and design (**B**) stress/strain diagrams for reinforcing steel in tension and compression, EC2 [2]

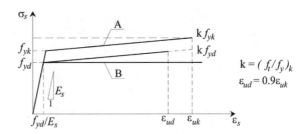

where n is an exponent, ε_{c2} is the concrete strain (reaching the maximum strength), and ε_{cu2} the ultimate strain defined in Table 2.1 in terms of concrete class. In Fig. 2.1 this equation is represented, for both characteristic and design values (compressive stress and compressive strain in absolute values, as in EC2 [2]). The tensile stresses acting in the concrete are ignored.

Eurocode 2 [2] allows the use of two stress–strain design diagrams for steel (continuous lines of Fig. 2.2):

(a)   Linear elastic diagram up to the stress, $f_{yd} = f_{yk}/\gamma_s$, followed by a sloped branch defined by a stress of kf_{yk}/γ_s for the maximum strain limit of ε_{uk}, but with a strain limit of ε_{ud}.

(b)   Linear elastic diagram up to the stress, $f_{yd} = f_{yk}/\gamma_s$, followed by an horizontal branch without strain limit.

2.3   Limit States of Bending and Axial Force

The ultimate limit states are defined in terms of maximum strains in both the concrete and the reinforcements and are attained when at least one of the maximum strains is reached. The range of possible strain distributions in the ultimate limit state are shown in Fig. 2.3a, where the strain limits are imposed at the levels of points A, B and C.

Point A is defined by the maximum strain limit of the steel, ε_{ud}, point B by the compression strain limit of the concrete, ε_{cu2}, and point C by the strain limit of the concrete under simple compression, ε_{c2}. In order to illustrate the relation between

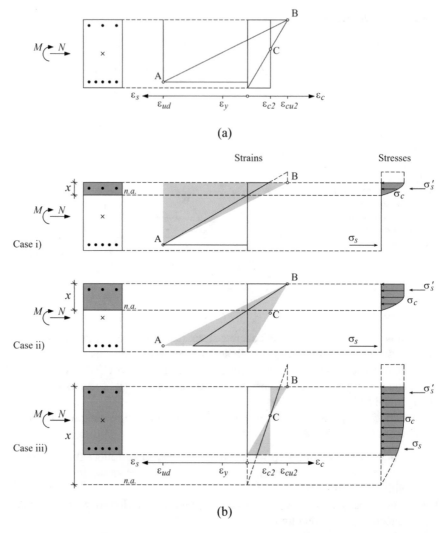

Fig. 2.3 **a** Range of possible strain distributions in a cross-section in the ultimate limit state; **b** Examples of stress and strain distributions

deformations and internal stresses, Fig. 2.3b shows the stresses corresponding to three strain distributions: Case (i) corresponds to a situation where the maximum steel strain (point A) is governing and the ultimate limit state is achieved without concrete crushing, that is, the maximum compressive strain is lower than ε_{cu2}; Case (ii) corresponds to the situation where the neutral axis (n.a.) is inside the cross-section and the ultimate limit ststate is reached when the maximum compressive strain is ε_{cu2}; Case (iii) corresponds to the situation where the compressive strain at point C is governing, which occurs when axial force eccentricity, M/N, is low and the neutral axis lies outside de cross-section.

The tables and design charts were prepared respecting the concrete strain limits at points B and C, with the corresponding strains taken from Table 2.1 according to the concrete class. For the steel, point A, law b) was used, as set forth in Sect. 2.2 wherein the horizontal branch was limited to a strain of 2.5%. In general, this limit has no practical significance in the resistance of the cross-section.

2.4 Behaviour of Sections Under Bending

The resulting forces of a reinforced concrete cross-section are computed integrating the stresses over the concrete and the steel. Thereby, they are function of the behaviour of the materials, or, by other words, they are function of the stress–strain relations of both concrete and steel. The stresses over the section are evaluated for all possible strain configurations defined in Sect. 2.3.

The calculation of the ultimate moment and axial force resistance of reinforced and prestressed concrete cross-sections is based on the following assumptions:

i.   Plane sections remain plane after deformation;
ii.   The strains in bonded reinforcement are the same as those in the surrounding concrete;
iii. The tensile strength of the concrete is ignored;
iv.   The compressive concrete stresses are given by the corresponding design stress–strain diagram given in Fig. 2.1;
v.   The stresses in the steel reinforcement are given by the design stress–strain diagram given in Fig. 2.2.

Given the reinforcement distribution, in the case of uniaxial bending and axial force, the problem entails three unknowns: the axial force (N) and bending moment (M) resistances, and the depth of the neutral axis. This requires three equations: translational and rotational equilibrium and the compatibility equation. The later is expressed using the premises (i) and (ii) listed above. The equilibrium equations involve the determination of the stress resultants via integration over cross-section area. In the general case of biaxial bending and axial force, the number of unknowns grows to five: the axial force, the two bending moment projections (M_x, M_y) and the depth and direction of the neutral axis. The analytical solution to this problem is intricated and either numerical methods or design tables and charts need to be adopted in practice.

The three dimensional interaction surface shown in Fig. 2.4 is the locus of the (N, M_x, M_y) coordinates corresponding to the ultimate limite state of the cross-section. The uniaxial bending case corresponds to the intersection of the interaction surface with the vertical plane $M_x = 0$ (or $M_y = 0$). This intsersection is the basis of the design Charts 1 to 3, or Tables 3 to 11, presented in this book. In the case of beams, where the axial force is either null or negligible, the solution can be found along the

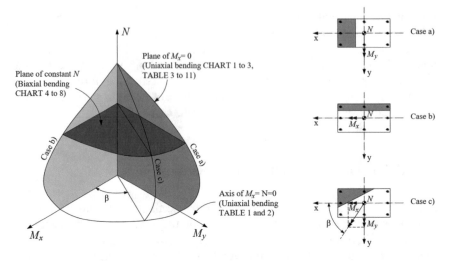

Fig. 2.4 Three dimensional interaction surface (left) of a reinforced concrete cross-section (right)

M_y (or M_x) axis. This situation is covered by the Tables 1 and 2. The general case of biaxial bending and axial force is treated using design charts that correspond to the intersection of the interaction surface with horizontal planes with constant N. This is the basis of the design Charts 4 to 8. If the cross-section is symmetrically reinforced with respect to the x- and y- axes, this intsersection becomes completely defined by the representation of a single quadrant.

In the case of Serviceability Limit States, the premises (i) to (iii) listed above were considered to obtain the stresses in concrete and steel in a cracked section under bending moment and axial load due to characteristic, frequent or quasi-permanent load combinations. In this case, the constitutive laws for concrete and steel are considered linear elastic with elasticity modulus $E_{c,eff}$ and E_s respectively. A given reinforced concrete section may be considered cracked whenever the strength of concrete under tension has been exceeded.

2.5 Summary

This chapter describes the assumptions for the design or safety assessment of reinforced concrete cross-sections subjected to bending moments and axial force. The design properties and constitutive relationships of concrete and steel were defined according to EC2 [2] recommendations, as well as the strain distribution in the cross-sections for the ultimate limit states under bending moment and axial force. The criteria defining the ultimate limit states were introduced. The three-dimensional interaction surface defining the locus of the (N, M_x, M_y) coordinates corresponding to the ultimate limite state of the cross-section was described.

Chapter 3
Implementation of Tables and Design Charts

3.1 Introduction

This chapter describes the model used in the evaluation of the stress resultants (axial force and bending moments) considering the different classes of concrete and steel in cross-sections of the structural members, as well as the reinforcement amount and layout. The tables are applied to rectangular cross-sections subjected to bending moment only, or bending moment and axial force. The design charts consider axial force and bending moment in rectangular and circular cross-sections and biaxial bending moments of rectangular sections with different distribution of the reinforcement along the sides. These tables and design charts are available for two different reinforcement concrete cover ratios. Some tables allow the direct reinforcement design by prescribing the position of the neutral axis depth.

3.2 Overview of the Resolution Model

The integration of the compression concrete stresses for the purposes of calculating the resultant axial and bending forces in the section is a laborious problem, bearing in mind the nonlinear constitutive law used in the ultimate limit state. In addition, there is an integral of area whose domain of integration is the region of the cross-section bounded by the neutral axis and where the concrete is in compression. In the present work this area is divided into triangles, in which the exact integration is performed, with the process described in [3]. Another important aspect is the imposition of the failure conditions that allow the definition of the strains in each point of the section. This aspect was addressed with Heaviside functions which allowed the definition of parametric functions in a single expression. The method is described in Barros et al. [4], being the one used in this work. The model was implemented in MAPLE, a mathematical manipulation program that solves the equations and compiles the tables and design charts. The diagrams for the evaluation of the concrete and steel stresses

© The Author(s), under exclusive license to Springer Nature Switzerland AG 2022
H. Barros et al., *Design of Reinforced Concrete Sections Under Bending and Axial Forces*,
https://doi.org/10.1007/978-3-030-80139-7_3

under Serviceability Limit States were also obtained with the MAPLE program. The integration of the stresses in the compression concrete and in the steel of a cracked section under service loads was also performed with this program. The results are presented in the form of diagrams allowing the evaluation of the maximum stresses in the steel and concrete and the neutral axis depth.

3.3   Tables and Design Charts

3.3.1   Organization of Tables and Design Charts

The tables and design charts compiled to assess the ultimate limit states of resistance of concrete cross-sections are grouped as follows:

- Tables for ultimate limit design of rectangular sections under simple bending

    - *Simply reinforced sections*: TABLE 1
    - *Doubly reinforced sections:* TABLE 2

- Tables for ultimate limit design of rectangular sections under bending and axial force

    - *Symmetric reinforcement*: TABLE 3
    - *Simply reinforced sections*: TABLE 4

- Tables for ultimate limit design of rectangular sections under bending and axial force with prescribed neutral axis depth, x/d: TABLE 5 to 8

- Tables for ultimate limit design of T-sections under bending and axial force: TABLE 9 to 11

- Charts for ultimate limit design of column cross-sections under bending and axial force

    - *Rectangular sections with symmetric reinforcement*: CHART 1
    - *Rectangular sections with non-symmetric reinforcement*: CHART 2
    - *Circular sections with reinforcement distributed along the contour*: CHART 3

- Charts for ultimate limit design of rectangular sections under biaxial bending moments and axial force

    - *Reinforcement at the section corners*: CHART 4
    - *Reinforcement distributed along the section contour*: CHART 5
    - *Different reinforcement ratios*: CHART 6 and 7
    - *Reinforcement in two opposite faces*: CHART 8

3.3.2   Organization of Diagrams

Calculation diagrams of allowable design stresses to verify the serviceability limit states are grouped as follows:

- Diagrams for rectangular sections under bending and axial forces:
  - *Simply reinforcement rectangular section*: DIAGRAM 1
  - *Doubly reinforced rectangular section with different A'/A*: DIAGRAM 2
  - *Symmetric reinforcement with imposed maximum concrete stress* $\sigma_{c,max}$: DIAGRAM 3
- Diagrams for simply reinforced T-sections under bending and axial force
  - *Flange web ratio $b/b_w = 4$ and variable h_f/d*: DIAGRAM 4
  - *Flange web ratio $b/b_w = 8$ and variable h_f/d*: DIAGRAM 5
  - *Diagrams for circular sections with uniformly distributed reinforcement along the contour*: DIAGRAM 6

3.4   Summary

This chapter describes, very concisely, the program used to obtain the tables and charts for ultimate limit design and the stress diagrams for allowable service design. The book content in terms of the tables and design charts applicable to the different cross section geometries and applied forces is also summarized.

Chapter 4
Application Examples

4.1 Introduction

The tables, charts and diagrams in this book can be used to design the longitudinal reinforcement in cross-sections subjected to axial forces and bending moments. The tables and charts refer to ultimate limit state conditions while the diagrams can be used for assessing the stresses in cracked sections in serviceability conditions where linear elastic behavior may be assumed. The examples presented in this section were prepared to guide the reader through the correct use of these tools.

4.2 Examples

Example 1 *Simple bending. Singly reinforced rectangular section.*

Design the flexural reinforcement for a rectangular 0.25×0.50 m^2 cross-section subjected to the design bending moment of 250 kNm, (Fig. 4.1). **Materials:** C30 concrete and S500 steel.

From Table 5.1 TABLE 1_Concrete Classes C12-C90_Steel Class S500 we get

$$\mu = \frac{M_{Rd}}{bd^2 f_{cd}} = \frac{250}{0.25 \times 0.45^2 \times 20{,}000} = 0.247;$$

from columns 1 and 2 it is concluded that: $\alpha = 0.359$; $\omega = 0.290$. The required reinforcement is then given by:

$$A_s = \omega bd \frac{f_{cd}}{f_{yd}} = 0.290 \times 0.25 \times 0.45 \frac{20}{435} = 0.0015\,\text{m}^2 = 15\,\text{cm}^2.$$

© The Author(s), under exclusive license to Springer Nature Switzerland AG 2022
H. Barros et al., *Design of Reinforced Concrete Sections Under Bending and Axial Forces*,
https://doi.org/10.1007/978-3-030-80139-7_4

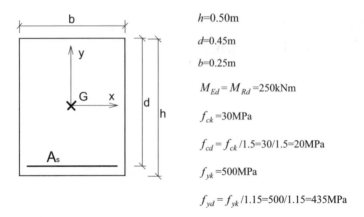

h=0.50m

d=0.45m

b=0.25m

$M_{Ed} = M_{Rd}$ =250kNm

f_{ck} =30MPa

$f_{cd} = f_{ck}/1.5=30/1.5=20$MPa

f_{yk} =500MPa

$f_{yd} = f_{yk}/1.15=500/1.15=435$MPa

Fig. 4.1 Schematic representation of the cross-section of the Example n°1

This is satisfied with 5Ø20 (15.7 cm^2), placed as follows, (see Fig. 4.2):

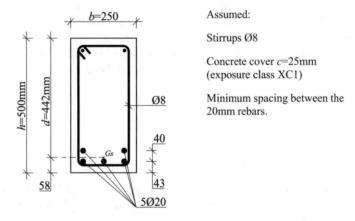

Assumed:

Stirrups Ø8

Concrete cover c=25mm
(exposure class XC1)

Minimum spacing between the
20mm rebars.

Fig. 4.2 Layout of the reinforcement in the cross-section of the Example n°1

It can be concluded that the effective depth $d = 0.442$ m is slightly lower than the 0.45 m adopted in the calculations. Once the difference is small and the provided reinforcement area (15.7 cm^2) is somewhat larger than that required (15.0 cm^2), it is likely that it is not necessary to change the layout. Verification:

$$\varpi = \frac{A_s f_{yd}}{bd f_{cd}} = \frac{15.7 \times 435}{25 \times 44.2 \times 20} = 0.309$$

From Table 5.1 TABLE 1_Concrete Classes C12-C90_Steel Class S500 we get $\mu = 0.260$, which corresponds to a design bending moment of:

$$M_{Rd} = \mu b d^2 f_{cd} = 0.26 \times 0.25 \times 0.442^2 \times 20{,}000 = 254 \text{ kNm} \geq 250 \text{ kNm } (OK).$$

It is confirmed that the designed cross-section fulfills the safety requirements.

Example 2 *Simple bending. Doubly reinforced rectangular section.*

Design the flexural reinforcement for a rectangular 0.30×0.60 m^2 cross-section subjected to the design bending moment of 800 kNm (see Fig. 4.3). **Materials:** C30 concrete and S500 steel.

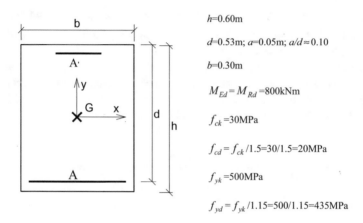

$h=0.60$m

$d=0.53$m; $a=0.05$m; $a/d \approx 0.10$

$b=0.30$m

$M_{Ed} = M_{Rd} = 800$kNm

$f_{ck} = 30$MPa

$f_{cd} = f_{ck}/1.5 = 30/1.5 = 20$MPa

$f_{yk} = 500$MPa

$f_{yd} = f_{yk}/1.15 = 500/1.15 = 435$MPa

Fig. 4.3 Schematic representation of the cross-section of the Example n°2

(a)   From Table 5.1 TABLE 2_Concrete Classes C12-C50_Steel Class S500_a/d $= 0.10$ we get:

$$\mu = \frac{M_{Rd}}{bd^2 f_{cd}} = \frac{800}{0.3 \times 0.53^2 \times 20{,}000} = 0.475;$$

$A'/A = 0$; has no solution

$$A'/A = 0.2; \alpha = 0.602; \quad \varpi = 0.609, \quad \varpi + 0.2 \ \varpi = 0.731$$

$$A'/A = 0.3; \quad \alpha = 0.498; \quad \varpi = 0.575; \quad \varpi + 0.3 \ \varpi = 0.748$$

It is concluded that the most economical solution is obtained with $A'/A = 0.2$ and so:

Fig. 4.4 Cross-section with the neutral axis depth and strain profile

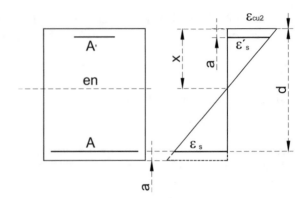

$A = \varpi bd \dfrac{f_{cd}}{f_{yd}} = 0.609 \times 30 \times 53 \times \dfrac{20}{435} = 44.5\,\text{cm}^2$ and $A'/A \times 0.2 = 44.5 \times 0.2 = 8.9\,\text{cm}^2$ and the position of the neutral axis given by: $x = \alpha\,d = 0.602 \times 53 = 31.9\,\text{cm}$.

Note that the solution $A'/A = 0.3$ with neutral axis position, $x = 0.498 \times 53 = 26.4$ cm, though with slightly higher steel consumption, may be preferrable due to its greater ductility, expressed by the lower value of x/d (see Fig. 4.4). To obtain the strains in the steel and concrete, it is deduced from the figure that:

$$\frac{\varepsilon_{cu2}}{x} = \frac{\varepsilon_s}{d-x} \ \text{with}\ \varepsilon_s = \frac{d-x}{x}\varepsilon_{cu2} = \frac{53-31.9}{31.9}3.5\%_{00} = 2.32\%_{00}.$$

and that:

$$\frac{\varepsilon_{cu2}}{x} = \frac{\varepsilon'_s}{x-a} \ \text{with}\ \varepsilon'_s = \frac{x/d - a/d}{x/d}\varepsilon_{cu2} = \frac{0.602-0.10}{0.602}3.5\%_{00} = 2.92\%_{00}.$$

In design values, the yield strain of grade S500 steel is:

$$f_{yd}/E_s = (f_{yk}/\gamma_s)/E_s = (500/1.15)/20,0000 = 2.17\%_{00}.$$

Whence, it can be concluded that both reinforcements are undergoing yielding ($\varepsilon_s > 2.17\%_{00}$ and $\varepsilon'_s > 2.17\%_{00}$), so with stress equal to yielding strength $\sigma_s = \sigma'_s = f_{yd} = 435$ MPa, and that the maximum compressive stress in the concrete upper fibers is $\sigma_c = f_{cd} = 20$ MPa.

(b) From Tab. 5.29 TABLE 7_Concrete Classes C12-C50_Steel Class S500, considering $\alpha=0.450$ and $a/d = 0.10$, comes: $N_{Ed} = 0 \rightarrow M_{Eds} = M_{Ed}$, leading to $\mu_s = \mu = 0.475$.

$$\varpi_{1,s} = 0.564 \rightarrow A_s = \frac{1}{f_{yd}}(\varpi_{1,s}bdf_{cd} + N_{Ed}) = \frac{1}{435,000}(0.564 \times 30 \times 53 \times 20,000 + 0) = 41.2\,\text{cm}^2$$

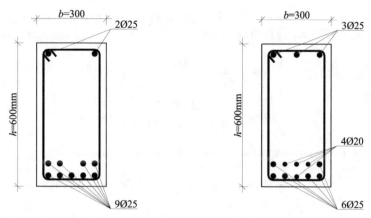

a) More economical solution: $\alpha=0.602$, $A_s = 54.0cm^2$.   b) More ductile solution: $\alpha=0.450$, $A_s = 56.8cm^2$.

Fig. 4.5 Two possible reinforcement layouts in the cross-section of Example n°2

$$\varpi_2 = 0.199 \rightarrow A_s' = \varpi_2 bd \frac{f_{cd}}{f_{yd}} = 0.199 \times 30 \times 63 \times \frac{20}{435} = 14.6 \, cm^2$$

Two possible reinforcement layouts corresponding to solutions a) e b) are the following (Fig. 4.5):

Example 3 _Bending and axial force._

Determine the reinforcement for a rectangular, $0.40 \times 0.60 \, m^2$ cross-section subjected to bending moment and axial force whose design values are respectively 500 kNm and 1000 kN (compressive), (Fig. 4.6). **Materials**: C35 concrete and S500 steel.

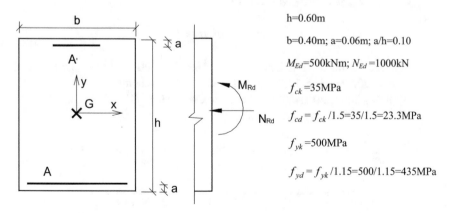

h=0.60m

b=0.40m; a=0.06m; a/h=0.10

M_{Ed}=500kNm; N_{Ed} =1000kN

f_{ck} =35MPa

$f_{cd} = f_{ck}/1.5$=35/1.5=23.3MPa

f_{yk} =500MPa

$f_{yd} = f_{yk}/1.15$=500/1.15=435MPa

Fig. 4.6 Schematic representation of the cross-section of the Example n°3

(a)   Applying   Table 5.11   TABLE 3_Concrete Classes C12-C50_Steel
      Class S500_a/h $= 0.10$ in which $A' = A$ (symmetric reinforcements) we get:

$$\mu = \frac{500}{0.40 \ \times 0.60^2 \times \ 23{,}300} = 0.149;$$

$$\nu = \frac{1000}{0.40 \times \ 0.60 \times \ 23{,}300} = 0.179;$$

The following results are obtained after sucessive interpolations:

$$\mu = 0.149; \ \nu = 0.1; \ \varpi = 0.2634; \ \alpha = 0.1746;$$

$$\mu = 0.149; \ \nu = 0.2; \ \varpi = 0.1746; \ \alpha = 0.2508;$$

$$\mu = 0.149; \ \nu = 0.179; \ \varpi = 0.1932; \alpha = 0.2348;$$

It   is   concluded   that   the   total   reinforcement   area A_s $=$
$\left(0.1932 \times \ 40 \ \times 60 \times \ \frac{23.3}{435}\right) = 24.8 \ \text{cm}^2$ and so $A = A' = 24.8/2 = 12.4 \ \text{cm}^2$.

(b)   Applying Fig. 6.2 CHART 1_C12-C50_S500_a/h $= 0.10$ in which $A' = A$
      (equal reinforcements) we get for $\mu = 0.149$ and $\nu = 0.179$:

$$\omega = 0.2; \ \alpha \approx 0.25.$$

The total reinforcement area $A_s = 0.2 \times 40 \times 60 \times \frac{23.3}{435} = 25.7 \ \text{cm}^2$ and so $A = A' = 25.7/2 = 12.85 \ \text{cm}^2$.
Note that the tables allow greater precision in the results than the charts.

(c)   Applying Fig. 6.4 CHART 2_C12-C50_S500_a/h $= 0.10$ in which $A' = 0.5A$
      we get for $\mu = 0.149$ and $\nu = 0.179$:

$$\varpi = 0.16; \ \alpha \approx 0.3.$$

The total reinforcement area is $A_s = 0.16 \times 40 \times 60 \times \frac{23.3}{435} = 20.52 \ \text{cm}^2$ and
so $A = 20.52/1.5 = 13.7 \ \text{cm}^2$ and $A' = 13.7/2 = 6.8 \text{cm}^2$.

(d)   Applying Tab. 5.20 TABLE 4_Concrete Classes C12-C50_Steel Class S500 in
      which $A' = 0$ (singly reinforced) we get:

$$M_{Eds} = M_{Ed} - N_{Ed}(h/2 - a) = 500 + 1{,}000 \times (0.60/2 - 0.06) = 740 \ \text{kNm}.$$

$$\mu_s = \frac{M_{Eds}}{bd^2 f_{cd}} = \frac{740}{0.40 \ \times 0.54^2 \ \times 23{,}300} = 0.272;$$

The following results are obtained after interpolation:

Fig. 4.7 Schematic representation of the cross-section with the forces acting at the centroid of the tensile reinforcement

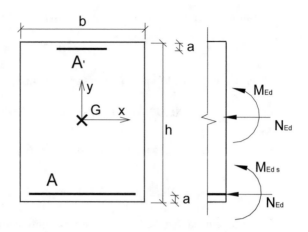

$$\mu_s = 0.272; \ \varpi_{1,s} = 0.3268; \ \alpha = 0.4038;$$

It is concluded that the required reinforcement is $A = 10{,}000 \times (0.3268 \times 0.40 \times 0.54 \times 23{,}300{-}1{,}000)/435000 = 14.8 \ \text{cm}^2$.

(e) Applying Tab. 5.23 TABLE 5_Concrete Classes C12-C50_Steel Class S500_x/d $= 0.250$ with prescribed $\alpha = 0.250$ gives (see Fig. 4.7):

$$M_{Eds} = M_{Ed} - N_{Ed}(h/2 - a) = 500 + 1{,}000 \times (0.60/2 - 0.06) = 740 \ \text{kNm}.$$

$$\mu_s = \frac{M_{Eds}}{bd^2 f_{cd}} = \frac{740}{0.40 \times 0.54^2 \times 23{,}300} = 0.272;$$

We get the following results after interpolation, bearing in mind that $a = 0.06$ m, $d = 0.60{-}0.06 = 0.54$ m and $a/d = 0.06/0.54 = 0.111$:

$$\mu_s = 0.272; \ a/d = 0.10; \ \varpi_{1,s} = 0.3032; \ \varpi_2 = 0.1042;$$

$$\mu_s = 0.272; \ a/d = 0.15; \ \varpi_{1,s} = 0.3092; \ \varpi_2 = 0.1656;$$

$$\mu_s = 0.272; \ a/d = 0.111; \ \varpi_{1,s} = 0.3045; \ \varpi_2 = 0.1177;$$

It is concluded that the lower and upper reinforcement requirements are respectively:
$A = 10{,}000 \times (0.3045 \times .40 \times .54 \times 23{,}300 - 1{,}000)/435{,}000 = 12.24 \ \text{cm}^2$,
and $A' = 10{,}000 \times (0.1177 \times .40 \times .54 \ \frac{23.3}{435}) = 13.6 \ \text{cm}^2$.

(f) Applying Table 5.26 TABLE 6_Concrete Classes C12-C50_Steel Class S500_x/d $= 0.350$ with prescribed $\alpha = 0.350$, and bearing in mind that statically equivalent moment applied at the centroid of the tensile reinforcement is equal to that obtained in point d) above, we get:

$$\mu_s = 0.272;$$

Interpolating, the following results are obtained:

$$\mu_s = 0.272; \ a/d = 0.10; \ \varpi_{1,s} = 0.305; \ \varpi_2 = 0.0312$$

$$\mu_s = 0.272; \ a/d = 0.15; \ \varpi_{1,s} = 0.3162; \ \varpi_2 = 0.0332$$

$$\mu_s = 0.272; \ a/d = 0.111; \ \varpi_{1,s} = 0.3075; \ \varpi_2 = 0.0316$$

It is concluded that the lower and upper reinforcement areas are respectively:

$$A = 10,000 \times (0.3075 \times 0.40 \times 0.54 \times \ 23,300 - 1,000)/435,000 = 12.6 \ cm^2$$

$$A' = 10,000 \times (0.0316 \times 0.40 \times 0.54 \ \frac{23.3}{435}) = 3.6 \ cm^2$$

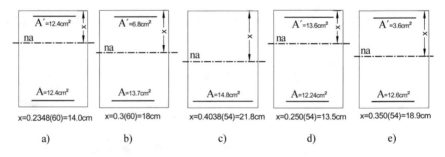

Fig. 4.8 Comparison of the different solutions obtained for the Example n°3

Figure 4.8 shows the solutions obtained in:

- points (a) and (b) with symmetric reinforcements $A' = A$ (Fig. 4.8a);
- point (c) with $A' = 0.5A$ (Fig. 4.8b);
- point (d) with a singly reinforced cross-section $A' = 0$ (Fig. 4.8c);
- point (e) with imposed neutral axis depth $x/d = 0.250$ (Fig. 4.8d);
- point (f) with imposed neutral axis depth $x/d = 0.350$ (Fig. 4.8e);

Note that in this case, and looking only at the specified pair bending and axial force, the singly reinforced cross-section appears as the most economical. However, in the case of a real column cross-section in which the acting bending moments may invert its sign in different load combinations, the solutions with $A' \neq 0$ become the most adequate. The detailing of the reinforcement corresponding to the different solutions is not discussed here.

Example 4 *Simple bending of a T-section.*

Design the longitudinal tensile reinforcement of a T-section with $b = 1.20$ m, $b_w = 0.30$ m; $h = 1.10$ m, $h_f = 0.15$ m and $d = 1.00$ m (Fig. 4.9a), and subjected to the design bending moment $M_{Ed} = 2,400$ kN/m. **Materials:** C20/25; S500.

The cross-section dimensions are such that $b/b_w = 1.20/0.30 = 4$ and $h_f/d = 0.15/1.0 = 0.15$ and the non-dimensional bending moment is:

$$\mu_s = \frac{2.4}{1.2 \times 1.0^2 \times 20/1.5} = 0.150$$

Applying Tab. 5.35 TABLE 9_Concrete Classes C12-C50_b/b$_w$ = 4.

$\mu_s = 0.15$, $h_f/d = 0.14$, $\varpi_{1,s} = 0.165$ $\alpha = 0.263$; $x = \alpha d = 0.263$m $> h_f$

$\mu_s = 0.15$, $h_f/d = 0.16$, $\varpi_{1,s} = 0.164$ (neutral axis falls outside flange)

$$\mu_s = 0.15, \quad h_f/d = 0.15, \quad \varpi_{1,s} = 0.1645$$

$$A_s = \frac{0.1645 \times 1.2 \times 1.0 \times 20/1.5}{435} = 60.5 \times 10^{-4} \text{ m}^2 = 60.5 \text{ cm}^2 \ (13\phi25).$$

See the reinforcement layout in Fig. 4.9b).

Example 5 Bending with axial force of a T-section.

Design the ordinary longitudinal tensile reinforcement of a T-section with the geometry described in Example 4, subjected to the design bending moment of M_{Sd}

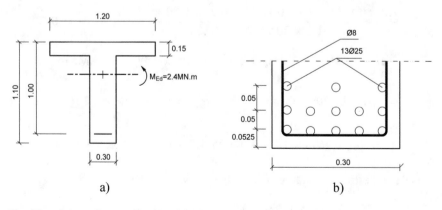

Fig. 4.9 a Schematic representation of the cross-section of Example n°4; **b** Detailing of the tensile reinforcement

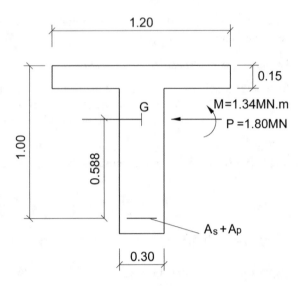

Fig. 4.10 Schematic representation of the cross-section of Example n°5

$= 1340$ kNm and to a compressive axial force (due to prestressing) $P = 1800$ kN. Consider that the effective depth corresponding to the ordinary reinforcement (d_s) and to the prestressing reinforcement (d_p) is the same $d = d_s = d_p = 1.0$ m. The prestressing cable is constituted by twelve 0.6 inch strands with total area $A_p = 1.4 \times 12 = 16.8$cm^2 (Fig. 4.10). **Materials:** C25/30; S400; Y1860 $(f_{puk} = 1860$ MPa).

Applying Table 5.35 TABLE 9_Concrete Classes C12-C50_b/b$_w$ = 4, we need to determine the bending moment M_{Eds} (forces applied at the centroid of the tensile reinforcement forces).

Distance from the geometric centre of the T-section to the tensile reinforcement, z_s:

$$z_s = \frac{1.2 \times 0.15 \times (1.10 - 0.15/2) + 0.95 \times 0.30 \times 0.95/2}{1.2 \times 0.15 + 0.95 \times 0.30} - 0.10 = 0.588\text{m}$$

$$M_{Eds} = M_{Ed} - N_{Ed}\,z_s = 1.34 - (-1.8 \times 0.588) = 2.4 \text{ MNm}$$

$$\mu_s = 0.150; \quad h_f/d = 0.15 \rightarrow \varpi_{1,s} = 0.1645; \quad \alpha = 0.262$$

$$A_s = \frac{1}{348}(0.1645 \times 1.2 \times 1.0 \times 25/1.5 - 1.8) = 42.82 \times 10^{-4}\text{m}^2.$$

Is it the prestressed reinforcement in the yielding domain?

$$\varepsilon_{yp} = \frac{1{,}860 \times 0.9}{1.15}\Big/(200 \times 10^3) = 7.3‰$$

Fig. 4.11 Schematic representation of the cross-section of Example n°6

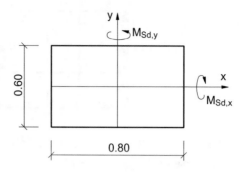

$$\varepsilon_p = \varepsilon_p^0 + \Delta\varepsilon_p = \frac{1.8}{16.8 \times 10^{-4} \times 200 \times 10^3} + 3.5 \times 10^{-3} \times \frac{1 - 0.262}{0.262}$$

$$= 5.3 \times 10^{-3} + 9.9 \times 10^{-3} = 15.2\%_{00} > 7.3\%_{00}$$

OK, it is yielding.

Ordinary reinforcement is required to check safety:

$$A_{s\,S400} = A_s - A_{pi} = 42.82 - \left(\left(\frac{1,860 \times 0.9}{1.15} - \frac{1.8}{16.8 \times 10^{-4}}\right)/348\right) \times 16.8$$

$$= 42.82 - (384/348) \times 16.8 = 24.27 \text{cm}^2 \ (8\phi 20)$$

Example 6 *Biaxial bending with axial force.*

Design the longitudinal reinforcement for a column with a rectangular section: $b = 0.60$ m, $h = 0.80$ m, subjected to the following design axial force and bending moments, $N_{Ed} = 3000$ kN; $M_{Edx} = 525$ kN/m; $M_{Edy} = 1516$ kN/m (Fig. 4.11). **Materials:** C30/37; S500.

(a) Choosing a solution with reinforcement equally distributed by the four sides of the cross-section, the Fig. 6.11 CHART 5_C12-C50_S500_1/4A$_S$ at side contour_ $a_1/h = c_1/b = 0.10$—Biaxial bending is selected considering $\frac{a_1}{h} = \frac{c_1}{b} = 0.10$:

$$v = \frac{3.0}{0.6 \times 0.8 \times 20} = 0.313$$

$$\mu_x = \frac{0.525}{0.8 \times 0.6 \times 0.6 \times 20} = 0.091$$

$$\mu_y = \frac{1.516}{0.6 \times 0.8 \times 0.8 \times 20} = 0.197$$

$$v = 0.2; \quad \mu_x = 0.091 \quad \mu_y = 0.197 \ \rightarrow \ \varpi = 0.50$$

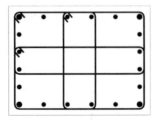 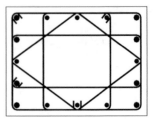

Longitudinal reinf.:	4Ø32 (corners)	8Ø32 (4 in each short side)
	16Ø25	8Ø25
Transverse reinf.:	Ø8//0.30	Ø8//.30

Fig. 4.12 Two alternative reinforcement layouts for the column cross-section of Example 6

$$v = 0.4; \quad \mu_x = 0.091 \quad \mu_y = 0.197 \ \rightarrow \ \varpi = 0.47$$

We get.

$$v = 0.313; \quad \mu_x = 0.091 \quad \mu_y = 0.197 \ \rightarrow \ \varpi \cong 0.483$$

$$A_s = 0.483 \times 0.6 \times 0.8 \times 20/435 = 106.6 \times 10^{-4} \mathrm{m}^2 = 106.6 \mathrm{cm}^2.$$

Possible solution:: 4Ø32 + 16Ø25 (110.7 cm²), distributed as indicated in Fig. 4.12a).

(b) Choosing a solution with 2/3 of reinforcement distributed by the two sides of the cross-section parallel to the largest non-dimensional moment vector, the Fig. 6.13 CHART 6_C12-C50_S500_A$_s$/3, A$_s$/6_a₁/h = c₁/b = 0.10— Biaxial bending is selected considering $\frac{a_1}{h} = \frac{c_1}{b} = 0.10$:

$$v = 0.2; \quad \mu_x = 0.091 \quad \mu_y = 0.197 \ \rightarrow \ \varpi = 0.47$$

$$v = 0.4; \quad \mu_x = 0.091 \quad \mu_y = 0.197 \ \rightarrow \ \varpi = 0.45$$

We get.

$$v = 0.313; \quad \mu_x = 0.091 \quad \mu_y = 0.197 \ \rightarrow \ \varpi \cong 0.459$$

$$A_s = 0.459 \times 0.6 \times 0.8 \times 20/435 = 101.3 \times 10^{-4} \mathrm{m}^2 = 101.3 \mathrm{cm}^2$$

Possible solution: 8Ø32 + 8Ø25 (103.6 cm²), distributed as indicated in Fig. 4.12b).

Fig. 4.13 Reinforcement
layout for the column
cross-section of Example 7

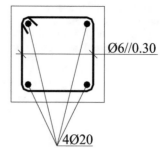

Ø6//0.30

4Ø20

Example 7 *Biaxial bending with axial force: column with cross-Section 0.30 ×
0.30 m².*

Design the longitudinal reinforcement for a column with a rectangular section: $b = h$
$= 0.30$ m, subjected to the following design axial force and bending moments: N_{Ed}
$= 900$ kN; $M_{Edx} = 45$k Nm; $M_{Edy} = 70$ kNm (Fig. 4.13). **Materials:** C25/30; S500.

Taking into account the small size of the cross-section, we chose reinforcement
layout constituted by one rebar on each corner. In this case we should use the Fig. 6.10
CHART 4_C12-C50_S500_1/4 A_s at section corners_$a_1/h = c_1/b = 0.15$—Biaxial
Bending, considering $\frac{a_1}{h} = \frac{c_1}{b} = 0.15$:

$$\nu = \frac{0.900}{0.3 \times 0.3 \times 16.667} = 0.600$$

$$\mu_x = \frac{0.045}{0.3 \times 0.3 \times 0.30 \times 16.667} = 0.100$$

$$\mu_y = \frac{0.070}{0.3 \times 0.3 \times 0.30 \times 16.667} = 0.156$$

We get.

$$\varpi = 0.36 \rightarrow A_s = 0.36 \times 0.3 \times 0.3 \times 16.667/435$$
$$= 12.4 \times 10^{-4}\text{m}^2 = 12.4\text{cm}^2 \ (4\text{Ø}20)$$

Example 8 *Biaxial bending with axial force: safety verification of a column subjected
to several load combinations.*

Check the safety with respect to the ultimate limite state of a column with cross-
section $b = 0.30$ m, $h = 0.40$ m, reinforced with 3Ø20 in each of the shorter sides as
represented in Fig. 4.14. The cross-section is subjected to the 3 pairs of axial force
and bending moment corresponding to different fundamental load combinations,
with the values indicated in KN and KNm. Consider the nominal concrete cover to
be $c_{nom} = 35$ mm. **Materials:** C30/37; S500

Fig. 4.14 Reinforcement layout and schematic representation of the column cross-section of Example 8

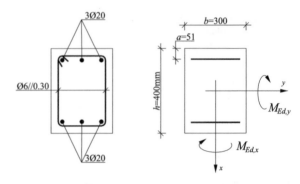

Combination	N_{Ed}	M_{Edx}	M_{Edy}
ULS1	1200	0	120
ULS 2	700	40	160
ULS3	700	90	60

Considering the nominal concrete $c_{nom} = 35$ mm and the stirrup diameter of 6 mm, the distance a from the surface to the center of the longitudinal reinforcement can be calculated at $a = 51$ mm, which leads to the ratio $a/h = 0.1275$. In this case it is conservative to use the tables and charts with $a/h = 0.15$. If increased accuracy is necessary, the results can be interpolated between those obtained with $a/h = 0.10$ and $a/h = 0.15$.

The combination ULS1 leads to uniaxial bending and axial force. Using the Tab. 5.12 TABLE 3_Concrete Classes C12-C50_Steel Class S500_a/h = 0.15:

$$v = \frac{1.200}{0.3 \times 0.4 \times 20} = 0.500$$

$$\mu = \frac{0.120}{0.3 \times 0.4 \times 0.4 \times 20} = 0.125$$

we get the required mechanical reinforcement ratio:

$$\varpi = 0.012.$$

The mechanical reinforcement corresponding to the existing reinforcement (A_s = A + A′ = 18.85 cm²) is:
$\varpi = \frac{18.85 \times 10^{-4} \times 435}{0.3 \times 0.4 \times 20} = 0.342 \geq 0.012$, OK, safety is verified in the combination ULS1.

The combination ULS2 leads to biaxial bending and axial force with the following non-dimensional values:

$$v = \frac{0.700}{0.3 \times 0.4 \times 20} = 0.292$$

$$\mu_x = \frac{0.040}{0.4 \times 0.3 \times 0.30 \times 20} = 0.056$$

$$\mu_y = \frac{0.160}{0.3 \times 0.4 \times 0.40 \times 20} = 0.167$$

Considering the provided reinforcement layout, we choose the Fig. 6.18 CHART 8_C12-C50_S500_opposite faces_$a_1/h = c_1/h = 0.15$ corresponding to uniformly distributed reinforcement along the two opposite shorter sides:

$$\nu = 0.2; \quad \mu_x = 0.056 \quad \mu_y = 0.167 \ \to \ \varpi = 0.33$$

$$\nu = 0.4; \quad \mu_x = 0.056 \quad \mu_y = 0.167 \ \to \ \varpi = 0.28$$

After interpolation we get the following required mechanical reinforcement ratio:

$$\nu = 0.292; \quad \mu_x = 0.056 \quad \mu_y = 0.167 \ \to \ \varpi \cong 0.31.$$

As the mechanical reinforcement ratio corresponding to the provided 6Ø20 rebars is $\varpi = 0.342 \geq 0.31$, safety is verified in the combination ULS2.

The combination ULS3 leads to biaxial bending and axial force with the following non-dimensional values:

$$\nu = \frac{0.700}{0.3 \times 0.4 \times 20} = 0.292$$

$$\mu_x = \frac{0.090}{0.4 \times 0.3 \times 0.30 \times 20} = 0.125$$

$$\mu_y = \frac{0.060}{0.3 \times 0.4 \times 0.40 \times 20} = 0.0625$$

In this case the maximum non-dimensional moment demands the cross-section side with less reinforcement, that is, $\mu_x > \mu_y$. This situation is not covered in any of the design charts provided in this book. Still, an approximate and conservative solution can be found by neglecting the contribution of the rebar paced at the centre of the two shorter opposite sides. In this case it is possible to use the Fig. 6.10 CHART 4_C12-C50_S500_1/4A_s at section corner_$a1/h = c1/h = 0.15$ corresponding to a reinforcement layout with 4 bars, one per corner:

$$\nu = 0.2; \quad \mu_x = 0.0625 \quad \mu_y = 0.125 \ \to \ \varpi = 0.20$$

$$\nu = 0.4; \quad \mu_x = 0.0625 \quad \mu_y = 0.125 \ \to \ \varpi = 0.12$$

After interpolation we get the following required mechanical reinforcement ratio:

$$v = 0.292; \quad \mu_x = 0.0625 \quad \mu_y = 0.125 \quad \rightarrow \quad \varpi \cong 0.163$$

As the rebars at the centre of the shorter sides are neglected, the provided mechanical reinforcement ratio corresponds to 4Ø20 only:

$\varpi = \frac{12.57 \times 10^{-4} \times 435}{0.3 \times 0.4 \times 20} = 0.228 \geq 0.163$ OK, safety is verified in the combination ULS3.

Example 9 *Bending and axial force in a circular cross-section.*

Design the longitudinal reinforcement of a circular column with diameter $2r = 0.60$ m subjected to the following forces: $N_{Ed} = 3,100$ kN; $M_{Edx} = 250$ kNm; $M_{Edy} = 475$ kNm. **Materials:** C30/37; S500.

In a circular cross-section with longitudinal reinforcement uniformly distributed along the perimeter bending is always uniaxial. Therefore, the resultant bending moment is:

$$M_{Ed} = \sqrt{M_{Edx}^2 + M_{Edy}^2} = 537 \text{ kNm}$$

The non-dimensional values of the axial force and bending moment are:

$$v = \frac{3.100}{\pi \times 0.3 \times 0.3 \times 20} = 0.548$$

$$\mu = \frac{0.537}{2 \times \pi \times 0.3 \times 0.3 \times 0.3 \times 20} = 0.158$$

Using the Fig. 6.6 CHART 3_C12-C50_S500_a/(2r) = 0.10, Bending with axial force with $\frac{a}{2r} = 0.10$, we get.

$\varpi = 0.30 \rightarrow A_s = 0.30 \times \pi \times 0.3 \times 0.3 \times 20/435 = 39.0 \times 10^{-4} \text{m}^2 = 39.0$ cm^2 (8Ø25) (Fig. 4.15).

Example 10 *Stresses for rectangular section in serviceability limit states.*

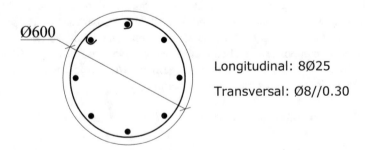

Longitudinal: 8Ø25

Transversal: Ø8//0.30

Fig. 4.15 Reinforcement layout for the column cross-section of Example 9

Calculate the stresses in the concrete (upper fiber) and in the tensile reinforcement of the doubly reinforced rectangular section, as described in Example 2, when subjected to a service bending moment, $M_{Ed} = 400$ kN/m. Take $\alpha_e = E_s/E_c = 10$.

Applying Fig. 7.3 DIAGRAM 2_A'/A = 0.2:

$$A'/A = 8.9/44.5 = 0.2; \, a/d = 0.1; \, \rho = \frac{44.5 \times 10^{-4}}{0.30 \times 0.53} = 0.028$$

$$\alpha_e \rho = 10 \times 0.028 = 0.28; \, M_{Eds} = M_{Ed} = 400 \text{kNm}$$

$$\frac{N_{Ed} \cdot d}{M_{Eds}} = 0 \rightarrow C_c = 4.0; \, \sigma_c = 4.0 \times \frac{0.400}{0.3 \times 0.53^2} = 19.0 \text{ MPa}$$

$$\rightarrow C_s \rho = 1.20; \, \sigma_s = \frac{1.2}{0.028} \times \frac{0.4}{0.3 \times 0.53^2} = 203.4 \text{ MPa}$$

Example 11 *Stresses for a prestressed T-section in serviceability limit states.*

Calculate the maximum compressive stresses in the concrete and tensile stresses in the steel of a T-section designed as described in Example 5, when loaded by the bending moment and axial force for the frequent combination of actions: $M_{Ed} = 600$ kN/m; $N_{Ed} = P = -1{,}800$ kN. Take $\alpha_e = Es/Ec = 15$.

Applying Fig. 7.12 DIAGRAM 4_b/bw = 4_ h$_f$/d = 0.14 and Fig. 7.13 DIAGRAM 4_b/b$_w$ = 4_ hf /d = 0.16 and interpolating for $h_f/d = 0.15$ we get:

$$M_{Eds} = 600 + 1{,}800 \times 0.588 = 1{,}658.4 \text{ kNm}$$

$$\frac{M_{Eds}}{1.2 \times 1.0^2} = 1.38; \, \rho = \frac{Ap + As}{bd} = \frac{(16.8 + 24.27) \times 10^{-4}}{1.2 \times 1.0} = 0.034$$

$$\alpha_e \rho = 0.051; \rightarrow N_{Ed} \times d/M_{Eds} = -1{,}800 \times 1.0/1{,}658.4 \cong -1.1$$

$$\alpha = 0.77$$

$$C_c = 6.0 \rightarrow \sigma_c = 6.0 \times 1.38 = 8.3 \text{MPa}.$$

$C_s \rho = 0.1 \rightarrow \sigma_s = \frac{0.1}{0.0034} \times 1.38 = 40.6 \text{MPa}$ (stress in the steel of the passive reinforcement).

Example 12 *Stresses for a rectangular section in serviceability limit states under bending with axial force.*

Confirm whether the stress in the concrete, in the most compressed fiber of a column designed as in Example 3 a) with symmetric reinforcements, is less than $0.45\,f_{ck}$. Consider that in the quasi-permanent load combination the moment and normal force are respectively 250kNm and 550 kN (compressive). Take $\alpha_e = E_s/E_c = 15$.

Applying Fig. 7.8 DIAGRAM $3\_\sigma_{c,max} = 0.45\,f_{ck}\_A'/A = 1.0$

$$A = A' = 12.4 \text{ cm}^2; b = 0.40\text{m}; h = 0.60\text{m}; a = 0.06 \text{ m}$$
$$d = h - a = 0.54 \text{ m}; a/d = 0.06/0.54 = 0.11 \cong 0.1$$

$$\mu_e = \frac{M_{Ed}}{bd^2 f_{ck}} = \frac{250}{0.40 \times 0.54^2 \times 35,000} = 0.061$$

$$v_e = \frac{N_{Ed}}{bd f_{ck}} = \frac{550}{0.40 \times 0.54 \times 35,000} = 0.073 \rightarrow \alpha_e \rho = 0.05$$

$$\rho = 0.05/15 = 0.0033; A = 0.0033 \times 40 \times 54 = 7.2 \text{ cm}^2.$$

Bearing in mind that the existing reinforcement $A = A' = 12.4 \text{ cm}^2$ is greater than 7.2 cm^2, it is concluded that the stress in the concrete in the most compressed fiber is less than $0.45\,f_{ck}$.

Example 13 *Stresses for a circular cross-section in serviceability limit states under bending with axial force.*

Determine the maximum compressive stress in the concrete and the maximum tensile stresses in the reinforcement of the circular cross-section of Example 9. Consider the acting axial force N = 1,000 kN and bending moment 200 kN/m. Take $\alpha_e = E_s/E_c = 15$.

Applying Fig. 7.24 DIAGRAM $6\_a/(2r) = 0.10$:

$$A_s = 39.3 \text{ cm}^2; \rho = \frac{A_s}{\pi r^2} = \frac{39.3}{\pi \times 30 \times 30} = 0.0139$$

$$\alpha \rho_e = 15 \times 0.0139 = 0.208; \frac{N_{Ed} r}{M_{Ed}} = \frac{-1,000 \times 0.30}{200} = -1.5$$

$$\alpha = 0.58$$

$$C_c \approx 1.65 \rightarrow \sigma_c = 1.65 \times \frac{0.200}{0.3 \times 0.3 \times 0.3} = 12.29 \text{ MPa}$$

$$C_s/\alpha_e \approx 0.90 \rightarrow C_s = 15 \times 0.90 = 13.5 \rightarrow \sigma_s$$
$$= 13.5 \times \frac{0.200}{0.3 \times 0.3 \times 0.3} = 100 \text{ MPa}.$$

4.3 Summary

A set of 13 examples were presented to illustrate the application of the tables, charts and diagrams in the longitudinal reinforcement design of reinforced concrete beams and columns subjeted to axial forces and bending moments. The examples cover both the design of reinforcement to fulfil safety in the ultimate limit state, as well as the calculation of stresses in cracked sections in serviceability conditions.

Chapter 5
Tables

5.1 Introduction

The tables presented in this chapter are auxiliary tools for professional engineers and architects, and students engaged in the reinforced and prestressed concrete design of buildings and other infrastructure concrete elements. The design tables, elaborated according to the EC2 [2] rules, are applied to the design and safety verification to the ultimate limit state of reinforced and prestressed concrete cross–sections of beams and columns subjected to bending and axial force. The advantage of the tables is the accuracy in the design and verification of structural concrete sections with the information about the ductility behavior, the economic and practical construction aspects, and the prompt availability of a set of solutions to decide on.

The internal forces acting on the cross–sections can be the simple bending moment or both bending moment and axial force, resulting from the specified load combination. The included section geometries are the rectangular and T–section, as they are currently used in beams, slabs and columns of the buildings. The most common steel reinforcement layouts are available. The mechanical reinforcement percentage is given in the tables as a function of the non–dimensional bending and axial forces, the non–dimensional concrete covers and the materials strength, whenever these variables influence the results.

The tables include solutions valid for the concrete strength classes C12–C90 with the properties described in EC2 [2]. The two steel strength classes S400 and S500 are available as reinforcement of C12–C50 cross–sections. However, for the strength classes C55–C90 only the S500 steel can be considered. A procedure to use S600 steel is also presented for simply reinforced rectangular sections under bending moment.

The nomenclature of the tables is denoted as in the following examples:

- *Tab. 5.15 TABLE 2_Concrete class C55_Steel class S500_a/d = 0.10*, means that table number 2 considers the concrete class C55, the steel class S500 and the nondimensional concrete cover a/d = 0.10;

Force	Reinforcement / Steel Class	Concrete Cover a or h/d	Neutral axis depth x/d	Concrete Class		
				C12-C50	C55	C60 / C70 / C80 / C90
Bending Moment	Simply / S400. S500. S600	Variable	Variable	TABLE 1 (Tab. 5.1)		
	Doubly / S500	0.10	Variable	TABLE 2 (Tab. 5.2)	TABLE 2 (Tab. 5.6)	TABLE 2(Tab. 5.7) / TABLE 2 (Tab. 5.8) / TABLE 2(Tab. 5.9) / TABLE 2(Tab. 5.10)
		0.15	Variable	TABLE 2 (Tab. 5.3)		
	Doubly / S400	0.10		TABLE 2 (Tab. 5.4)		
		0.15		TABLE 2 (Tab. 5.5)		
Bending Moment and Axial Force	Simply / S500	Variable	Variable	TABLE 4 (Tab. 5.20)	TABLE 4 (Tab. 5.22)	
	Simply / S400			TABLE 4 (Tab. 5.21)		
	Doubly / S500	Variable (0.05 to 0.20)	0.250	TABLE 5 (Tab. 5.23)	TABLE 5 (Tab. 5.25)	
			0.350	TABLE 6 (Tab. 5.26)	TABLE 6 (Tab. 5.28)	
			0.450	TABLE 7 (Tab. 5.29)	TABLE 7 (Tab. 5.31)	
			0.617	TABLE 8 (Tab. 5.32)	TABLE 8 (Tab. 5.34)	
	Doubly / S400	Variable (0.05 to 0.20)	0.250	TABLE 5 (Tab. 5.24)		
			0.350	TABLE 6 (Tab. 5.27)		
			0.450	TABLE 7 (Tab. 5.30)		
			0.617	TABLE 8 (Tab. 5.33)		
	Symmetrical / S500	0.10	Variable	TABLE 3 (Tab. 5.11)	TABLE 3 (Tab. 5.15)	TABLE 3(Tab. 5.16) / TABLE 3(Tab. 5.17) / TABLE 3(Tab. 5.18) / TABLE 3(Tab. 5.19)
		0.15	Variable	TABLE 3 (Tab. 5.12)		
	Symmetrical / S400	0.10		TABLE 3 (Tab. 5.13)		
		0.15		TABLE 3 (Tab. 5.14)		

Figure. 5.1 Design tables for rectangular sections

Force	Reinforcement / Steel Class	b/b$_w$	Concrete Class	
			C12-C50	C55
Bending Moment and Axial Force	Simply /S400;S500;S600	4	TABLE 9 (Tab. 5.35)	TABLE 9 (Tab. 5.36)
		8	TABLE 10 (Tab. 5.37)	TABLE 10 (Tab. 5.38)
		16	TABLE 11 (Tab. 5.39)	TABLE 11 (Tab. 5.40)

Figure. 5.2 Design tables for T–sections

- *Tab. 5.32 TABLE 8_Concrete classes C12–C50_Steel class S500_x/d = 0.617,* means that this table is available for concrete classes from C12 to C50, the steel class S500 and the neutral axis depth x/d = 0.617.

To streamline the choice of the appropriate table, the schedule of the tables content is summarized in Figs. 5.1 and 5.2.

5.2 Simple Bending of Rectangular Sections

Purpose of the Tables in This Section

The tables in this section serve to determine the reinforcement amount for rectangular sections under bending moment. Both simply reinforced, TABLE 1, or doubly reinforced, TABLE 2, cross sections are included. For a given design case, the results from both tables can be obtained and compared. It is possible to choose the best solution in terms of the minimum reinforcing steel area or of the envisaged cross–section ductility. The comparison of the different solutions allows a quick optimization of the section. Increasing of the reinforcing amount in the compression zone increases the ductility, which can be assessed by the neutral axis position, also delivered in the tables.

How to Use the Tables in This Section
The examples regarding directly these tables are n° 1 and n° 2, where there is no applied axial force (see Chap. 4). In TABLE 2, there are two possible concrete covers (a/d = 0.10 and a/d = 0.15) allowing interpolation for intermediate values.

Cautions When Using the Tables in This Section
The tables are valid with the definitions presented in the figures attatched to each table and the corresponding evaluation of non–dimensional values.

5.2.1 Simply Reinforced Sections—TABLE 1

Tab. 5.1

TABLE 1_Concrete classes C12-C90_Steel class S500
Simply reinforced sections under bending moment [1]

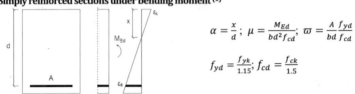

$$\alpha = \frac{x}{d}; \quad \mu = \frac{M_{Ed}}{bd^2 f_{cd}}; \quad \varpi = \frac{A\, f_{yd}}{bd\, f_{cd}}$$

$$f_{yd} = \frac{f_{yk}}{1.15}; \quad f_{cd} = \frac{f_{ck}}{1.5}$$

	C12-C50		C55		C60		C70		C80		C90	
μ	ϖ	α	ϖ	α	ϖ	α	ϖ	α	ϖ	α	ϖ	α
0.005	0.005	0.021	0.005	0.023	0.005	0.024	0.005	0.026	0.005	0.027	0.005	0.027
0.010	0.010	0.030	0.010	0.033	0.010	0.035	0.010	0.037	0.010	0.038	0.010	0.039
0.015	0.015	0.037	0.015	0.041	0.015	0.043	0.015	0.046	0.015	0.047	0.015	0.048
0.020	0.020	0.044	0.020	0.048	0.020	0.050	0.020	0.053	0.020	0.055	0.020	0.055
0.025	0.026	0.050	0.025	0.054	0.026	0.057	0.026	0.060	0.026	0.062	0.025	0.062
0.030	0.031	0.055	0.031	0.060	0.031	0.063	0.031	0.066	0.031	0.068	0.030	0.069
0.035	0.036	0.061	0.036	0.065	0.036	0.068	0.036	0.072	0.036	0.074	0.036	0.075
0.040	0.041	0.066	0.041	0.071	0.041	0.074	0.041	0.077	0.041	0.079	0.041	0.080
0.045	0.046	0.071	0.046	0.076	0.046	0.079	0.046	0.082	0.046	0.084	0.046	0.086
0.050	0.052	0.076	0.052	0.081	0.052	0.084	0.052	0.087	0.052	0.090	0.051	0.091
0.055	0.057	0.081	0.057	0.086	0.057	0.089	0.057	0.092	0.057	0.095	0.057	0.097
0.060	0.062	0.087	0.062	0.091	0.062	0.094	0.062	0.098	0.062	0.104	0.062	0.106
0.065	0.067	0.092	0.067	0.096	0.068	0.099	0.068	0.106	0.068	0.113	0.067	0.115
0.070	0.073	0.097	0.073	0.102	0.073	0.105	0.073	0.115	0.073	0.122	0.073	0.125
0.075	0.078	0.102	0.078	0.107	0.078	0.113	0.078	0.123	0.079	0.131	0.078	0.134
0.080	0.084	0.107	0.084	0.113	0.084	0.121	0.084	0.132	0.084	0.140	0.084	0.144
0.085	0.089	0.113	0.089	0.120	0.089	0.128	0.090	0.140	0.090	0.150	0.090	0.153
0.090	0.095	0.118	0.095	0.128	0.095	0.137	0.095	0.149	0.095	0.159	0.095	0.163
0.095	0.100	0.124	0.100	0.135	0.100	0.144	0.101	0.158	0.101	0.168	0.101	0.173
0.100	0.106	0.131	0.106	0.143	0.106	0.153	0.106	0.167	0.107	0.178	0.107	0.183
0.105	0.111	0.137	0.111	0.150	0.112	0.161	0.112	0.176	0.112	0.188	0.112	0.192
0.110	0.117	0.144	0.117	0.158	0.117	0.169	0.118	0.185	0.118	0.197	0.118	0.202
0.115	0.123	0.151	0.123	0.166	0.123	0.177	0.124	0.194	0.124	0.207	0.124	0.212
0.120	0.128	0.159	0.129	0.173	0.129	0.186	0.129	0.203	0.130	0.217	0.130	0.223
0.125	0.134	0.166	0.135	0.181	0.135	0.194	0.135	0.212	0.136	0.227	0.136	0.233
0.130	0.140	0.173	0.140	0.189	0.141	0.202	0.141	0.222	0.142	0.237	0.142	0.243
0.135	0.146	0.180	0.146	0.197	0.147	0.211	0.147	0.231	0.148	0.247	0.148	0.254
0.140	0.152	0.188	0.152	0.205	0.153	0.220	0.153	0.241	0.154	0.257	0.154	0.264
0.145	0.158	0.195	0.158	0.213	0.159	0.228	0.159	0.250	0.160	0.267	0.160	0.275
0.150	0.164	0.202	0.164	0.221	0.165	0.237	0.165	0.260	0.166	0.277	0.166	0.285
0.155	0.170	0.210	0.170	0.229	0.171	0.246	0.172	0.269	0.173	0.288	0.173	0.296
0.160	0.176	0.217	0.176	0.238	0.177	0.255	0.178	0.279	0.179	0.298	0.179	0.307
0.165	0.182	0.225	0.183	0.246	0.183	0.264	0.184	0.289	0.185	0.309	0.185	0.318
0.170	0.188	0.232	0.189	0.254	0.189	0.272	0.191	0.299	0.192	0.320	0.192	0.329
0.175	0.194	0.240	0.195	0.263	0.196	0.282	0.197	0.309	0.198	0.331	0.199	0.340
0.180	0.201	0.248	0.201	0.271	0.202	0.291	0.203	0.319	0.205	0.342	0.205	0.352
0.185	0.207	0.256	0.208	0.280	0.209	0.300	0.210	0.330	0.211	0.353	0.212	0.363
0.190	0.213	0.264	0.214	0.289	0.215	0.309	0.217	0.340	0.218	0.364	0.219	0.375
0.195	0.220	0.271	0.221	0.297	0.222	0.319	0.223	0.350	0.225	0.375	0.225	0.386
0.200	0.226	0.280	0.227	0.306	0.228	0.328	0.230	0.361	0.232	0.387	0.232	0.398
0.205	0.233	0.288	0.234	0.315	0.235	0.338	0.237	0.372	0.239	0.398	0.239	0.410
0.210	0.239	0.296	0.241	0.324	0.242	0.348	0.244	0.382	0.246	0.410	0.246	0.422
0.215	0.246	0.304	0.247	0.333	0.248	0.357	0.251	0.393	0.253	0.422	0.254	0.435
0.220	0.253	0.312	0.254	0.342	0.255	0.367	0.258	0.404	0.260	0.434	0.261	0.447
0.225	0.260	0.321	0.261	0.352	0.262	0.377	0.265	0.415	0.267	0.446	0.268	0.460
0.230	0.266	0.329	0.268	0.361	0.269	0.387	0.272	0.427	0.275	0.458	0.276	0.472
0.235	0.273	0.338	0.275	0.370	0.276	0.398	0.279	0.438	0.282	0.470	0.283	0.485
0.240	0.280	0.346	0.282	0.380	0.284	0.408	0.287	0.450	0.290	0.483	0.291	0.499
0.245	0.287	0.355	0.289	0.390	0.291	0.419	0.294	0.461	0.297	0.496	0.299	0.512
0.250	0.295	0.364	0.296	0.399	0.298	0.429	0.302	0.473	0.305	0.509	0.306	0.525

Note: the minimum reinforcement mechanical percentage in beams is the greater of $\varpi_{min} = 0.226 f_{ctm}/f_{cd}$ and
$\varpi_{min} = 0.226 f_{yd}/f_{cd}$ (section 9.2.1.1 of EC2).

TABLE 1_Concrete classes C12-C90_Steel class S500 (cont.)
Simply reinforced sections under bending moment [1]

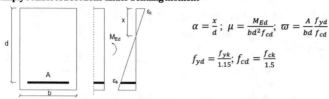

$$\alpha = \frac{x}{d}\ ;\ \mu = \frac{M_{Ed}}{bd^2 f_{cd}};\ \varpi = \frac{A\,f_{yd}}{bd\,f_{cd}}$$

$$f_{yd} = \frac{f_{yk}}{1.15};\ f_{cd} = \frac{f_{ck}}{1.5}$$

μ	C12-C50 ϖ	α	C55 ϖ	α	C60 ϖ	α	C70 ϖ	α	C80 ϖ	α	C90 ϖ	α
0.255	0.302	0.373	0.304	0.409	0.306	0.440	0.309	0.485	0.313	0.522	0.315	0.539
0.260	0.309	0.382	0.311	0.419	0.313	0.451	0.317	0.498	0.321	0.535	0.334	0.553
0.265	0.316	0.391	0.319	0.429	0.321	0.462	0.325	0.510	0.335	0.549	0.363	0.567
0.270	0.324	0.400	0.326	0.440	0.329	0.473	0.333	0.522	0.363	0.563	0.394	0.582
0.275	0.331	0.409	0.334	0.450	0.336	0.484	0.341	0.535	0.394	0.577	0.430	0.596
0.280	0.339	0.419	0.342	0.460	0.344	0.495	0.349	0.548	0.428	0.591	0.469	0.611
0.285	0.347	0.428	0.349	0.471	0.352	0.507	0.368	0.561	0.465	0.605	0.512	0.626
0.290	0.355	0.438	0.357	0.482	0.360	0.519	0.398	0.574	0.508	0.620	0.561	0.642
0.295	0.362	0.448	0.366	0.493	0.369	0.531	0.431	0.588	0.554	0.635	0.617	0.658
0.300	0.370	0.458	0.374	0.504	0.377	0.543	0.467	0.602	0.607	0.651	0.680	0.674
0.305	0.379	0.468	0.382	0.515	0.386	0.555	0.507	0.616	0.667	0.666	0.751	0.690
0.310	0.387	0.478	0.391	0.526	0.394	0.567	0.551	0.630	0.734	0.682	0.834	0.707
0.315	0.395	0.488	0.399	0.538	0.418	0.580	0.600	0.645	0.812	0.699	0.931	0.725
0.320	0.404	0.499	0.408	0.550	0.450	0.593	0.656	0.660	0.901	0.715	1.044	0.742
0.325	0.412	0.509	0.417	0.562	0.486	0.606	0.718	0.675	1.005	0.732	1.180	0.761
0.330	0.421	0.520	0.426	0.574	0.526	0.620	0.789	0.690	1.128	0.750	1.344	0.779
0.335	0.430	0.531	0.435	0.586	0.569	0.633	0.871	0.706	1.276	0.768	1.547	0.799
0.340	0.439	0.542	0.465	0.599	0.618	0.647	0.964	0.722	1.455	0.787	1.804	0.819
0.345	0.448	0.554	0.501	0.611	0.673	0.661	1.074	0.739	1.678	0.806	2.143	0.840
0.350	0.458	0.565	0.540	0.624	0.734	0.676	1.202	0.756	1.962	0.826	2.601	0.861
0.355	0.467	0.577	0.585	0.638	0.803	0.691	1.357	0.774	2.335	0.846	3.260	0.883
0.360	0.477	0.589	0.634	0.651	0.883	0.706	1.545	0.792	2.849	0.868	4.287	0.907
0.365	0.487	0.601	0.688	0.665	0.973	0.721	1.779	0.811	3.598	0.890	6.111	0.931
0.370	0.497	0.613	0.750	0.680	1.079	0.737	2.075	0.830	4.791	0.913		
0.375	0.528	0.626	0.820	0.694	1.203	0.754	2.468	0.850	7.006	0.937		
0.380	0.570	0.639	0.900	0.709	1.350	0.771	3.006	0.871				
0.385	0.617	0.653	0.992	0.725	1.529	0.788	3.795	0.892				
0.390	0.670	0.666	1.099	0.740	1.748	0.806	5.056	0.915				
0.395	0.729	0.680	1.224	0.757	2.026	0.825	7.391	0.939				
0.400	0.796	0.695	1.375	0.774	2.389	0.844						
0.405	0.873	0.710	1.556	0.791	2.879	0.865						
0.410	0.961	0.725	1.783	0.809	3.583	0.886						
0.415	1.065	0.741	2.069	0.828	4.678	0.908						
0.420	1.188	0.757	2.448	0.847	6.618	0.932						
0.425	1.335	0.774	2.970	0.868								
0.430	1.516	0.792	3.730	0.890								
0.435	1.743	0.811	4.955	0.913								
0.440	2.036	0.830	7.255	0.937								
0.445	2.431	0.850										
0.450	2.995	0.872										
0.455	3.865	0.896										
0.460	5.384	0.921										
0.465	8.800	0.949										

[1] **Note:** The table is valid for S400, S500 and S600 steel up to the continuous line (upper part of the table). From this line on the procedure is the following:
S400 use the mechanical percentages in the table between the continuous and the dashed lines. From the dashed line on the mechanical percentage is obtained by multiplying the value obtained in the table by a factor $f_{yk}/500=400/500=0.8$;
S500 the table is valid for all values;
S600 From the continuous line on the mechanical percentage is obtained by multiplying the value obtained in the table by a factor $f_{yk}/500=600/500=1.2$.

5.2.2 Doubly Reinforced Sections—TABLE 2

Tab. 5.2

TABLE 2_Concrete classes C12-C50_Steel class S500_a/d=0.10

Doubly reinforced sections under bending moment

$$\alpha = \frac{x}{d}; \quad \mu = \frac{M_{Ed}}{bd^2 f_{cd}}; \quad \varpi = \frac{A}{bd}\frac{f_{yd}}{f_{cd}}$$

$$f_{yd} = \frac{f_{yk}}{1.15}; \quad f_{cd} = \frac{f_{ck}}{1.5}; \quad \frac{a}{d} = 0.10$$

μ	$A'/A=0.0$		$A'/A=0.2$		$A'/A=0.3$		$A'/A=0.4$		$A'/A=0.5$		$A'/A=1.0$	
	ϖ	α	ϖ	α	ϖ	α	ϖ	α	ϖ	α	ϖ	α
0.005	0.005	0.021	0.005	0.022	0.005	0.023	0.005	0.024	0.005	0.025	0.004	0.027
0.010	0.010	0.030	0.010	0.032	0.010	0.033	0.010	0.034	0.010	0.035	0.009	0.039
0.015	0.015	0.037	0.015	0.040	0.015	0.041	0.015	0.042	0.015	0.043	0.014	0.047
0.020	0.020	0.044	0.020	0.047	0.020	0.048	0.020	0.049	0.020	0.050	0.019	0.055
0.025	0.026	0.050	0.025	0.053	0.025	0.054	0.025	0.055	0.025	0.057	0.024	0.061
0.030	0.031	0.055	0.030	0.058	0.030	0.060	0.030	0.061	0.030	0.062	0.030	0.067
0.035	0.036	0.061	0.036	0.064	0.036	0.065	0.035	0.066	0.036	0.068	0.035	0.072
0.040	0.041	0.066	0.041	0.069	0.041	0.070	0.041	0.071	0.041	0.072	0.040	0.077
0.045	0.046	0.071	0.046	0.074	0.046	0.075	0.046	0.076	0.046	0.077	0.045	0.081
0.050	0.052	0.076	0.051	0.079	0.051	0.080	0.051	0.081	0.051	0.082	0.051	0.085
0.055	0.057	0.081	0.057	0.083	0.057	0.085	0.057	0.085	0.057	0.086	0.056	0.089
0.060	0.062	0.087	0.062	0.088	0.062	0.089	0.062	0.089	0.062	0.090	0.062	0.092
0.065	0.067	0.092	0.067	0.093	0.067	0.093	0.067	0.094	0.068	0.094	0.067	0.095
0.070	0.073	0.097	0.073	0.097	0.073	0.098	0.073	0.098	0.073	0.098	0.073	0.098
0.075	0.078	0.102	0.078	0.102	0.079	0.102	0.078	0.102	0.078	0.102	0.078	0.101
0.080	0.084	0.107	0.084	0.106	0.084	0.106	0.084	0.105	0.084	0.105	0.084	0.104
0.085	0.089	0.113	0.089	0.110	0.089	0.110	0.089	0.109	0.089	0.108	0.089	0.106
0.090	0.095	0.118	0.095	0.115	0.095	0.113	0.095	0.112	0.095	0.111	0.095	0.108
0.095	0.100	0.124	0.100	0.119	0.100	0.117	0.100	0.116	0.100	0.115	0.100	0.110
0.100	0.106	0.131	0.106	0.123	0.106	0.121	0.106	0.119	0.106	0.118	0.106	0.112
0.105	0.111	0.137	0.111	0.128	0.111	0.124	0.111	0.122	0.112	0.120	0.111	0.114
0.110	0.117	0.144	0.117	0.133	0.117	0.129	0.117	0.126	0.117	0.123	0.117	0.116
0.115	0.123	0.151	0.122	0.138	0.123	0.133	0.123	0.129	0.122	0.126	0.122	0.118
0.120	0.128	0.159	0.128	0.143	0.128	0.137	0.128	0.133	0.128	0.129	0.128	0.119
0.125	0.134	0.166	0.134	0.148	0.134	0.142	0.134	0.137	0.133	0.132	0.133	0.121
0.130	0.140	0.173	0.139	0.153	0.139	0.146	0.139	0.140	0.139	0.136	0.139	0.122
0.135	0.146	0.180	0.145	0.158	0.145	0.150	0.145	0.144	0.145	0.139	0.144	0.124
0.140	0.152	0.188	0.151	0.163	0.151	0.154	0.151	0.148	0.150	0.142	0.150	0.125
0.145	0.158	0.195	0.157	0.168	0.156	0.159	0.156	0.151	0.156	0.145	0.155	0.127
0.150	0.164	0.202	0.162	0.173	0.162	0.163	0.162	0.154	0.162	0.148	0.161	0.128
0.155	0.170	0.210	0.168	0.178	0.168	0.167	0.167	0.158	0.168	0.151	0.167	0.130
0.160	0.176	0.217	0.174	0.183	0.173	0.171	0.173	0.162	0.173	0.154	0.172	0.131
0.165	0.182	0.225	0.180	0.188	0.179	0.175	0.179	0.165	0.179	0.157	0.178	0.133
0.170	0.188	0.232	0.185	0.193	0.185	0.179	0.185	0.168	0.184	0.159	0.183	0.134
0.175	0.194	0.240	0.191	0.199	0.190	0.184	0.190	0.172	0.190	0.162	0.189	0.135
0.180	0.201	0.248	0.197	0.204	0.197	0.188	0.196	0.175	0.195	0.165	0.194	0.137
0.185	0.207	0.256	0.203	0.209	0.202	0.192	0.202	0.179	0.201	0.168	0.200	0.138
0.190	0.213	0.264	0.209	0.214	0.208	0.196	0.207	0.182	0.207	0.170	0.206	0.139
0.195	0.220	0.271	0.215	0.219	0.214	0.200	0.213	0.185	0.212	0.173	0.211	0.140
0.200	0.226	0.280	0.221	0.224	0.220	0.204	0.219	0.189	0.218	0.176	0.217	0.141
0.205	0.233	0.288	0.227	0.229	0.226	0.209	0.224	0.192	0.224	0.179	0.222	0.143
0.210	0.239	0.296	0.233	0.234	0.231	0.213	0.230	0.195	0.230	0.181	0.228	0.144
0.215	0.246	0.304	0.239	0.240	0.237	0.217	0.236	0.198	0.235	0.184	0.233	0.145
0.220	0.253	0.312	0.245	0.245	0.243	0.221	0.242	0.202	0.241	0.187	0.239	0.146
0.225	0.260	0.321	0.251	0.250	0.249	0.225	0.248	0.205	0.246	0.189	0.245	0.147
0.230	0.266	0.329	0.257	0.255	0.255	0.229	0.253	0.208	0.252	0.192	0.250	0.148
0.235	0.273	0.338	0.263	0.260	0.260	0.233	0.259	0.211	0.258	0.194	0.256	0.149
0.240	0.280	0.346	0.269	0.266	0.266	0.237	0.265	0.215	0.264	0.197	0.261	0.150
0.245	0.287	0.355	0.275	0.272	0.272	0.241	0.271	0.218	0.269	0.199	0.267	0.151
0.250	0.295	0.364	0.282	0.278	0.278	0.245	0.276	0.221	0.275	0.202	0.273	0.152

TABLE 2_Concrete classes C12-C50_Steel class S500_a/d=0.10 (cont.)
Doubly reinforced sections under bending moment

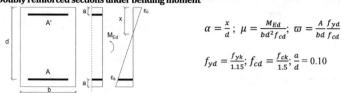

$$\alpha = \frac{x}{d}; \quad \mu = \frac{M_{Ed}}{bd^2 f_{cd}}; \quad \varpi = \frac{A\, f_{yd}}{bd\, f_{cd}}$$

$$f_{yd} = \frac{f_{yk}}{1.15}; \quad f_{cd} = \frac{f_{ck}}{1.5}; \quad \frac{a}{d} = 0.10$$

	$A'/A=0.0$		$A'/A=0.2$		$A'/A=0.3$		$A'/A=0.4$		$A'/A=0.5$		$A'/A=1.0$	
μ	ϖ	α	ϖ	α	ϖ	α	ϖ	α	ϖ	α	ϖ	α
0.255	0.302	0.373	0.288	0.285	0.284	0.250	0.282	0.224	0.281	0.204	0.278	0.153
0.260	0.309	0.382	0.294	0.291	0.290	0.254	0.288	0.227	0.287	0.207	0.284	0.154
0.265	0.316	0.391	0.301	0.297	0.296	0.258	0.294	0.231	0.293	0.209	0.289	0.155
0.270	0.324	0.400	0.307	0.303	0.302	0.262	0.299	0.234	0.298	0.212	0.295	0.155
0.275	0.331	0.409	0.314	0.310	0.308	0.267	0.305	0.237	0.303	0.214	0.301	0.156
0.280	0.339	0.419	0.320	0.316	0.314	0.272	0.311	0.240	0.309	0.217	0.306	0.157
0.285	0.347	0.428	0.327	0.323	0.320	0.277	0.317	0.243	0.315	0.219	0.312	0.158
0.290	0.355	0.438	0.333	0.329	0.327	0.283	0.323	0.246	0.321	0.222	0.317	0.159
0.295	0.362	0.448	0.340	0.336	0.333	0.288	0.329	0.250	0.326	0.224	0.323	0.159
0.300	0.370	0.458	0.346	0.342	0.339	0.293	0.335	0.253	0.332	0.226	0.328	0.160
0.310	0.387	0.478	0.360	0.355	0.352	0.304	0.347	0.259	0.344	0.231	0.340	0.162
0.320	0.404	0.499	0.373	0.369	0.364	0.315	0.358	0.266	0.355	0.236	0.351	0.163
0.330	0.421	0.520	0.387	0.382	0.377	0.326	0.370	0.274	0.367	0.240	0.362	0.165
0.340	0.439	0.542	0.401	0.396	0.390	0.337	0.382	0.283	0.379	0.245	0.373	0.166
0.350	0.458	0.565	0.415	0.410	0.403	0.349	0.395	0.292	0.390	0.250	0.384	0.167
0.360	0.477	0.589	0.429	0.424	0.416	0.360	0.407	0.302	0.402	0.254	0.396	0.169
0.370	0.497	0.613	0.444	0.438	0.429	0.371	0.419	0.311	0.413	0.258	0.406	0.170
0.380	0.570	0.639	0.458	0.453	0.443	0.383	0.432	0.320	0.425	0.263	0.418	0.171
0.390	0.670	0.666	0.473	0.468	0.456	0.395	0.444	0.329	0.437	0.270	0.429	0.172
0.400	0.796	0.695	0.488	0.482	0.470	0.406	0.457	0.339	0.449	0.277	0.440	0.173
0.410	0.961	0.725	0.503	0.497	0.483	0.418	0.470	0.348	0.460	0.284	0.451	0.174
0.420	1.188	0.757	0.519	0.513	0.497	0.430	0.483	0.358	0.473	0.292	0.462	0.176
0.430	1.516	0.792	0.535	0.528	0.511	0.442	0.495	0.367	0.485	0.299	0.474	0.177
0.440	2.036	0.830	0.551	0.544	0.525	0.454	0.508	0.377	0.497	0.307	0.485	0.178
0.450	2.995	0.872	0.567	0.560	0.540	0.467	0.521	0.386	0.509	0.314	0.496	0.179
0.460	5.384	0.921	0.584	0.577	0.554	0.479	0.534	0.396	0.521	0.322	0.507	0.180
0.470			0.601	0.594	0.568	0.492	0.547	0.406	0.533	0.329	0.518	0.181
0.480			0.618	0.611	0.583	0.504	0.561	0.416	0.546	0.337	0.529	0.181
0.490			0.648	0.622	0.598	0.517	0.574	0.425	0.558	0.345	0.540	0.182
0.500			0.686	0.630	0.613	0.530	0.588	0.435	0.571	0.352	0.552	0.183
0.510			0.725	0.638	0.628	0.543	0.601	0.445	0.583	0.360	0.563	0.184
0.520			0.765	0.646	0.644	0.556	0.615	0.456	0.596	0.368	0.574	0.185
0.530			0.805	0.653	0.659	0.570	0.628	0.466	0.608	0.376	0.585	0.186
0.540			0.847	0.660	0.675	0.584	0.642	0.476	0.620	0.383	0.596	0.187
0.550			0.890	0.666	0.691	0.597	0.656	0.486	0.633	0.391	0.607	0.187
0.560			0.933	0.673	0.707	0.611	0.670	0.497	0.646	0.399	0.618	0.188
0.570			0.976	0.679	0.732	0.620	0.684	0.507	0.659	0.407	0.630	0.189
0.580			1.021	0.684	0.762	0.626	0.698	0.517	0.672	0.415	0.641	0.190
0.590			1.066	0.690	0.791	0.630	0.713	0.528	0.685	0.423	0.652	0.190
0.600			1.111	0.695	0.823	0.635	0.727	0.539	0.697	0.431	0.663	0.191
0.610			1.158	0.700	0.852	0.640	0.742	0.550	0.710	0.439	0.674	0.192
0.620			1.204	0.705	0.883	0.644	0.756	0.560	0.724	0.447	0.686	0.192
0.630			1.250	0.710	0.915	0.648	0.771	0.571	0.737	0.455	0.697	0.193
0.640			1.299	0.714	0.947	0.652	0.786	0.582	0.750	0.463	0.708	0.194
0.650			1.346	0.718	0.978	0.656	0.801	0.593	0.763	0.471	0.719	0.194
0.660			1.395	0.722	1.010	0.660	0.816	0.605	0.776	0.479	0.730	0.195
0.670			1.443	0.726	1.042	0.664	0.831	0.616	0.790	0.488	0.741	0.196
0.680			1.492	0.730	1.074	0.667	0.855	0.620	0.803	0.496	0.752	0.196
0.690			1.541	0.733	1.106	0.671	0.880	0.623	0.817	0.505	0.764	0.197
0.700			1.591	0.737	1.138	0.674	0.904	0.626	0.830	0.513	0.774	0.197

Tab. 5.3

TABLE 2_Concrete classes C12-C50_Steel class S500_a/d=0.15

Doubly reinforced sections under bending moment

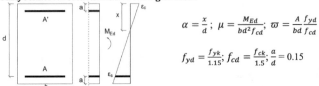

$$\alpha = \frac{x}{d}\;;\;\; \mu = \frac{M_{Ed}}{bd^2 f_{cd}};\;\; \varpi = \frac{A\,f_{yd}}{bd\,f_{cd}}$$

$$f_{yd} = \frac{f_{yk}}{1.15};\; f_{cd} = \frac{f_{ck}}{1.5};\; \frac{a}{d} = 0.15$$

	$A'/A=0.0$		$A'/A=0.2$		$A'/A=0.3$		$A'/A=0.4$		$A'/A=0.5$		$A'/A=1.0$	
μ	ϖ	α	ϖ	α	ϖ	α	ϖ	α	ϖ	α	ϖ	α
0.005	0.005	0.021	0.005	0.022	0.005	0.023	0.005	0.024	0.005	0.025	0.004	0.027
0.010	0.010	0.030	0.010	0.033	0.010	0.034	0.010	0.035	0.009	0.036	0.009	0.040
0.015	0.015	0.037	0.015	0.041	0.015	0.042	0.015	0.044	0.014	0.045	0.013	0.051
0.020	0.020	0.044	0.020	0.048	0.020	0.050	0.019	0.052	0.019	0.053	0.018	0.061
0.025	0.025	0.050	0.025	0.055	0.025	0.057	0.024	0.059	0.024	0.061	0.023	0.070
0.030	0.031	0.055	0.030	0.061	0.030	0.063	0.029	0.066	0.029	0.068	0.028	0.077
0.035	0.036	0.061	0.035	0.067	0.035	0.070	0.034	0.072	0.034	0.075	0.033	0.084
0.040	0.041	0.066	0.040	0.073	0.040	0.075	0.040	0.078	0.039	0.081	0.038	0.091
0.045	0.046	0.071	0.045	0.078	0.045	0.081	0.045	0.084	0.045	0.087	0.043	0.097
0.050	0.051	0.076	0.051	0.084	0.050	0.087	0.050	0.090	0.050	0.092	0.049	0.102
0.055	0.057	0.081	0.056	0.089	0.056	0.092	0.055	0.095	0.055	0.098	0.054	0.107
0.060	0.062	0.086	0.061	0.094	0.061	0.097	0.061	0.100	0.061	0.103	0.060	0.112
0.065	0.067	0.092	0.067	0.099	0.066	0.102	0.066	0.105	0.066	0.108	0.065	0.117
0.070	0.073	0.097	0.072	0.104	0.072	0.108	0.072	0.110	0.072	0.113	0.071	0.121
0.075	0.078	0.102	0.078	0.110	0.078	0.113	0.077	0.115	0.077	0.117	0.076	0.125
0.080	0.084	0.107	0.083	0.114	0.083	0.117	0.083	0.119	0.083	0.121	0.082	0.129
0.085	0.089	0.113	0.089	0.119	0.089	0.122	0.088	0.124	0.089	0.126	0.088	0.132
0.090	0.095	0.118	0.094	0.124	0.094	0.127	0.094	0.129	0.094	0.130	0.094	0.135
0.095	0.100	0.124	0.100	0.130	0.100	0.132	0.100	0.133	0.100	0.135	0.099	0.139
0.100	0.106	0.130	0.106	0.135	0.106	0.137	0.106	0.138	0.105	0.139	0.105	0.142
0.105	0.111	0.138	0.111	0.140	0.111	0.142	0.112	0.142	0.111	0.143	0.111	0.145
0.110	0.117	0.144	0.117	0.146	0.117	0.146	0.117	0.147	0.117	0.147	0.117	0.148
0.115	0.123	0.151	0.123	0.151	0.123	0.151	0.123	0.151	0.123	0.151	0.122	0.151
0.120	0.128	0.159	0.128	0.157	0.129	0.156	0.129	0.155	0.129	0.155	0.128	0.153
0.125	0.134	0.166	0.134	0.162	0.135	0.161	0.135	0.160	0.135	0.159	0.134	0.156
0.130	0.140	0.173	0.140	0.167	0.140	0.165	0.140	0.164	0.140	0.163	0.140	0.158
0.135	0.146	0.180	0.146	0.173	0.146	0.170	0.146	0.168	0.146	0.166	0.146	0.161
0.140	0.152	0.188	0.152	0.178	0.152	0.175	0.152	0.172	0.152	0.170	0.152	0.163
0.145	0.158	0.195	0.158	0.183	0.158	0.180	0.158	0.176	0.158	0.174	0.158	0.165
0.150	0.164	0.202	0.164	0.189	0.164	0.184	0.164	0.180	0.164	0.177	0.163	0.168
0.155	0.170	0.210	0.169	0.194	0.170	0.189	0.170	0.184	0.169	0.181	0.169	0.170
0.160	0.176	0.217	0.176	0.199	0.175	0.193	0.176	0.188	0.176	0.184	0.175	0.172
0.165	0.182	0.225	0.181	0.205	0.181	0.198	0.182	0.193	0.182	0.188	0.181	0.174
0.170	0.188	0.232	0.187	0.210	0.187	0.203	0.187	0.196	0.187	0.191	0.187	0.176
0.175	0.194	0.240	0.193	0.216	0.193	0.207	0.193	0.200	0.193	0.195	0.193	0.178
0.180	0.201	0.248	0.199	0.221	0.199	0.212	0.199	0.204	0.199	0.198	0.199	0.180
0.185	0.207	0.256	0.206	0.226	0.205	0.216	0.205	0.208	0.205	0.201	0.205	0.182
0.190	0.213	0.263	0.212	0.232	0.211	0.221	0.211	0.212	0.211	0.205	0.210	0.184
0.195	0.220	0.272	0.218	0.237	0.217	0.225	0.217	0.216	0.217	0.208	0.216	0.185
0.200	0.226	0.279	0.224	0.243	0.224	0.230	0.223	0.220	0.223	0.211	0.222	0.187
0.205	0.233	0.288	0.230	0.248	0.229	0.234	0.229	0.223	0.229	0.214	0.228	0.189
0.210	0.239	0.296	0.236	0.254	0.236	0.239	0.235	0.227	0.235	0.217	0.234	0.190
0.215	0.246	0.304	0.243	0.259	0.242	0.243	0.241	0.231	0.241	0.221	0.240	0.192
0.220	0.253	0.312	0.249	0.264	0.248	0.248	0.247	0.234	0.247	0.224	0.246	0.194
0.225	0.260	0.321	0.255	0.270	0.254	0.252	0.253	0.238	0.253	0.227	0.252	0.195
0.230	0.266	0.329	0.261	0.275	0.260	0.257	0.259	0.242	0.259	0.230	0.258	0.197
0.235	0.273	0.338	0.267	0.281	0.266	0.261	0.265	0.246	0.265	0.233	0.263	0.198
0.240	0.280	0.346	0.274	0.286	0.272	0.266	0.272	0.249	0.271	0.236	0.269	0.200
0.245	0.287	0.355	0.280	0.292	0.279	0.270	0.278	0.253	0.277	0.239	0.275	0.201
0.250	0.295	0.364	0.287	0.298	0.285	0.275	0.284	0.257	0.283	0.242	0.281	0.202

TABLE 2_Concrete classes C12-C50_Steel class S500_a/d=0.15 (cont.)
Doubly reinforced sections under bending moment

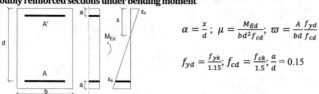

$$\alpha = \frac{x}{d}; \quad \mu = \frac{M_{Ed}}{bd^2 f_{cd}}; \quad \varpi = \frac{A\,f_{yd}}{bd\,f_{cd}}$$

$$f_{yd} = \frac{f_{yk}}{1.15}; \quad f_{cd} = \frac{f_{ck}}{1.5}; \quad \frac{a}{d} = 0.15$$

	$A'/A=0.0$		$A'/A=0.2$		$A'/A=0.3$		$A'/A=0.4$		$A'/A=0.5$		$A'/A=1.0$	
μ	ϖ	α	ϖ	α	ϖ	α	ϖ	α	ϖ	α	ϖ	α
0.255	0.302	0.373	0.293	0.303	0.291	0.279	0.290	0.260	0.289	0.245	0.287	0.204
0.260	0.309	0.382	0.300	0.309	0.297	0.284	0.296	0.264	0.295	0.248	0.293	0.205
0.265	0.316	0.391	0.306	0.314	0.304	0.288	0.302	0.268	0.301	0.251	0.299	0.207
0.270	0.324	0.400	0.313	0.320	0.310	0.293	0.308	0.271	0.307	0.254	0.305	0.208
0.275	0.331	0.409	0.319	0.326	0.316	0.297	0.314	0.275	0.313	0.257	0.311	0.209
0.280	0.339	0.419	0.326	0.331	0.323	0.302	0.321	0.278	0.319	0.260	0.317	0.210
0.285	0.347	0.428	0.332	0.337	0.329	0.306	0.327	0.282	0.325	0.263	0.323	0.212
0.290	0.355	0.438	0.339	0.343	0.335	0.311	0.333	0.286	0.331	0.266	0.329	0.213
0.295	0.363	0.448	0.345	0.348	0.341	0.315	0.339	0.289	0.337	0.269	0.334	0.214
0.300	0.371	0.458	0.352	0.354	0.348	0.320	0.345	0.293	0.344	0.272	0.340	0.215
0.310	0.387	0.478	0.365	0.366	0.361	0.328	0.358	0.300	0.356	0.277	0.352	0.218
0.320	0.404	0.499	0.379	0.377	0.374	0.338	0.370	0.307	0.368	0.283	0.364	0.220
0.330	0.421	0.520	0.393	0.389	0.386	0.347	0.383	0.314	0.380	0.288	0.376	0.222
0.340	0.439	0.542	0.407	0.402	0.399	0.356	0.395	0.321	0.392	0.294	0.388	0.224
0.350	0.458	0.565	0.421	0.416	0.413	0.365	0.407	0.328	0.405	0.299	0.400	0.226
0.360	0.477	0.589	0.435	0.430	0.425	0.374	0.420	0.335	0.417	0.305	0.411	0.228
0.370	0.497	0.613	0.450	0.445	0.439	0.383	0.433	0.342	0.429	0.310	0.423	0.230
0.380	0.570	0.639	0.465	0.460	0.452	0.392	0.446	0.349	0.441	0.315	0.435	0.232
0.390	0.670	0.666	0.480	0.475	0.466	0.403	0.458	0.356	0.454	0.321	0.447	0.234
0.400	0.796	0.695	0.496	0.490	0.480	0.415	0.471	0.362	0.466	0.326	0.459	0.236
0.410	0.961	0.725	0.511	0.505	0.494	0.427	0.484	0.369	0.479	0.331	0.471	0.237
0.420	1.188	0.757	0.527	0.521	0.508	0.439	0.496	0.376	0.491	0.336	0.482	0.239
0.430	1.516	0.792	0.543	0.537	0.522	0.452	0.509	0.383	0.504	0.341	0.494	0.241
0.440	2.036	0.830	0.560	0.553	0.537	0.464	0.522	0.390	0.516	0.347	0.506	0.242
0.450	2.995	0.872	0.577	0.570	0.551	0.477	0.535	0.397	0.528	0.351	0.518	0.244
0.460	5.384	0.921	0.594	0.587	0.566	0.490	0.549	0.407	0.541	0.357	0.530	0.245
0.470			0.611	0.604	0.581	0.503	0.562	0.417	0.553	0.362	0.542	0.247
0.480			0.635	0.619	0.596	0.516	0.576	0.427	0.566	0.367	0.554	0.248
0.490			0.673	0.628	0.612	0.529	0.590	0.437	0.579	0.372	0.565	0.250
0.500			0.714	0.636	0.627	0.542	0.604	0.448	0.591	0.377	0.577	0.251
0.510			0.755	0.644	0.643	0.556	0.618	0.458	0.603	0.381	0.589	0.252
0.520			0.797	0.651	0.659	0.570	0.632	0.468	0.616	0.386	0.601	0.254
0.530			0.840	0.659	0.675	0.584	0.646	0.479	0.629	0.391	0.613	0.255
0.540			0.885	0.666	0.691	0.598	0.660	0.489	0.641	0.396	0.625	0.256
0.550			0.930	0.672	0.708	0.612	0.675	0.500	0.654	0.404	0.636	0.258
0.560			0.976	0.679	0.736	0.621	0.689	0.511	0.667	0.412	0.648	0.259
0.570			1.022	0.685	0.766	0.626	0.704	0.522	0.681	0.420	0.660	0.260
0.580			1.068	0.690	0.797	0.631	0.719	0.533	0.694	0.429	0.672	0.261
0.590			1.117	0.696	0.829	0.636	0.734	0.544	0.707	0.437	0.684	0.262
0.600			1.165	0.701	0.862	0.641	0.749	0.555	0.721	0.445	0.696	0.264
0.610			1.214	0.706	0.895	0.646	0.764	0.566	0.734	0.454	0.708	0.265
0.620			1.263	0.711	0.927	0.650	0.779	0.577	0.748	0.462	0.719	0.266
0.630			1.313	0.715	0.961	0.654	0.795	0.589	0.762	0.471	0.731	0.267
0.640			1.365	0.720	0.994	0.658	0.810	0.600	0.776	0.479	0.743	0.268
0.650			1.415	0.724	1.026	0.662	0.826	0.612	0.789	0.487	0.755	0.269
0.660			1.467	0.728	1.062	0.666	0.849	0.619	0.803	0.496	0.767	0.270
0.670			1.519	0.732	1.095	0.669	0.873	0.622	0.817	0.505	0.778	0.271
0.680			1.570	0.735	1.128	0.673	0.900	0.626	0.831	0.513	0.790	0.272
0.690			1.623	0.739	1.164	0.676	0.925	0.629	0.845	0.522	0.802	0.273
0.700			1.674	0.742	1.200	0.680	0.950	0.632	0.860	0.531	0.814	0.274

Tab. 5.4

TABLE 2_Concrete classes C12-C50_Steel class S400_a/d=0.10

Doubly Reinforced Sections under Bending Moment

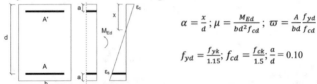

$$\alpha = \frac{x}{d}; \mu = \frac{M_{Ed}}{bd^2 f_{cd}}; \varpi = \frac{A}{bd}\frac{f_{yd}}{f_{cd}}$$

$$f_{yd} = \frac{f_{yk}}{1.15}; f_{cd} = \frac{f_{ck}}{1.5}; \frac{a}{d} = 0.10$$

	A'/A=0.0		A'/A=0.2		A'/A=0.3		A'/A=0.4		A'/A=0.5		A'/A=1.0	
μ	ϖ	α	ϖ	α	ϖ	α	ϖ	α	ϖ	α	ϖ	α
0.005	0.005	0.021	0.005	0.023	0.005	0.024	0.005	0.024	0.005	0.025	0.004	0.028
0.010	0.010	0.030	0.010	0.033	0.010	0.034	0.010	0.035	0.010	0.036	0.009	0.040
0.015	0.015	0.037	0.015	0.041	0.015	0.042	0.015	0.043	0.015	0.044	0.014	0.049
0.020	0.020	0.044	0.020	0.047	0.020	0.049	0.020	0.050	0.020	0.052	0.019	0.057
0.025	0.026	0.050	0.025	0.053	0.025	0.055	0.025	0.056	0.025	0.058	0.024	0.063
0.030	0.031	0.055	0.030	0.059	0.030	0.061	0.030	0.062	0.030	0.063	0.029	0.069
0.035	0.036	0.061	0.036	0.064	0.035	0.066	0.036	0.067	0.035	0.069	0.035	0.074
0.040	0.041	0.066	0.041	0.069	0.041	0.071	0.041	0.072	0.041	0.074	0.040	0.078
0.045	0.046	0.071	0.046	0.074	0.046	0.076	0.046	0.077	0.046	0.078	0.045	0.082
0.050	0.052	0.076	0.051	0.079	0.051	0.080	0.051	0.082	0.051	0.083	0.051	0.086
0.055	0.057	0.081	0.057	0.084	0.057	0.085	0.057	0.086	0.057	0.087	0.056	0.090
0.060	0.062	0.087	0.062	0.089	0.062	0.090	0.062	0.090	0.062	0.091	0.062	0.093
0.065	0.067	0.092	0.067	0.093	0.067	0.094	0.068	0.094	0.067	0.094	0.067	0.096
0.070	0.073	0.097	0.073	0.098	0.073	0.098	0.073	0.098	0.073	0.098	0.073	0.098
0.075	0.078	0.102	0.078	0.102	0.078	0.102	0.078	0.102	0.078	0.101	0.078	0.101
0.080	0.084	0.107	0.084	0.106	0.084	0.105	0.084	0.105	0.084	0.105	0.084	0.103
0.085	0.089	0.113	0.089	0.110	0.089	0.109	0.089	0.108	0.089	0.108	0.089	0.105
0.090	0.095	0.118	0.095	0.114	0.095	0.113	0.095	0.112	0.095	0.110	0.095	0.107
0.095	0.100	0.124	0.100	0.118	0.101	0.116	0.100	0.115	0.100	0.113	0.100	0.109
0.100	0.106	0.131	0.106	0.122	0.106	0.119	0.106	0.118	0.106	0.116	0.106	0.111
0.105	0.111	0.137	0.111	0.126	0.111	0.122	0.112	0.120	0.112	0.118	0.111	0.112
0.110	0.117	0.144	0.117	0.131	0.117	0.126	0.117	0.123	0.117	0.121	0.117	0.114
0.115	0.123	0.151	0.122	0.135	0.123	0.130	0.122	0.126	0.123	0.123	0.122	0.115
0.120	0.128	0.159	0.128	0.140	0.128	0.134	0.128	0.129	0.128	0.126	0.128	0.117
0.125	0.134	0.166	0.134	0.145	0.134	0.138	0.133	0.132	0.133	0.128	0.133	0.118
0.130	0.140	0.173	0.139	0.149	0.139	0.141	0.139	0.136	0.139	0.131	0.139	0.119
0.135	0.146	0.180	0.145	0.154	0.145	0.145	0.145	0.139	0.145	0.134	0.144	0.120
0.140	0.152	0.188	0.151	0.159	0.151	0.149	0.150	0.142	0.151	0.136	0.150	0.121
0.145	0.158	0.195	0.156	0.163	0.156	0.153	0.156	0.145	0.156	0.139	0.155	0.122
0.150	0.164	0.202	0.162	0.168	0.162	0.156	0.162	0.148	0.162	0.141	0.161	0.123
0.155	0.170	0.210	0.168	0.172	0.168	0.160	0.168	0.151	0.168	0.144	0.167	0.124
0.160	0.176	0.217	0.174	0.177	0.173	0.164	0.173	0.154	0.173	0.146	0.172	0.126
0.165	0.182	0.225	0.179	0.181	0.179	0.167	0.179	0.157	0.179	0.148	0.178	0.127
0.170	0.188	0.232	0.185	0.186	0.185	0.171	0.184	0.159	0.184	0.151	0.183	0.128
0.175	0.194	0.240	0.191	0.191	0.190	0.174	0.190	0.162	0.189	0.153	0.189	0.129
0.180	0.201	0.248	0.197	0.195	0.196	0.178	0.195	0.165	0.195	0.155	0.194	0.130
0.185	0.207	0.256	0.203	0.200	0.202	0.182	0.201	0.168	0.201	0.157	0.200	0.131
0.190	0.213	0.264	0.208	0.206	0.208	0.185	0.207	0.170	0.206	0.159	0.206	0.132
0.195	0.220	0.271	0.214	0.212	0.213	0.188	0.212	0.173	0.212	0.162	0.211	0.132
0.200	0.226	0.280	0.220	0.218	0.219	0.192	0.218	0.176	0.218	0.164	0.217	0.133
0.205	0.233	0.288	0.226	0.224	0.225	0.196	0.224	0.179	0.223	0.166	0.222	0.134
0.210	0.239	0.296	0.232	0.230	0.230	0.199	0.230	0.181	0.229	0.168	0.228	0.135
0.215	0.246	0.304	0.238	0.236	0.236	0.204	0.235	0.184	0.234	0.170	0.234	0.136
0.220	0.253	0.312	0.245	0.242	0.242	0.209	0.241	0.187	0.241	0.172	0.239	0.136
0.225	0.260	0.321	0.251	0.248	0.248	0.214	0.247	0.189	0.246	0.174	0.245	0.137
0.230	0.266	0.329	0.257	0.254	0.254	0.220	0.252	0.192	0.251	0.176	0.250	0.138
0.235	0.273	0.338	0.263	0.260	0.260	0.225	0.258	0.194	0.257	0.178	0.256	0.139
0.240	0.280	0.346	0.269	0.266	0.266	0.230	0.264	0.197	0.263	0.180	0.261	0.139
0.245	0.287	0.355	0.275	0.272	0.272	0.235	0.269	0.200	0.268	0.182	0.267	0.140
0.250	0.295	0.364	0.282	0.278	0.278	0.240	0.275	0.204	0.274	0.184	0.273	0.141

TABLE 2_Concrete classes C12-C50_Steel class S400_a/d=0.10 (cont.)
Doubly reinforced sections under bending moment

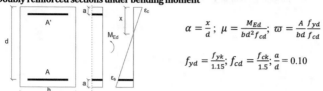

$$\alpha = \frac{x}{d}; \quad \mu = \frac{M_{Ed}}{bd^2 f_{cd}}; \quad \varpi = \frac{A\,f_{yd}}{bd\,f_{cd}}$$

$$f_{yd} = \frac{f_{yk}}{1.15}; \quad f_{cd} = \frac{f_{ck}}{1.5}; \quad \frac{a}{d} = 0.10$$

μ	A'/A=0.0 ϖ	α	A'/A=0.2 ϖ	α	A'/A=0.3 ϖ	α	A'/A=0.4 ϖ	α	A'/A=0.5 ϖ	α	A'/A=1.0 ϖ	α
0.255	0.302	0.373	0.288	0.285	0.284	0.245	0.281	0.208	0.280	0.186	0.278	0.141
0.260	0.309	0.382	0.294	0.291	0.290	0.251	0.287	0.212	0.286	0.187	0.284	0.142
0.265	0.316	0.391	0.301	0.297	0.296	0.256	0.293	0.217	0.291	0.189	0.289	0.143
0.270	0.324	0.400	0.307	0.303	0.302	0.261	0.299	0.221	0.297	0.191	0.295	0.143
0.275	0.331	0.409	0.314	0.310	0.308	0.267	0.304	0.226	0.302	0.193	0.300	0.144
0.280	0.339	0.419	0.320	0.316	0.314	0.272	0.310	0.230	0.308	0.195	0.306	0.144
0.285	0.347	0.428	0.327	0.323	0.320	0.277	0.316	0.234	0.314	0.196	0.312	0.145
0.290	0.355	0.438	0.333	0.329	0.327	0.283	0.322	0.239	0.320	0.198	0.317	0.145
0.295	0.362	0.448	0.340	0.336	0.333	0.288	0.328	0.243	0.325	0.201	0.323	0.146
0.300	0.370	0.458	0.346	0.342	0.339	0.293	0.334	0.248	0.331	0.204	0.328	0.147
0.310	0.387	0.478	0.360	0.355	0.352	0.304	0.346	0.256	0.342	0.211	0.339	0.148
0.320	0.404	0.499	0.373	0.369	0.364	0.315	0.358	0.266	0.354	0.219	0.351	0.149
0.330	0.421	0.520	0.387	0.382	0.377	0.326	0.370	0.274	0.366	0.226	0.362	0.149
0.340	0.439	0.542	0.401	0.396	0.390	0.337	0.382	0.283	0.377	0.233	0.373	0.150
0.350	0.458	0.565	0.415	0.410	0.403	0.349	0.395	0.292	0.389	0.240	0.384	0.151
0.360	0.477	0.589	0.429	0.424	0.416	0.360	0.407	0.302	0.401	0.248	0.395	0.152
0.370	0.497	0.613	0.444	0.438	0.429	0.371	0.419	0.311	0.413	0.255	0.406	0.153
0.380	0.518	0.639	0.458	0.453	0.443	0.383	0.432	0.320	0.425	0.262	0.417	0.154
0.390	0.540	0.666	0.473	0.468	0.456	0.395	0.444	0.329	0.437	0.270	0.428	0.154
0.400	0.637	0.695	0.488	0.482	0.470	0.406	0.457	0.339	0.449	0.277	0.440	0.155
0.410	0.769	0.725	0.503	0.497	0.483	0.418	0.470	0.348	0.460	0.284	0.451	0.156
0.420	0.951	0.757	0.519	0.513	0.497	0.430	0.483	0.358	0.473	0.292	0.462	0.156
0.430	1.213	0.792	0.535	0.528	0.511	0.442	0.495	0.367	0.485	0.299	0.473	0.157
0.440	1.629	0.830	0.551	0.544	0.525	0.454	0.508	0.377	0.497	0.307	0.484	0.158
0.450	2.396	0.872	0.567	0.560	0.540	0.467	0.521	0.386	0.509	0.314	0.495	0.158
0.460	4.307	0.921	0.584	0.577	0.554	0.479	0.534	0.396	0.521	0.322	0.507	0.159
0.470			0.601	0.594	0.568	0.492	0.547	0.406	0.533	0.329	0.517	0.160
0.480			0.618	0.611	0.583	0.504	0.561	0.416	0.546	0.337	0.529	0.160
0.490			0.635	0.628	0.598	0.517	0.574	0.425	0.558	0.345	0.539	0.161
0.500			0.653	0.646	0.613	0.530	0.588	0.435	0.571	0.352	0.551	0.161
0.510			0.672	0.664	0.628	0.543	0.601	0.445	0.583	0.360	0.562	0.162
0.520			0.707	0.674	0.644	0.556	0.615	0.456	0.596	0.368	0.573	0.162
0.530			0.748	0.682	0.659	0.570	0.628	0.466	0.608	0.376	0.585	0.163
0.540			0.789	0.689	0.675	0.584	0.642	0.476	0.620	0.383	0.596	0.163
0.550			0.832	0.696	0.691	0.597	0.656	0.486	0.633	0.391	0.607	0.164
0.560			0.874	0.703	0.707	0.611	0.670	0.497	0.646	0.399	0.617	0.164
0.570			0.918	0.709	0.724	0.626	0.684	0.507	0.659	0.407	0.628	0.164
0.580			0.963	0.715	0.740	0.640	0.698	0.517	0.672	0.415	0.640	0.165
0.590			1.008	0.721	0.757	0.655	0.713	0.528	0.685	0.423	0.651	0.165
0.600			1.054	0.726	0.776	0.669	0.727	0.539	0.697	0.431	0.662	0.166
0.610			1.100	0.732	0.806	0.673	0.742	0.550	0.710	0.439	0.673	0.166
0.620			1.147	0.737	0.837	0.678	0.756	0.560	0.724	0.447	0.685	0.166
0.630			1.194	0.741	0.870	0.683	0.771	0.571	0.737	0.455	0.696	0.167
0.640			1.242	0.746	0.900	0.687	0.786	0.582	0.750	0.463	0.707	0.167
0.650			1.290	0.750	0.932	0.691	0.801	0.593	0.763	0.471	0.718	0.168
0.660			1.339	0.754	0.965	0.695	0.816	0.605	0.776	0.479	0.729	0.168
0.670			1.389	0.758	0.995	0.699	0.831	0.616	0.790	0.488	0.740	0.168
0.680			1.437	0.762	1.029	0.703	0.847	0.627	0.803	0.496	0.752	0.169
0.690			1.487	0.765	1.063	0.706	0.862	0.639	0.817	0.505	0.763	0.169
0.700			1.537	0.769	1.094	0.709	0.878	0.651	0.830	0.513	0.773	0.169

Tab. 5.5

TABLE 2_Concrete classes C12-C50_Steel class S400_a/d=0.15

Doubly reinforced sections under bending moment

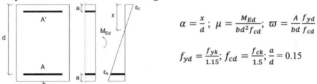

$$\alpha = \frac{x}{d}\ ;\ \mu = \frac{M_{Ed}}{bd^2 f_{cd}};\ \varpi = \frac{A\ f_{yd}}{bd\ f_{cd}}$$

$$f_{yd} = \frac{f_{yk}}{1.15};\ f_{cd} = \frac{f_{ck}}{1.5};\ \frac{a}{d} = 0.15$$

	$A'/A=0.0$		$A'/A=0.2$		$A'/A=0.3$		$A'/A=0.4$		$A'/A=0.5$		$A'/A=1.0$	
μ	ϖ	α	ϖ	α	ϖ	α	ϖ	α	ϖ	α	ϖ	α
0.005	0.005	0.021	0.005	0.022	0.005	0.023	0.005	0.024	0.005	0.025	0.004	0.027
0.010	0.010	0.030	0.010	0.033	0.010	0.034	0.010	0.035	0.009	0.036	0.009	0.040
0.015	0.015	0.037	0.015	0.041	0.015	0.042	0.015	0.044	0.014	0.045	0.013	0.051
0.020	0.020	0.044	0.020	0.048	0.020	0.050	0.019	0.052	0.019	0.053	0.018	0.061
0.025	0.025	0.050	0.025	0.055	0.025	0.057	0.024	0.059	0.024	0.061	0.023	0.070
0.030	0.031	0.055	0.030	0.061	0.030	0.063	0.029	0.066	0.029	0.068	0.027	0.079
0.035	0.036	0.061	0.035	0.067	0.035	0.070	0.034	0.073	0.034	0.075	0.032	0.088
0.040	0.041	0.066	0.040	0.073	0.040	0.076	0.039	0.079	0.039	0.083	0.037	0.095
0.045	0.046	0.071	0.045	0.079	0.045	0.083	0.045	0.087	0.044	0.090	0.043	0.101
0.050	0.051	0.076	0.050	0.085	0.050	0.089	0.050	0.092	0.050	0.095	0.048	0.106
0.055	0.057	0.081	0.056	0.091	0.056	0.094	0.055	0.098	0.055	0.100	0.054	0.111
0.060	0.062	0.086	0.061	0.096	0.061	0.100	0.061	0.103	0.060	0.106	0.059	0.116
0.065	0.067	0.092	0.067	0.101	0.066	0.105	0.066	0.108	0.066	0.111	0.065	0.120
0.070	0.073	0.097	0.072	0.106	0.072	0.110	0.072	0.113	0.071	0.115	0.071	0.124
0.075	0.078	0.102	0.078	0.111	0.077	0.114	0.077	0.117	0.077	0.119	0.076	0.128
0.080	0.084	0.107	0.083	0.116	0.083	0.119	0.083	0.121	0.083	0.124	0.082	0.131
0.085	0.089	0.113	0.089	0.121	0.089	0.123	0.088	0.126	0.088	0.128	0.088	0.134
0.090	0.095	0.118	0.094	0.125	0.094	0.128	0.094	0.130	0.094	0.132	0.094	0.137
0.095	0.100	0.124	0.100	0.131	0.100	0.133	0.100	0.135	0.100	0.136	0.099	0.140
0.100	0.106	0.130	0.106	0.136	0.106	0.138	0.105	0.139	0.105	0.140	0.105	0.143
0.105	0.111	0.138	0.111	0.141	0.111	0.142	0.111	0.143	0.111	0.144	0.111	0.145
0.110	0.117	0.144	0.117	0.146	0.117	0.147	0.117	0.147	0.117	0.147	0.117	0.148
0.115	0.123	0.151	0.123	0.151	0.123	0.151	0.123	0.151	0.123	0.151	0.122	0.150
0.120	0.128	0.159	0.129	0.156	0.128	0.155	0.129	0.155	0.129	0.155	0.128	0.153
0.125	0.134	0.166	0.134	0.161	0.135	0.160	0.135	0.159	0.135	0.158	0.134	0.155
0.130	0.140	0.173	0.140	0.166	0.140	0.164	0.140	0.163	0.140	0.161	0.140	0.157
0.135	0.146	0.180	0.146	0.171	0.146	0.169	0.146	0.166	0.146	0.165	0.146	0.159
0.140	0.152	0.188	0.152	0.176	0.152	0.173	0.152	0.170	0.152	0.168	0.152	0.161
0.145	0.158	0.195	0.158	0.181	0.158	0.177	0.158	0.174	0.158	0.171	0.158	0.163
0.150	0.164	0.202	0.164	0.186	0.164	0.181	0.164	0.177	0.164	0.174	0.163	0.165
0.155	0.170	0.210	0.169	0.191	0.169	0.185	0.169	0.181	0.170	0.177	0.169	0.167
0.160	0.176	0.217	0.175	0.196	0.176	0.190	0.176	0.184	0.175	0.180	0.175	0.169
0.165	0.182	0.225	0.181	0.201	0.181	0.194	0.182	0.188	0.181	0.183	0.181	0.170
0.170	0.188	0.232	0.187	0.206	0.187	0.198	0.187	0.191	0.187	0.186	0.187	0.172
0.175	0.194	0.240	0.193	0.211	0.193	0.202	0.193	0.195	0.193	0.189	0.193	0.173
0.180	0.201	0.248	0.199	0.216	0.199	0.206	0.199	0.198	0.199	0.192	0.199	0.175
0.185	0.207	0.256	0.205	0.221	0.205	0.210	0.205	0.201	0.205	0.195	0.205	0.176
0.190	0.213	0.263	0.212	0.226	0.211	0.214	0.211	0.205	0.211	0.198	0.210	0.178
0.195	0.220	0.272	0.218	0.231	0.217	0.218	0.217	0.208	0.217	0.200	0.216	0.179
0.200	0.226	0.279	0.224	0.236	0.223	0.222	0.223	0.211	0.223	0.203	0.222	0.181
0.205	0.233	0.288	0.230	0.241	0.229	0.226	0.229	0.214	0.229	0.206	0.228	0.182
0.210	0.239	0.296	0.236	0.246	0.235	0.230	0.235	0.218	0.235	0.208	0.234	0.183
0.215	0.246	0.304	0.242	0.251	0.241	0.234	0.241	0.221	0.241	0.211	0.240	0.185
0.220	0.253	0.312	0.248	0.256	0.247	0.238	0.247	0.224	0.247	0.213	0.246	0.186
0.225	0.260	0.321	0.254	0.261	0.253	0.241	0.253	0.227	0.253	0.216	0.252	0.187
0.230	0.266	0.329	0.261	0.265	0.260	0.245	0.259	0.230	0.259	0.219	0.258	0.188
0.235	0.273	0.338	0.267	0.271	0.265	0.249	0.265	0.233	0.264	0.221	0.263	0.189
0.240	0.280	0.346	0.273	0.275	0.272	0.253	0.271	0.236	0.270	0.224	0.269	0.190
0.245	0.287	0.355	0.279	0.280	0.278	0.257	0.277	0.239	0.277	0.226	0.275	0.192
0.250	0.295	0.364	0.286	0.285	0.284	0.261	0.283	0.242	0.282	0.228	0.281	0.193

TABLE 2_Concrete classes C12-C50_Steel class S400_a/d=0.15 (cont.)
Doubly reinforced sections under bending moment

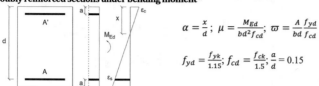

$$\alpha = \frac{x}{d}; \quad \mu = \frac{M_{Ed}}{bd^2 f_{cd}}; \quad \varpi = \frac{A\, f_{yd}}{bd\, f_{cd}}$$

$$f_{yd} = \frac{f_{yk}}{1.15}; \quad f_{cd} = \frac{f_{ck}}{1.5}; \quad \frac{a}{d} = 0.15$$

μ	$A'/A=0.0$ ϖ	α	$A'/A=0.2$ ϖ	α	$A'/A=0.3$ ϖ	α	$A'/A=0.4$ ϖ	α	$A'/A=0.5$ ϖ	α	$A'/A=1.0$ ϖ	α
0.255	0.302	0.373	0.292	0.290	0.290	0.265	0.289	0.245	0.288	0.231	0.287	0.194
0.260	0.309	0.382	0.298	0.295	0.296	0.268	0.295	0.248	0.295	0.233	0.293	0.195
0.265	0.316	0.391	0.305	0.301	0.302	0.272	0.301	0.251	0.301	0.236	0.299	0.196
0.270	0.324	0.400	0.311	0.308	0.309	0.276	0.307	0.254	0.307	0.238	0.305	0.197
0.275	0.331	0.409	0.318	0.314	0.315	0.280	0.313	0.257	0.313	0.240	0.311	0.198
0.280	0.339	0.419	0.324	0.320	0.321	0.284	0.319	0.260	0.319	0.243	0.317	0.199
0.285	0.347	0.428	0.331	0.327	0.327	0.287	0.325	0.263	0.325	0.245	0.322	0.200
0.290	0.355	0.438	0.338	0.334	0.333	0.291	0.331	0.266	0.330	0.247	0.328	0.200
0.295	0.363	0.448	0.344	0.340	0.339	0.295	0.337	0.269	0.336	0.249	0.334	0.201
0.300	0.371	0.458	0.351	0.347	0.346	0.299	0.344	0.272	0.343	0.251	0.340	0.202
0.310	0.387	0.478	0.365	0.360	0.359	0.310	0.356	0.277	0.354	0.256	0.352	0.204
0.320	0.404	0.499	0.379	0.374	0.372	0.321	0.368	0.283	0.367	0.260	0.364	0.205
0.330	0.421	0.520	0.392	0.388	0.385	0.333	0.380	0.288	0.378	0.264	0.375	0.207
0.340	0.439	0.542	0.407	0.402	0.398	0.344	0.392	0.294	0.391	0.268	0.387	0.209
0.350	0.458	0.565	0.421	0.416	0.411	0.356	0.405	0.300	0.402	0.272	0.399	0.210
0.360	0.477	0.589	0.435	0.430	0.424	0.367	0.417	0.309	0.415	0.276	0.411	0.211
0.370	0.497	0.613	0.450	0.445	0.438	0.379	0.430	0.319	0.427	0.280	0.423	0.213
0.380	0.518	0.639	0.465	0.460	0.452	0.391	0.443	0.328	0.439	0.285	0.435	0.214
0.390	0.540	0.666	0.480	0.475	0.466	0.403	0.456	0.338	0.451	0.288	0.446	0.215
0.400	0.637	0.695	0.496	0.490	0.480	0.415	0.469	0.347	0.463	0.292	0.458	0.216
0.410	0.769	0.725	0.511	0.505	0.494	0.427	0.482	0.357	0.475	0.296	0.470	0.218
0.420	0.951	0.757	0.527	0.521	0.508	0.439	0.495	0.367	0.487	0.301	0.482	0.219
0.430	1.213	0.792	0.543	0.537	0.522	0.452	0.508	0.377	0.500	0.309	0.494	0.220
0.440	1.629	0.830	0.560	0.553	0.537	0.464	0.522	0.387	0.512	0.316	0.505	0.221
0.450	2.396	0.872	0.577	0.570	0.551	0.477	0.535	0.397	0.525	0.324	0.517	0.222
0.460	4.307	0.921	0.594	0.587	0.566	0.490	0.549	0.407	0.537	0.332	0.529	0.223
0.470			0.611	0.604	0.581	0.503	0.562	0.417	0.550	0.340	0.541	0.224
0.480			0.629	0.621	0.596	0.516	0.576	0.427	0.563	0.348	0.553	0.225
0.490			0.647	0.639	0.612	0.529	0.590	0.437	0.576	0.356	0.565	0.226
0.500			0.666	0.658	0.627	0.542	0.604	0.448	0.589	0.364	0.576	0.227
0.510			0.695	0.672	0.643	0.556	0.618	0.458	0.602	0.372	0.588	0.228
0.520			0.736	0.680	0.659	0.570	0.632	0.468	0.615	0.380	0.600	0.229
0.530			0.780	0.688	0.675	0.584	0.646	0.479	0.628	0.388	0.612	0.229
0.540			0.824	0.695	0.691	0.598	0.660	0.489	0.641	0.396	0.624	0.230
0.550			0.869	0.702	0.708	0.612	0.675	0.500	0.654	0.404	0.635	0.231
0.560			0.914	0.709	0.725	0.627	0.689	0.511	0.667	0.412	0.647	0.232
0.570			0.961	0.715	0.742	0.642	0.704	0.522	0.681	0.420	0.659	0.233
0.580			1.009	0.721	0.759	0.657	0.719	0.533	0.694	0.429	0.671	0.233
0.590			1.056	0.727	0.781	0.669	0.734	0.544	0.707	0.437	0.682	0.234
0.600			1.105	0.732	0.814	0.675	0.749	0.555	0.721	0.445	0.694	0.235
0.610			1.154	0.737	0.845	0.679	0.764	0.566	0.734	0.454	0.706	0.235
0.620			1.203	0.742	0.878	0.684	0.779	0.577	0.748	0.462	0.718	0.236
0.630			1.254	0.747	0.912	0.688	0.795	0.589	0.762	0.471	0.729	0.237
0.640			1.305	0.751	0.946	0.693	0.810	0.600	0.776	0.479	0.741	0.237
0.650			1.357	0.756	0.979	0.697	0.826	0.612	0.789	0.487	0.754	0.238
0.660			1.409	0.760	1.014	0.701	0.842	0.624	0.803	0.496	0.765	0.239
0.670			1.461	0.764	1.048	0.705	0.857	0.636	0.817	0.505	0.777	0.239
0.680			1.514	0.767	1.081	0.708	0.874	0.648	0.831	0.513	0.788	0.240
0.690			1.565	0.771	1.118	0.712	0.890	0.660	0.845	0.522	0.800	0.240
0.700			1.620	0.774	1.151	0.715	0.910	0.669	0.860	0.531	0.813	0.241

Tab. 5.6

TABLE 2_Concrete class C55_Steel class S500_a/d=0.10

Doubly reinforced sections under bending moment

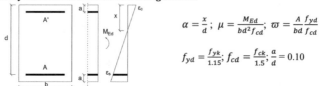

$$\alpha = \frac{x}{d}\,;\ \mu = \frac{M_{Ed}}{bd^2 f_{cd}}\,;\ \varpi = \frac{A}{bd}\frac{f_{yd}}{f_{cd}}$$

$$f_{yd} = \frac{f_{yk}}{1.15}\,;\ f_{cd} = \frac{f_{ck}}{1.5}\,;\ \frac{a}{d} = 0.10$$

	$A'/A=0.0$		$A'/A=0.2$		$A'/A=0.3$		$A'/A=0.4$			$A'/A=0.5$		$A'/A=1.0$
μ	ϖ	α	ϖ	α	ϖ	α	μ	ϖ	α	ϖ	α	ϖ
0.005	0.005	0.023	0.005	0.025	0.005	0.026	0.005	0.026	0.005	0.027	0.004	0.030
0.010	0.010	0.033	0.010	0.035	0.010	0.036	0.010	0.038	0.010	0.038	0.009	0.042
0.015	0.015	0.041	0.015	0.044	0.015	0.045	0.015	0.046	0.015	0.047	0.014	0.051
0.020	0.020	0.048	0.020	0.051	0.020	0.052	0.020	0.053	0.020	0.054	0.019	0.058
0.025	0.025	0.054	0.025	0.057	0.025	0.058	0.025	0.059	0.025	0.060	0.025	0.065
0.030	0.031	0.060	0.031	0.063	0.030	0.064	0.030	0.065	0.030	0.066	0.030	0.070
0.035	0.036	0.065	0.036	0.068	0.036	0.069	0.035	0.070	0.035	0.071	0.035	0.075
0.040	0.041	0.071	0.041	0.073	0.041	0.074	0.041	0.075	0.041	0.076	0.040	0.080
0.045	0.046	0.076	0.046	0.078	0.046	0.079	0.046	0.080	0.046	0.081	0.046	0.084
0.050	0.052	0.081	0.051	0.083	0.052	0.084	0.051	0.085	0.051	0.085	0.051	0.088
0.055	0.057	0.086	0.057	0.088	0.057	0.088	0.057	0.089	0.057	0.090	0.056	0.092
0.060	0.062	0.091	0.062	0.092	0.062	0.093	0.062	0.093	0.062	0.094	0.062	0.095
0.065	0.067	0.096	0.068	0.097	0.068	0.097	0.068	0.097	0.068	0.098	0.067	0.098
0.070	0.073	0.102	0.073	0.101	0.073	0.101	0.073	0.101	0.073	0.101	0.073	0.101
0.075	0.078	0.107	0.078	0.106	0.079	0.105	0.078	0.105	0.078	0.105	0.078	0.103
0.080	0.084	0.113	0.084	0.110	0.084	0.109	0.084	0.109	0.084	0.108	0.084	0.106
0.085	0.089	0.120	0.089	0.116	0.089	0.114	0.089	0.113	0.089	0.111	0.089	0.108
0.090	0.095	0.128	0.095	0.121	0.095	0.119	0.095	0.117	0.095	0.116	0.095	0.110
0.095	0.100	0.135	0.100	0.127	0.100	0.124	0.101	0.122	0.100	0.119	0.100	0.113
0.100	0.106	0.143	0.106	0.133	0.106	0.129	0.106	0.126	0.106	0.124	0.106	0.115
0.105	0.111	0.150	0.111	0.138	0.112	0.134	0.111	0.130	0.112	0.127	0.111	0.118
0.110	0.117	0.158	0.117	0.144	0.117	0.139	0.117	0.135	0.117	0.131	0.117	0.120
0.115	0.123	0.166	0.123	0.150	0.123	0.144	0.123	0.139	0.123	0.135	0.122	0.122
0.120	0.129	0.173	0.128	0.155	0.128	0.149	0.128	0.143	0.128	0.138	0.128	0.124
0.125	0.135	0.181	0.134	0.161	0.134	0.154	0.134	0.147	0.134	0.142	0.134	0.126
0.130	0.140	0.189	0.140	0.167	0.140	0.159	0.140	0.152	0.139	0.146	0.139	0.128
0.135	0.146	0.197	0.145	0.172	0.145	0.163	0.145	0.156	0.145	0.149	0.145	0.130
0.140	0.152	0.205	0.151	0.178	0.151	0.168	0.151	0.160	0.151	0.153	0.150	0.132
0.145	0.158	0.213	0.157	0.184	0.156	0.173	0.157	0.164	0.156	0.156	0.156	0.134
0.150	0.164	0.221	0.163	0.190	0.162	0.178	0.162	0.168	0.162	0.160	0.161	0.136
0.155	0.170	0.229	0.168	0.195	0.168	0.183	0.168	0.172	0.168	0.163	0.167	0.137
0.160	0.176	0.238	0.174	0.201	0.174	0.187	0.173	0.176	0.173	0.167	0.172	0.139
0.165	0.183	0.246	0.180	0.207	0.180	0.192	0.179	0.180	0.179	0.170	0.178	0.141
0.170	0.189	0.254	0.186	0.213	0.185	0.197	0.185	0.184	0.184	0.173	0.184	0.142
0.175	0.195	0.263	0.192	0.219	0.191	0.202	0.191	0.188	0.190	0.177	0.189	0.144
0.180	0.201	0.271	0.198	0.224	0.197	0.207	0.196	0.192	0.196	0.180	0.195	0.145
0.185	0.208	0.280	0.204	0.230	0.203	0.212	0.202	0.196	0.202	0.184	0.200	0.147
0.190	0.214	0.289	0.210	0.236	0.208	0.216	0.208	0.200	0.207	0.187	0.206	0.148
0.195	0.221	0.297	0.216	0.242	0.214	0.221	0.213	0.204	0.213	0.190	0.211	0.150
0.200	0.227	0.306	0.222	0.248	0.220	0.226	0.219	0.208	0.218	0.193	0.217	0.151
0.205	0.234	0.315	0.228	0.254	0.226	0.231	0.225	0.212	0.224	0.196	0.223	0.153
0.210	0.241	0.324	0.234	0.260	0.232	0.236	0.231	0.216	0.230	0.199	0.228	0.154
0.215	0.247	0.333	0.240	0.266	0.238	0.240	0.236	0.220	0.236	0.203	0.234	0.155
0.220	0.254	0.342	0.246	0.272	0.244	0.245	0.242	0.224	0.241	0.206	0.239	0.156
0.225	0.261	0.352	0.252	0.278	0.250	0.250	0.248	0.228	0.247	0.209	0.245	0.158
0.230	0.268	0.361	0.258	0.284	0.256	0.255	0.254	0.231	0.253	0.212	0.250	0.159
0.235	0.275	0.370	0.264	0.290	0.261	0.260	0.260	0.235	0.259	0.215	0.256	0.160
0.240	0.282	0.380	0.270	0.296	0.268	0.265	0.266	0.239	0.264	0.218	0.262	0.161
0.245	0.289	0.390	0.277	0.302	0.273	0.269	0.271	0.243	0.270	0.221	0.267	0.163
0.250	0.296	0.399	0.283	0.308	0.279	0.274	0.277	0.247	0.276	0.225	0.273	0.164

TABLE 2_Concrete class C55_Steel class S500_a/d=0.10 (cont.)
Doubly reinforced sections under bending moment

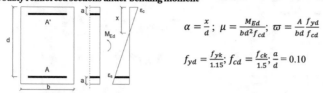

$$\alpha = \frac{x}{d}; \quad \mu = \frac{M_{Ed}}{bd^2 f_{cd}}; \quad \varpi = \frac{A\, f_{yd}}{bd\, f_{cd}}$$

$$f_{yd} = \frac{f_{yk}}{1.15}; \quad f_{cd} = \frac{f_{ck}}{1.5}; \quad \frac{a}{d} = 0.10$$

μ	$A'/A=0.0$		$A'/A=0.2$		$A'/A=0.3$		$A'/A=0.4$			$A'/A=0.5$		$A'/A=1.0$	
	ϖ	α	ϖ	α	ϖ	α	μ	ϖ	α	ϖ	α	ϖ	
0.255	0.304	0.409	0.289	0.314	0.286	0.279	0.283	0.251	0.281	0.228	0.278	0.165	
0.260	0.311	0.419	0.296	0.320	0.292	0.284	0.289	0.255	0.287	0.231	0.284	0.166	
0.265	0.319	0.429	0.302	0.326	0.298	0.289	0.295	0.258	0.293	0.234	0.290	0.167	
0.270	0.326	0.440	0.308	0.333	0.304	0.294	0.301	0.262	0.299	0.237	0.295	0.168	
0.275	0.334	0.450	0.315	0.339	0.310	0.299	0.307	0.266	0.305	0.240	0.301	0.169	
0.280	0.342	0.460	0.321	0.347	0.316	0.304	0.312	0.270	0.311	0.243	0.306	0.170	
0.285	0.349	0.471	0.328	0.354	0.322	0.309	0.318	0.274	0.316	0.246	0.312	0.171	
0.290	0.357	0.482	0.335	0.361	0.328	0.313	0.324	0.278	0.322	0.249	0.318	0.172	
0.295	0.366	0.493	0.341	0.368	0.334	0.318	0.330	0.281	0.328	0.252	0.323	0.173	
0.300	0.374	0.504	0.348	0.375	0.340	0.323	0.336	0.285	0.333	0.255	0.329	0.174	
0.310	0.391	0.526	0.361	0.390	0.353	0.333	0.348	0.293	0.345	0.261	0.340	0.176	
0.320	0.408	0.550	0.375	0.404	0.366	0.345	0.360	0.301	0.357	0.267	0.351	0.178	
0.330	0.426	0.574	0.389	0.419	0.379	0.357	0.372	0.308	0.368	0.272	0.362	0.180	
0.340	0.465	0.599	0.403	0.434	0.391	0.369	0.384	0.316	0.380	0.278	0.374	0.181	
0.350	0.540	0.624	0.417	0.450	0.404	0.382	0.396	0.324	0.391	0.284	0.385	0.183	
0.360	0.634	0.651	0.432	0.465	0.418	0.394	0.408	0.331	0.403	0.290	0.396	0.185	
0.370	0.750	0.680	0.446	0.481	0.431	0.407	0.420	0.340	0.415	0.296	0.407	0.186	
0.380	0.900	0.709	0.461	0.497	0.445	0.420	0.433	0.350	0.427	0.301	0.418	0.188	
0.390	1.099	0.740	0.476	0.513	0.458	0.432	0.446	0.360	0.439	0.307	0.429	0.189	
0.400	1.375	0.774	0.491	0.530	0.472	0.445	0.458	0.371	0.451	0.313	0.441	0.191	
0.410	1.783	0.809	0.507	0.547	0.486	0.458	0.471	0.381	0.462	0.318	0.452	0.192	
0.420	2.448	0.847	0.523	0.564	0.500	0.472	0.484	0.391	0.474	0.324	0.463	0.194	
0.430	3.730	0.890	0.539	0.581	0.514	0.485	0.497	0.402	0.486	0.330	0.474	0.195	
0.440			0.567	0.594	0.528	0.498	0.510	0.412	0.498	0.335	0.486	0.196	
0.450			0.603	0.602	0.543	0.512	0.523	0.423	0.510	0.344	0.497	0.198	
0.460			0.640	0.611	0.557	0.526	0.536	0.434	0.522	0.352	0.507	0.199	
0.470			0.678	0.619	0.572	0.539	0.549	0.444	0.534	0.360	0.518	0.200	
0.480			0.717	0.627	0.587	0.553	0.563	0.455	0.547	0.369	0.530	0.201	
0.490			0.757	0.634	0.602	0.568	0.576	0.466	0.559	0.377	0.541	0.203	
0.500			0.798	0.642	0.617	0.582	0.590	0.477	0.572	0.385	0.552	0.204	
0.510			0.839	0.648	0.640	0.591	0.603	0.488	0.585	0.394	0.563	0.205	
0.520			0.882	0.655	0.669	0.597	0.617	0.499	0.597	0.402	0.575	0.206	
0.530			0.926	0.661	0.698	0.603	0.631	0.510	0.610	0.411	0.586	0.207	
0.540			0.969	0.667	0.728	0.608	0.645	0.522	0.623	0.420	0.597	0.208	
0.550			1.013	0.673	0.758	0.613	0.659	0.533	0.635	0.428	0.608	0.209	
0.560			1.059	0.679	0.789	0.618	0.673	0.544	0.648	0.437	0.619	0.210	
0.570			1.104	0.684	0.819	0.623	0.687	0.556	0.661	0.445	0.630	0.211	
0.580			1.149	0.689	0.849	0.627	0.702	0.567	0.674	0.454	0.641	0.212	
0.590			1.196	0.694	0.881	0.632	0.716	0.579	0.687	0.463	0.653	0.213	
0.600			1.244	0.698	0.911	0.636	0.734	0.589	0.700	0.472	0.664	0.214	
0.610			1.290	0.703	0.944	0.640	0.757	0.593	0.713	0.480	0.675	0.215	
0.620			1.338	0.707	0.975	0.644	0.781	0.596	0.726	0.489	0.686	0.216	
0.630			1.386	0.711	1.006	0.647	0.806	0.600	0.739	0.498	0.698	0.217	
0.640			1.436	0.715	1.040	0.651	0.830	0.603	0.753	0.507	0.708	0.218	
0.650			1.485	0.718	1.071	0.654	0.855	0.606	0.766	0.516	0.720	0.219	
0.660			1.534	0.722	1.103	0.658	0.878	0.609	0.779	0.525	0.731	0.220	
0.670			1.582	0.725	1.136	0.661	0.903	0.612	0.793	0.534	0.742	0.221	
0.680			1.632	0.729	1.171	0.664	0.929	0.615	0.806	0.543	0.753	0.221	
0.690			1.682	0.732	1.204	0.667	0.953	0.618	0.820	0.553	0.765	0.222	
0.700			1.732	0.735	1.236	0.670	0.978	0.621	0.833	0.562	0.775	0.223	

Tab. 5.7

TABLE 2_Concrete class C60_Steel class S500_a/d=0.10

Doubly reinforced sections under bending moment

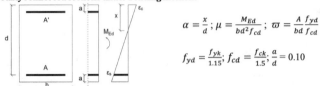

$$\alpha = \frac{x}{d} \; ; \mu = \frac{M_{Ed}}{bd^2 f_{cd}} \; ; \; \varpi = \frac{A\, f_{yd}}{bd\, f_{cd}}$$

$$f_{yd} = \frac{f_{yk}}{1.15}; f_{cd} = \frac{f_{ck}}{1.5}; \frac{a}{d} = 0.10$$

	$A'/A=0.0$		$A'/A=0.2$		$A'/A=0.3$		$A'/A=0.4$		$A'/A=0.5$		$A'/A=1.0$	
μ	ϖ	α	ϖ	α	ϖ	α	μ	ϖ	α	ϖ	α	ϖ
0.005	0.005	0.024	0.005	0.026	0.005	0.027	0.005	0.028	0.005	0.029	0.004	0.031
0.010	0.010	0.035	0.010	0.037	0.010	0.038	0.010	0.039	0.010	0.040	0.009	0.044
0.015	0.015	0.043	0.015	0.046	0.015	0.047	0.015	0.048	0.015	0.049	0.014	0.053
0.020	0.020	0.050	0.020	0.053	0.020	0.054	0.020	0.055	0.020	0.057	0.020	0.061
0.025	0.026	0.057	0.025	0.060	0.025	0.061	0.025	0.062	0.025	0.063	0.025	0.067
0.030	0.031	0.063	0.030	0.065	0.031	0.067	0.030	0.068	0.030	0.069	0.030	0.073
0.035	0.036	0.068	0.036	0.071	0.036	0.072	0.036	0.073	0.035	0.074	0.035	0.078
0.040	0.041	0.074	0.041	0.076	0.041	0.077	0.041	0.078	0.041	0.079	0.040	0.082
0.045	0.046	0.079	0.046	0.081	0.046	0.082	0.046	0.083	0.046	0.083	0.046	0.086
0.050	0.052	0.084	0.051	0.086	0.052	0.087	0.051	0.087	0.052	0.088	0.051	0.090
0.055	0.057	0.089	0.057	0.090	0.057	0.091	0.057	0.091	0.057	0.092	0.056	0.093
0.060	0.062	0.094	0.062	0.095	0.062	0.095	0.062	0.096	0.062	0.096	0.062	0.097
0.065	0.068	0.099	0.067	0.099	0.068	0.100	0.068	0.100	0.068	0.100	0.067	0.100
0.070	0.073	0.105	0.073	0.104	0.073	0.104	0.073	0.104	0.073	0.103	0.073	0.102
0.075	0.078	0.113	0.078	0.110	0.078	0.109	0.078	0.108	0.079	0.108	0.078	0.105
0.080	0.084	0.121	0.084	0.116	0.084	0.115	0.084	0.113	0.084	0.112	0.084	0.108
0.085	0.089	0.128	0.089	0.122	0.089	0.120	0.089	0.118	0.089	0.116	0.089	0.111
0.090	0.095	0.137	0.095	0.128	0.095	0.126	0.095	0.123	0.095	0.121	0.095	0.114
0.095	0.100	0.144	0.100	0.135	0.101	0.131	0.100	0.128	0.101	0.125	0.100	0.117
0.100	0.106	0.153	0.106	0.141	0.106	0.136	0.106	0.133	0.106	0.129	0.106	0.119
0.105	0.112	0.161	0.112	0.147	0.112	0.142	0.112	0.137	0.112	0.134	0.111	0.122
0.110	0.117	0.169	0.117	0.153	0.117	0.147	0.117	0.142	0.117	0.138	0.117	0.124
0.115	0.123	0.177	0.123	0.159	0.123	0.152	0.123	0.147	0.123	0.142	0.122	0.127
0.120	0.129	0.186	0.128	0.165	0.128	0.158	0.128	0.151	0.128	0.146	0.128	0.129
0.125	0.135	0.194	0.134	0.172	0.134	0.163	0.134	0.156	0.134	0.150	0.134	0.131
0.130	0.141	0.202	0.140	0.178	0.140	0.168	0.140	0.160	0.139	0.154	0.139	0.133
0.135	0.147	0.211	0.146	0.184	0.145	0.174	0.145	0.165	0.145	0.158	0.145	0.135
0.140	0.153	0.220	0.151	0.190	0.151	0.179	0.151	0.170	0.151	0.162	0.150	0.138
0.145	0.159	0.228	0.157	0.196	0.157	0.184	0.157	0.174	0.157	0.166	0.156	0.139
0.150	0.165	0.237	0.163	0.203	0.162	0.190	0.162	0.179	0.162	0.170	0.161	0.141
0.155	0.171	0.246	0.169	0.209	0.168	0.195	0.168	0.183	0.168	0.173	0.167	0.143
0.160	0.177	0.255	0.175	0.215	0.174	0.200	0.174	0.188	0.173	0.177	0.173	0.145
0.165	0.183	0.264	0.180	0.222	0.180	0.206	0.179	0.192	0.179	0.181	0.178	0.147
0.170	0.189	0.272	0.186	0.228	0.185	0.211	0.185	0.197	0.185	0.185	0.184	0.149
0.175	0.196	0.282	0.192	0.234	0.191	0.216	0.191	0.201	0.190	0.188	0.189	0.150
0.180	0.202	0.291	0.198	0.241	0.197	0.221	0.196	0.205	0.196	0.192	0.195	0.152
0.185	0.209	0.300	0.204	0.247	0.203	0.227	0.202	0.210	0.202	0.196	0.200	0.154
0.190	0.215	0.309	0.210	0.254	0.209	0.232	0.208	0.214	0.207	0.199	0.206	0.155
0.195	0.222	0.319	0.216	0.260	0.215	0.237	0.214	0.219	0.213	0.203	0.212	0.157
0.200	0.228	0.328	0.222	0.267	0.221	0.243	0.219	0.223	0.219	0.207	0.217	0.159
0.205	0.235	0.338	0.228	0.273	0.227	0.248	0.225	0.227	0.225	0.210	0.223	0.160
0.210	0.242	0.348	0.234	0.280	0.232	0.253	0.231	0.232	0.230	0.214	0.228	0.162
0.215	0.248	0.357	0.241	0.286	0.238	0.259	0.237	0.236	0.236	0.217	0.234	0.163
0.220	0.255	0.367	0.247	0.293	0.244	0.264	0.243	0.241	0.242	0.221	0.239	0.164
0.225	0.262	0.377	0.253	0.299	0.250	0.270	0.249	0.245	0.248	0.225	0.245	0.166
0.230	0.269	0.387	0.259	0.306	0.256	0.275	0.255	0.249	0.253	0.228	0.251	0.167
0.235	0.276	0.398	0.265	0.313	0.262	0.280	0.261	0.254	0.259	0.231	0.256	0.169
0.240	0.284	0.408	0.272	0.319	0.269	0.286	0.266	0.258	0.265	0.235	0.262	0.170
0.245	0.291	0.419	0.278	0.326	0.274	0.291	0.272	0.262	0.271	0.239	0.268	0.171
0.250	0.298	0.429	0.284	0.333	0.281	0.297	0.278	0.267	0.276	0.242	0.273	0.173

TABLE 2_Concrete class C60_Steel class S500_a/d=0.10 (cont.)
Doubly reinforced sections under bending moment

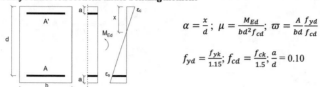

$$\alpha = \frac{x}{d}; \quad \mu = \frac{M_{Ed}}{bd^2 f_{cd}}; \quad \varpi = \frac{A\,f_{yd}}{bd\,f_{cd}}$$

$$f_{yd} = \frac{f_{yk}}{1.15}; \quad f_{cd} = \frac{f_{ck}}{1.5}; \quad \frac{a}{d} = 0.10$$

	A'/A=0.0		A'/A=0.2		A'/A=0.3		A'/A=0.4		A'/A=0.5		A'/A=1.0	
μ	ϖ	α	ϖ	α	ϖ	α	ϖ (μ)	α	ϖ	α	α	ϖ
0.255	0.306	0.440	0.291	0.340	0.287	0.302	0.284	0.271	0.282	0.245	0.279	0.174
0.260	0.313	0.451	0.297	0.346	0.293	0.308	0.290	0.275	0.288	0.249	0.284	0.175
0.265	0.321	0.462	0,303	0.353	0.299	0.313	0.296	0.280	0.294	0.252	0.290	0.176
0.270	0.329	0.473	0.310	0.360	0.305	0.318	0.302	0.284	0.300	0.256	0.296	0.178
0.275	0.336	0.484	0.316	0.367	0.311	0.324	0.308	0.288	0.305	0.259	0.301	0.179
0.280	0.344	0.495	0.323	0.374	0.317	0.329	0.314	0.293	0.311	0.263	0.306	0.180
0.285	0.352	0.507	0.330	0.381	0.323	0.335	0.320	0.297	0.317	0.266	0.312	0.181
0.290	0.360	0.519	0.336	0.388	0.330	0.340	0.325	0.301	0.323	0.270	0.318	0.182
0.295	0.369	0.531	0.343	0.395	0.336	0.346	0.331	0.306	0.329	0.273	0.324	0.183
0.300	0.377	0.543	0.349	0.402	0.342	0.351	0.337	0.310	0.334	0.276	0.329	0.184
0.310	0.394	0.567	0.363	0.418	0.355	0.363	0.350	0.319	0.346	0.283	0.340	0.187
0.320	0.450	0.593	0.377	0.434	0.368	0.374	0.362	0.327	0.358	0.290	0.351	0.189
0.330	0.526	0.620	0.391	0.450	0.380	0.385	0.374	0.336	0.369	0.297	0.363	0.191
0.340	0.618	0.647	0.405	0.466	0.393	0.396	0.386	0.345	0.382	0.303	0.374	0.193
0.350	0.734	0.676	0.420	0.483	0.406	0.409	0.398	0.354	0.393	0.310	0.385	0.195
0.360	0.883	0.706	0.434	0.500	0.420	0.423	0.410	0.362	0.405	0.317	0.396	0.197
0.370	1.079	0.737	0.449	0.517	0.433	0.436	0.423	0.371	0.417	0.323	0.407	0.198
0.380	1.350	0.771	0.464	0.534	0.447	0.450	0.435	0.380	0.429	0.330	0.418	0.200
0.390	1.748	0.806	0.479	0.552	0.460	0.464	0.448	0.389	0.441	0.337	0.430	0.202
0.400	2.389	0.844	0.495	0.570	0.474	0.478	0.460	0.398	0.453	0.343	0.441	0.204
0.410	3.583	0.886	0.528	0.580	0.488	0.492	0.473	0.408	0.464	0.350	0.452	0.205
0.420			0.563	0.590	0.502	0.506	0.486	0.419	0.476	0.357	0.463	0.207
0.430			0.599	0.599	0.516	0.520	0.499	0.431	0.488	0.363	0.474	0.209
0.440			0.637	0.607	0.531	0.535	0.512	0.442	0.500	0.370	0.486	0.210
0.450			0.675	0.615	0.546	0.550	0.525	0.453	0.513	0.376	0.497	0.212
0.460			0.715	0.623	0.560	0.564	0.538	0.465	0.524	0.383	0.508	0.213
0.470			0.755	0.631	0.582	0.575	0.552	0.476	0.536	0.389	0.519	0.215
0.480			0.796	0.638	0.611	0.581	0.565	0.488	0.549	0.396	0.530	0.216
0.490			0.838	0.645	0.639	0.587	0.579	0.500	0.561	0.404	0.542	0.218
0.500			0.881	0.651	0.668	0.593	0.592	0.511	0.574	0.413	0.553	0.219
0.510			0.924	0.658	0.698	0.598	0.606	0.523	0.586	0.422	0.564	0.220
0.520			0.968	0.664	0.727	0.604	0.620	0.535	0.599	0.431	0.575	0.222
0.530			1.013	0.669	0.757	0.608	0.634	0.547	0.611	0.440	0.586	0.223
0.540			1.058	0.675	0.788	0.613	0.648	0.560	0.624	0.449	0.598	0.224
0.550			1.104	0.680	0.819	0.618	0.662	0.572	0.637	0.458	0.608	0.226
0.560			1.150	0.685	0.850	0.622	0.686	0.576	0.650	0.468	0.620	0.227
0.570			1.197	0.690	0.880	0.627	0.709	0.580	0.663	0.477	0.631	0.228
0.580			1.245	0.694	0.911	0.631	0.734	0.583	0.676	0.486	0.642	0.229
0.590			1.291	0.699	0.944	0.635	0.757	0.587	0.689	0.496	0.653	0.230
0.600			1.341	0.703	0.976	0.639	0.782	0.591	0.702	0.505	0.665	0.232
0.610			1.389	0.707	1.007	0.642	0.805	0.594	0.715	0.515	0.675	0.233
0.620			1.436	0.710	1.040	0.646	0.830	0.597	0.728	0.524	0.687	0.234
0.630			1.486	0.714	1.074	0.649	0.856	0.600	0.742	0.534	0.698	0.235
0.640			1.536	0.718	1.107	0.653	0.880	0.604	0.755	0.543	0.710	0.236
0.650			1.585	0.721	1.138	0.656	0.903	0.606	0.769	0.553	0.720	0.237
0.660			1.636	0.724	1.171	0.659	0.929	0.609	0.782	0.563	0.732	0.238
0.670			1.685	0.727	1.204	0.662	0.954	0.612	0.797	0.572	0.743	0.239
0.680			1.736	0.730	1.240	0.665	0.980	0.615	0.817	0.574	0.754	0.240
0.690			1.786	0.733	1.273	0.668	1.004	0.617	0.837	0.577	0.765	0.241
0.700			1.837	0.736	1.304	0.670	1.029	0.620	0.859	0.579	0.776	0.242

Tab. 5.8

TABLE 2_Concrete class C70_Steel class S500_a/d=0.10

Doubly reinforced sections under bending moment

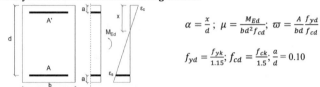

$$\alpha = \frac{x}{d}; \quad \mu = \frac{M_{Ed}}{bd^2 f_{cd}}; \quad \varpi = \frac{A}{bd}\frac{f_{yd}}{f_{cd}}$$

$$f_{yd} = \frac{f_{yk}}{1.15}; \quad f_{cd} = \frac{f_{ck}}{1.5}; \quad \frac{a}{d} = 0.10$$

| | $A'/A = 0.0$ | | $A'/A = 0.2$ | | $A'/A = 0.3$ | | $A'/A = 0.4$ | | $A'/A = 0.5$ | | $A'/A = 1.0$ | |
μ	ϖ	α	ϖ	α	ϖ	α	ϖ	α	ϖ	α	α	ϖ
0.005	0.005	0.026	0.005	0.028	0.005	0.029	0.005	0.030	0.005	0.030	0.004	0.033
0.010	0.010	0.037	0.010	0.039	0.010	0.041	0.010	0.042	0.010	0.043	0.009	0.046
0.015	0.015	0.046	0.015	0.048	0.015	0.050	0.015	0.051	0.015	0.052	0.014	0.056
0.020	0.020	0.053	0.020	0.056	0.020	0.057	0.020	0.058	0.020	0.059	0.020	0.063
0.025	0.026	0.060	0.025	0.062	0.025	0.064	0.025	0.065	0.025	0.066	0.025	0.070
0.030	0.031	0.066	0.031	0.068	0.030	0.069	0.030	0.071	0.030	0.071	0.030	0.075
0.035	0.036	0.072	0.036	0.074	0.036	0.075	0.036	0.076	0.036	0.077	0.035	0.080
0.040	0.041	0.077	0.041	0.079	0.041	0.080	0.041	0.081	0.041	0.082	0.040	0.084
0.045	0.046	0.082	0.046	0.084	0.046	0.085	0.046	0.085	0.046	0.086	0.046	0.088
0.050	0.052	0.087	0.052	0.089	0.051	0.089	0.051	0.090	0.052	0.090	0.051	0.092
0.055	0.057	0.092	0.057	0.093	0.057	0.094	0.057	0.094	0.057	0.094	0.057	0.095
0.060	0.062	0.098	0.062	0.098	0.062	0.098	0.062	0.098	0.062	0.099	0.062	0.099
0.065	0.068	0.106	0.068	0.105	0.068	0.104	0.068	0.104	0.068	0.104	0.067	0.103
0.070	0.073	0.115	0.073	0.112	0.073	0.111	0.073	0.110	0.073	0.109	0.073	0.106
0.075	0.078	0.123	0.078	0.118	0.079	0.117	0.079	0.115	0.079	0.114	0.079	0.110
0.080	0.084	0.132	0.084	0.125	0.084	0.123	0.084	0.121	0.084	0.119	0.084	0.113
0.085	0.090	0.140	0.090	0.132	0.090	0.129	0.090	0.126	0.090	0.124	0.090	0.116
0.090	0.095	0.149	0.095	0.139	0.095	0.135	0.095	0.132	0.095	0.129	0.095	0.119
0.095	0.101	0.158	0.101	0.146	0.101	0.141	0.101	0.137	0.101	0.133	0.100	0.122
0.100	0.106	0.167	0.106	0.152	0.106	0.147	0.106	0.142	0.106	0.138	0.106	0.125
0.105	0.112	0.176	0.112	0.159	0.112	0.153	0.112	0.148	0.112	0.143	0.112	0.128
0.110	0.118	0.185	0.117	0.166	0.117	0.159	0.118	0.153	0.117	0.148	0.117	0.130
0.115	0.124	0.194	0.123	0.173	0.123	0.165	0.123	0.158	0.123	0.152	0.123	0.133
0.120	0.129	0.203	0.129	0.180	0.129	0.171	0.129	0.163	0.129	0.157	0.128	0.136
0.125	0.135	0.212	0.135	0.187	0.134	0.177	0.134	0.168	0.134	0.161	0.134	0.138
0.130	0.141	0.222	0.140	0.194	0.140	0.183	0.140	0.173	0.140	0.165	0.139	0.140
0.135	0.147	0.231	0.146	0.201	0.146	0.189	0.146	0.179	0.145	0.170	0.145	0.143
0.140	0.153	0.241	0.152	0.208	0.152	0.195	0.151	0.184	0.151	0.174	0.151	0.145
0.145	0.159	0.250	0.158	0.215	0.157	0.201	0.157	0.189	0.157	0.179	0.156	0.147
0.150	0.165	0.260	0.164	0.222	0.163	0.207	0.163	0.194	0.162	0.183	0.162	0.149
0.155	0.172	0.269	0.169	0.229	0.169	0.213	0.168	0.199	0.168	0.187	0.167	0.152
0.160	0.178	0.279	0.175	0.236	0.174	0.218	0.174	0.204	0.174	0.192	0.173	0.154
0.165	0.184	0.289	0.181	0.243	0.180	0.225	0.180	0.209	0.179	0.196	0.178	0.156
0.170	0.191	0.299	0.187	0.250	0.186	0.231	0.186	0.214	0.185	0.200	0.184	0.158
0.175	0.197	0.309	0.193	0.257	0.192	0.237	0.191	0.219	0.191	0.204	0.190	0.160
0.180	0.203	0.319	0.199	0.264	0.198	0.243	0.197	0.224	0.196	0.209	0.195	0.161
0.185	0.210	0.330	0.205	0.272	0.204	0.249	0.203	0.229	0.202	0.213	0.201	0.163
0.190	0.217	0.340	0.211	0.279	0.210	0.255	0.209	0.234	0.208	0.217	0.206	0.165
0.195	0.223	0.350	0.217	0.286	0.216	0.261	0.215	0.239	0.214	0.221	0.212	0.167
0.200	0.230	0.361	0.223	0.293	0.222	0.267	0.220	0.244	0.219	0.225	0.217	0.169
0.205	0.237	0.372	0.230	0.301	0.228	0.273	0.226	0.249	0.225	0.229	0.223	0.170
0.210	0.244	0.382	0.236	0.308	0.234	0.279	0.232	0.254	0.231	0.234	0.229	0.172
0.215	0.251	0.393	0.242	0.315	0.240	0.285	0.238	0.259	0.237	0.238	0.234	0.174
0.220	0.258	0.404	0.248	0.323	0.246	0.291	0.244	0.264	0.242	0.242	0.240	0.175
0.225	0.265	0.415	0.255	0.330	0.251	0.297	0.250	0.269	0.248	0.246	0.245	0.177
0.230	0.272	0.427	0.261	0.338	0.258	0.303	0.255	0.274	0.254	0.250	0.251	0.179
0.235	0.279	0.438	0.267	0.345	0.264	0.309	0.261	0.279	0.260	0.254	0.257	0.180
0.240	0.287	0.450	0.274	0.353	0.270	0.316	0.267	0.285	0.265	0.258	0.262	0.182
0.245	0.294	0.461	0.280	0.360	0.276	0.322	0.273	0.289	0.271	0.262	0.268	0.183
0.250	0.302	0.473	0.286	0.368	0.282	0.328	0.279	0.294	0.277	0.266	0.273	0.185

TABLE 2_Concrete class C70_Steel class S500_a/d=0.10 (cont.)
Doubly reinforced sections under bending moment

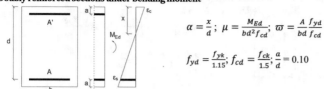

$$\alpha = \frac{x}{d}; \qquad \mu = \frac{M_{Ed}}{bd^2 f_{cd}}; \qquad \varpi = \frac{A\,f_{yd}}{bd\,f_{cd}}$$

$$f_{yd} = \frac{f_{yk}}{1.15}; \qquad f_{cd} = \frac{f_{ck}}{1.5}; \qquad \frac{a}{d} = 0.10$$

μ	$A'/A=0.0$		$A'/A=0.2$		$A'/A=0.3$		μ	$A'/A=0.4$		$A'/A=0.5$		$A'/A=1.0$	
	ϖ	α	ϖ	α	ϖ	α		ϖ	α	ϖ	α	ϖ	α
0.255	0.309	0.485	0.293	0.376	0.288	0.334	0.255	0.285	0.299	0.283	0.270	0.279	0.186
0.260	0.317	0.498	0.299	0.384	0.295	0.341	0.260	0.291	0.305	0.289	0.274	0.285	0.188
0.265	0.325	0.510	0.306	0.391	0.301	0.347	0.265	0.297	0.310	0.295	0.278	0.290	0.189
0.270	0.333	0.522	0.313	0.399	0.307	0.353	0.270	0.303	0.314	0.301	0.283	0.296	0.190
0.275	0.341	0.535	0.319	0.407	0.313	0.359	0.275	0.309	0.320	0.306	0.286	0.301	0.192
0.280	0.349	0.548	0.326	0.415	0.319	0.366	0.280	0.315	0.325	0.312	0.291	0.307	0.193
0.285	0.368	0.561	0.332	0.423	0.326	0.372	0.285	0.321	0.330	0.318	0.295	0.312	0.195
0.290	0.398	0.574	0.339	0.431	0.332	0.378	0.290	0.327	0.335	0.324	0.299	0.318	0.196
0.295	0.431	0.588	0.346	0.439	0.339	0.385	0.295	0.333	0.340	0.330	0.303	0.324	0.197
0.300	0.467	0.602	0.353	0.447	0.345	0.391	0.300	0.339	0.345	0.336	0.307	0.329	0.198
0.310	0.551	0.630	0.367	0.463	0.358	0.404	0.310	0.352	0.355	0.348	0.314	0.340	0.201
0.320	0.656	0.660	0.381	0.480	0.371	0.417	0.320	0.364	0.365	0.359	0.322	0.352	0.203
0.330	0.789	0.690	0.395	0.497	0.384	0.430	0.330	0.376	0.375	0.371	0.330	0.363	0.206
0.340	0.964	0.722	0.409	0.513	0.397	0.443	0.340	0.389	0.385	0.383	0.338	0.374	0.208
0.350	1.202	0.756	0.424	0.532	0.410	0.456	0.350	0.401	0.396	0.395	0.346	0.385	0.211
0.360	1.545	0.792	0.439	0.551	0.423	0.470	0.360	0.414	0.406	0.407	0.354	0.396	0.213
0.370	2.075	0.830	0.469	0.563	0.437	0.483	0.370	0.426	0.416	0.419	0.362	0.408	0.215
0.380	3.006	0.871	0.504	0.573	0.450	0.497	0.380	0.439	0.427	0.431	0.370	0.419	0.217
0.390			0.539	0.583	0.464	0.510	0.390	0.452	0.437	0.443	0.378	0.430	0.219
0.400			0.575	0.592	0.478	0.525	0.400	0.464	0.447	0.456	0.386	0.442	0.221
0.410			0.613	0.601	0.492	0.541	0.410	0.477	0.458	0.468	0.394	0.453	0.223
0.420			0.652	0.610	0.510	0.555	0.420	0.490	0.468	0.480	0.402	0.464	0.225
0.430			0.692	0.618	0.537	0.562	0.430	0.503	0.479	0.492	0.410	0.475	0.227
0.440			0.732	0.625	0.564	0.569	0.440	0.516	0.489	0.504	0.418	0.486	0.229
0.450			0.773	0.633	0.592	0.575	0.450	0.529	0.500	0.517	0.426	0.497	0.231
0.460			0.816	0.640	0.621	0.581	0.460	0.542	0.511	0.529	0.433	0.508	0.233
0.470			0.859	0.646	0.650	0.587	0.470	0.555	0.523	0.541	0.441	0.520	0.235
0.480			0.902	0.653	0.680	0.592	0.480	0.569	0.536	0.553	0.449	0.531	0.236
0.490			0.947	0.659	0.711	0.598	0.490	0.583	0.549	0.566	0.457	0.542	0.238
0.500			0.992	0.665	0.740	0.603	0.500	0.602	0.557	0.578	0.465	0.553	0.240
0.510			1.037	0.670	0.771	0.608	0.510	0.626	0.561	0.591	0.473	0.565	0.241
0.520			1.083	0.675	0.802	0.612	0.520	0.649	0.565	0.603	0.481	0.576	0.243
0.530			1.130	0.680	0.832	0.617	0.530	0.673	0.569	0.615	0.489	0.587	0.245
0.540			1.176	0.685	0.864	0.621	0.540	0.696	0.573	0.628	0.497	0.598	0.246
0.550			1.225	0.690	0.896	0.625	0.550	0.721	0.577	0.641	0.505	0.609	0.248
0.560			1.271	0.694	0.929	0.629	0.560	0.744	0.581	0.653	0.513	0.621	0.249
0.570			1.321	0.698	0.962	0.633	0.570	0.769	0.584	0.666	0.523	0.632	0.251
0.580			1.369	0.702	0.993	0.637	0.580	0.793	0.588	0.680	0.533	0.643	0.252
0.590			1.419	0.706	1.026	0.640	0.590	0.818	0.591	0.693	0.544	0.654	0.254
0.600			1.467	0.710	1.058	0.644	0.600	0.842	0.594	0.708	0.554	0.665	0.255
0.610			1.517	0.713	1.091	0.647	0.610	0.867	0.597	0.728	0.557	0.677	0.257
0.620			1.566	0.717	1.125	0.650	0.620	0.892	0.600	0.748	0.560	0.688	0.258
0.630			1.618	0.720	1.158	0.653	0.630	0.917	0.603	0.767	0.562	0.699	0.260
0.640			1.667	0.723	1.190	0.656	0.640	0.942	0.605	0.787	0.565	0.710	0.261
0.650			1.718	0.726	1.226	0.659	0.650	0.968	0.608	0.807	0.567	0.721	0.262
0.660			1.767	0.729	1.259	0.662	0.660	0.993	0.611	0.828	0.570	0.732	0.264
0.670			1.818	0.731	1.291	0.664	0.670	1.018	0.613	0.848	0.572	0.744	0.265
0.680			1.872	0.734	1.328	0.667	0.680	1.044	0.616	0.868	0.574	0.755	0.266
0.690			1.922	0.737	1.362	0.670	0.690	1.068	0.618	0.888	0.576	0.766	0.268
0.700			1.974	0.739	1.393	0.672	0.700	1.093	0.620	0.910	0.578	0.777	0.269

Tab. 5.9

TABLE 2_Concrete class C80_Steel class S500_a/d=0.10

Doubly reinforced sections under bending moment

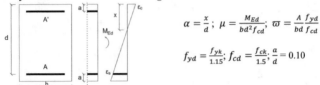

$$\alpha = \frac{x}{d}; \quad \mu = \frac{M_{Ed}}{bd^2 f_{cd}}; \quad \varpi = \frac{A}{bd}\frac{f_{yd}}{f_{cd}}$$

$$f_{yd} = \frac{f_{yk}}{1.15}; \quad f_{cd} = \frac{f_{ck}}{1.5}; \quad \frac{a}{d} = 0.10$$

	$A'/A=0.0$		$A'/A=0.2$		$A'/A=0.3$		$A'/A=0.4$		$A'/A=0.5$		$A'/A=1.0$	
μ	ϖ	α	ϖ	α	ϖ	α	μ	ϖ	α	ϖ	α	ϖ
0.005	0.005	0.027	0.005	0.029	0.005	0.030	0.005	0.031	0.005	0.032	0.005	0.034
0.010	0.010	0.038	0.010	0.041	0.010	0.042	0.010	0.043	0.010	0.044	0.009	0.048
0.015	0.015	0.047	0.015	0.050	0.015	0.051	0.015	0.052	0.015	0.053	0.015	0.057
0.020	0.020	0.055	0.020	0.057	0.020	0.059	0.020	0.060	0.020	0.061	0.020	0.065
0.025	0.026	0.062	0.025	0.064	0.025	0.065	0.025	0.066	0.025	0.067	0.025	0.071
0.030	0.031	0.068	0.031	0.070	0.031	0.071	0.030	0.072	0.031	0.073	0.030	0.077
0.035	0.036	0.074	0.036	0.076	0.036	0.077	0.036	0.078	0.036	0.079	0.035	0.081
0.040	0.041	0.079	0.041	0.081	0.041	0.082	0.041	0.083	0.041	0.083	0.041	0.086
0.045	0.046	0.084	0.046	0.086	0.046	0.087	0.046	0.087	0.046	0.088	0.046	0.090
0.050	0.052	0.090	0.052	0.091	0.052	0.091	0.052	0.092	0.052	0.092	0.051	0.093
0.055	0.057	0.095	0.057	0.096	0.057	0.096	0.057	0.097	0.057	0.097	0.057	0.097
0.060	0.062	0.104	0.062	0.103	0.062	0.103	0.062	0.103	0.062	0.102	0.062	0.102
0.065	0.068	0.113	0.068	0.110	0.068	0.110	0.068	0.109	0.068	0.108	0.068	0.106
0.070	0.073	0.122	0.073	0.118	0.073	0.116	0.073	0.115	0.073	0.114	0.073	0.109
0.075	0.079	0.131	0.079	0.125	0.079	0.123	0.079	0.121	0.079	0.119	0.079	0.113
0.080	0.084	0.140	0.084	0.132	0.084	0.129	0.084	0.127	0.084	0.124	0.084	0.117
0.085	0.090	0.150	0.090	0.140	0.090	0.136	0.090	0.132	0.090	0.129	0.090	0.120
0.090	0.095	0.159	0.095	0.147	0.095	0.142	0.095	0.138	0.095	0.135	0.095	0.123
0.095	0.101	0.168	0.101	0.154	0.101	0.149	0.101	0.144	0.101	0.140	0.101	0.126
0.100	0.107	0.178	0.106	0.161	0.106	0.155	0.106	0.149	0.106	0.145	0.106	0.129
0.105	0.112	0.188	0.112	0.169	0.112	0.162	0.112	0.155	0.112	0.150	0.112	0.132
0.110	0.118	0.197	0.118	0.176	0.118	0.168	0.118	0.161	0.118	0.155	0.117	0.135
0.115	0.124	0.207	0.123	0.184	0.123	0.174	0.123	0.167	0.123	0.160	0.123	0.138
0.120	0.130	0.217	0.129	0.191	0.129	0.181	0.129	0.172	0.129	0.165	0.128	0.140
0.125	0.136	0.227	0.135	0.198	0.135	0.187	0.135	0.178	0.134	0.169	0.134	0.143
0.130	0.142	0.237	0.141	0.206	0.140	0.194	0.140	0.183	0.140	0.174	0.140	0.146
0.135	0.148	0.247	0.147	0.213	0.146	0.200	0.146	0.189	0.146	0.179	0.145	0.148
0.140	0.154	0.257	0.152	0.221	0.152	0.207	0.152	0.194	0.151	0.184	0.151	0.150
0.145	0.160	0.267	0.158	0.228	0.158	0.213	0.157	0.200	0.157	0.188	0.156	0.153
0.150	0.166	0.277	0.164	0.236	0.164	0.220	0.163	0.205	0.163	0.193	0.162	0.155
0.155	0.173	0.288	0.170	0.244	0.169	0.226	0.169	0.211	0.168	0.198	0.167	0.157
0.160	0.179	0.298	0.176	0.251	0.175	0.232	0.175	0.216	0.174	0.203	0.173	0.160
0.165	0.185	0.309	0.182	0.259	0.181	0.239	0.180	0.222	0.180	0.207	0.179	0.162
0.170	0.192	0.320	0.188	0.267	0.187	0.245	0.186	0.227	0.186	0.212	0.184	0.164
0.175	0.198	0.331	0.194	0.274	0.193	0.252	0.192	0.233	0.191	0.216	0.190	0.166
0.180	0.205	0.342	0.200	0.282	0.198	0.258	0.198	0.238	0.197	0.221	0.195	0.168
0.185	0.211	0.353	0.206	0.290	0.204	0.265	0.204	0.244	0.203	0.226	0.201	0.170
0.190	0.218	0.364	0.212	0.298	0.210	0.272	0.209	0.249	0.208	0.230	0.206	0.172
0.195	0.225	0.375	0.218	0.306	0.216	0.278	0.215	0.255	0.214	0.235	0.212	0.174
0.200	0.232	0.387	0.225	0.314	0.222	0.285	0.221	0.260	0.220	0.239	0.218	0.176
0.205	0.239	0.398	0.231	0.322	0.228	0.291	0.227	0.266	0.226	0.244	0.223	0.178
0.210	0.246	0.410	0.237	0.330	0.234	0.298	0.233	0.271	0.231	0.248	0.229	0.180
0.215	0.253	0.422	0.243	0.338	0.240	0.305	0.239	0.277	0.237	0.253	0.234	0.181
0.220	0.260	0.434	0.250	0.346	0.246	0.311	0.245	0.282	0.243	0.257	0.240	0.183
0.225	0.267	0.446	0.256	0.354	0.253	0.318	0.250	0.288	0.249	0.262	0.246	0.185
0.230	0.275	0.458	0.262	0.362	0.259	0.325	0.256	0.293	0.255	0.266	0.251	0.187
0.235	0.282	0.470	0.269	0.370	0.265	0.331	0.262	0.299	0.260	0.271	0.257	0.188
0.240	0.290	0.483	0.275	0.378	0.271	0.338	0.268	0.304	0.266	0.275	0.262	0.190
0.245	0.297	0.496	0.282	0.387	0.277	0.345	0.274	0.310	0.272	0.280	0.268	0.192
0.250	0.305	0.509	0.288	0.395	0.283	0.352	0.280	0.315	0.278	0.284	0.274	0.193

TABLE 2_Concrete class C80_Steel class S500_a/d=0.10 (cont.)
Doubly reinforced sections under bending moment

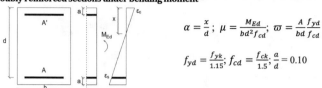

$$\alpha = \frac{x}{d}; \quad \mu = \frac{M_{Ed}}{bd^2 f_{cd}}; \quad \varpi = \frac{A\, f_{yd}}{bd\, f_{cd}}$$

$$f_{yd} = \frac{f_{yk}}{1.15}; \quad f_{cd} = \frac{f_{ck}}{1.5}; \quad \frac{a}{d} = 0.10$$

μ	$A'/A=0.0$		$A'/A=0.2$		$A'/A=0.3$		$A'/A=0.4$		$A'/A=0.5$		$A'/A=1.0$	
	ϖ	α	ϖ	α	ϖ	α	μ	ϖ	α	ϖ	α	ϖ
0.255	0.313	0.522	0.295	0.403	0.290	0.358	0.286	0.321	0.284	0.289	0.279	0.195
0.260	0.321	0.535	0.302	0.412	0.296	0.365	0.292	0.326	0.290	0.293	0.285	0.196
0.265	0.335	0.549	0.308	0.420	0.302	0.372	0.298	0.332	0.296	0.297	0.290	0.198
0.270	0.363	0.563	0.315	0.429	0.309	0.379	0.304	0.337	0.301	0.302	0.296	0.199
0.275	0.394	0.577	0.322	0.438	0.315	0.386	0.310	0.343	0.307	0.306	0.302	0.201
0.280	0.428	0.591	0.328	0.446	0.321	0.393	0.317	0.348	0.314	0.311	0.307	0.202
0.285	0.465	0.605	0.335	0.455	0.328	0.400	0.323	0.354	0.319	0.315	0.313	0.204
0.290	0.508	0.620	0.342	0.464	0.334	0.407	0.329	0.359	0.325	0.320	0.318	0.205
0.295	0.554	0.635	0.349	0.472	0.341	0.414	0.335	0.365	0.331	0.324	0.324	0.207
0.300	0.607	0.651	0.356	0.481	0.347	0.421	0.341	0.370	0.337	0.329	0.330	0.208
0.310	0.734	0.682	0.370	0.499	0.360	0.435	0.353	0.382	0.349	0.337	0.341	0.211
0.320	0.901	0.715	0.384	0.517	0.373	0.449	0.366	0.393	0.361	0.346	0.352	0.214
0.330	1.128	0.750	0.399	0.536	0.387	0.464	0.378	0.404	0.373	0.355	0.363	0.216
0.340	1.455	0.787	0.423	0.551	0.400	0.478	0.391	0.415	0.385	0.364	0.374	0.219
0.350	1.962	0.826	0.456	0.561	0.414	0.493	0.404	0.427	0.397	0.373	0.386	0.221
0.360	2.849	0.868	0.490	0.571	0.427	0.507	0.416	0.438	0.409	0.381	0.397	0.224
0.370			0.525	0.581	0.441	0.522	0.429	0.450	0.421	0.390	0.408	0.226
0.380			0.561	0.590	0.455	0.537	0.442	0.461	0.434	0.399	0.419	0.229
0.390			0.598	0.599	0.475	0.548	0.455	0.473	0.446	0.408	0.431	0.231
0.400			0.636	0.608	0.502	0.555	0.468	0.484	0.458	0.417	0.442	0.233
0.410			0.675	0.616	0.528	0.562	0.481	0.496	0.470	0.426	0.453	0.236
0.420			0.717	0.624	0.557	0.569	0.494	0.507	0.482	0.434	0.464	0.238
0.430			0.758	0.631	0.585	0.575	0.507	0.519	0.495	0.443	0.475	0.240
0.440			0.801	0.638	0.613	0.580	0.520	0.531	0.507	0.452	0.487	0.242
0.450			0.844	0.645	0.642	0.586	0.533	0.543	0.520	0.461	0.498	0.244
0.460			0.888	0.651	0.671	0.591	0.555	0.549	0.532	0.470	0.509	0.246
0.470			0.933	0.658	0.701	0.597	0.577	0.553	0.544	0.479	0.520	0.248
0.480			0.977	0.663	0.731	0.602	0.600	0.558	0.557	0.488	0.531	0.250
0.490			1.024	0.669	0.761	0.606	0.623	0.562	0.570	0.497	0.542	0.252
0.500			1.070	0.674	0.791	0.611	0.647	0.566	0.582	0.506	0.554	0.254
0.510			1.117	0.679	0.822	0.615	0.671	0.570	0.595	0.514	0.565	0.256
0.520			1.165	0.684	0.854	0.619	0.693	0.573	0.608	0.523	0.576	0.258
0.530			1.212	0.689	0.885	0.624	0.717	0.577	0.621	0.533	0.587	0.260
0.540			1.261	0.693	0.919	0.628	0.742	0.580	0.633	0.541	0.599	0.261
0.550			1.309	0.697	0.951	0.632	0.766	0.584	0.650	0.547	0.610	0.263
0.560			1.358	0.701	0.983	0.635	0.789	0.587	0.671	0.550	0.621	0.265
0.570			1.407	0.705	1.016	0.639	0.813	0.590	0.690	0.552	0.633	0.267
0.580			1.456	0.709	1.048	0.642	0.838	0.593	0.710	0.555	0.643	0.268
0.590			1.507	0.712	1.082	0.645	0.861	0.596	0.729	0.558	0.655	0.270
0.600			1.557	0.716	1.117	0.649	0.886	0.598	0.751	0.560	0.666	0.272
0.610			1.606	0.719	1.150	0.652	0.911	0.601	0.769	0.563	0.678	0.273
0.620			1.656	0.722	1.182	0.654	0.935	0.604	0.790	0.565	0.688	0.275
0.630			1.708	0.725	1.215	0.657	0.960	0.606	0.809	0.567	0.700	0.276
0.640			1.758	0.728	1.249	0.660	0.986	0.608	0.829	0.569	0.711	0.278
0.650			1.810	0.730	1.285	0.663	1.009	0.611	0.850	0.572	0.722	0.280
0.660			1.861	0.733	1.318	0.665	1.036	0.613	0.871	0.574	0.734	0.281
0.670			1.912	0.735	1.353	0.668	1.060	0.615	0.891	0.576	0.745	0.283
0.680			1.965	0.738	1.385	0.670	1.089	0.618	0.911	0.578	0.756	0.284
0.690			2.017	0.740	1.423	0.672	1.112	0.620	0.932	0.580	0.767	0.285
0.700			2.068	0.742	1.457	0.674	1.139	0.622	0.953	0.582	0.778	0.287

Tab. 5.10

TABLE 2_Concrete class C90_Steel class S500_a/d=0.10

Doubly reinforced sections under bending moment

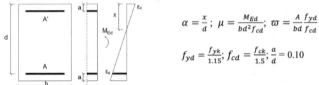

$$\alpha = \frac{x}{d}; \quad \mu = \frac{M_{Ed}}{bd^2 f_{cd}}; \quad \varpi = \frac{A}{bd}\frac{f_{yd}}{f_{cd}}$$

$$f_{yd} = \frac{f_{yk}}{1.15}; \quad f_{cd} = \frac{f_{ck}}{1.5}; \quad \frac{a}{d} = 0.10$$

	$A'/A=0.0$		$A'/A=0.2$		$A'/A=0.3$		$A'/A=0.4$			$A'/A=0.5$		$A'/A=1.0$
μ	ϖ	α	ϖ	α	ϖ	α	μ	ϖ	α	ϖ	α	ϖ
0.005	0.005	0.027	0.005	0.029	0.005	0.030	0.005	0.031		0.005	0.032	0.005	0.035
0.010	0.010	0.039	0.010	0.042	0.010	0.043	0.010	0.044		0.010	0.045	0.009	0.048
0.015	0.015	0.048	0.015	0.051	0.015	0.052	0.015	0.053		0.015	0.054	0.015	0.058
0.020	0.020	0.056	0.020	0.058	0.020	0.060	0.020	0.061		0.020	0.062	0.020	0.066
0.025	0.026	0.063	0.025	0.065	0.025	0.066	0.025	0.067		0.025	0.068	0.025	0.072
0.030	0.031	0.069	0.031	0.071	0.031	0.072	0.030	0.073		0.031	0.074	0.030	0.078
0.035	0.036	0.075	0.036	0.077	0.036	0.078	0.036	0.079		0.036	0.080	0.035	0.082
0.040	0.041	0.081	0.041	0.082	0.041	0.083	0.041	0.084		0.041	0.085	0.041	0.087
0.045	0.046	0.086	0.046	0.087	0.046	0.088	0.046	0.088		0.046	0.089	0.046	0.091
0.050	0.052	0.091	0.052	0.092	0.052	0.093	0.052	0.093		0.052	0.093	0.051	0.094
0.055	0.057	0.098	0.057	0.098	0.057	0.098	0.057	0.098		0.057	0.099	0.057	0.099
0.060	0.062	0.107	0.062	0.105	0.062	0.105	0.062	0.105		0.062	0.104	0.062	0.103
0.065	0.068	0.116	0.068	0.113	0.068	0.112	0.068	0.111		0.068	0.110	0.068	0.107
0.070	0.073	0.126	0.073	0.120	0.073	0.119	0.073	0.117		0.073	0.116	0.073	0.111
0.075	0.079	0.135	0.079	0.128	0.079	0.125	0.079	0.123		0.079	0.121	0.079	0.114
0.080	0.084	0.144	0.084	0.135	0.084	0.132	0.084	0.129		0.085	0.127	0.084	0.118
0.085	0.090	0.154	0.090	0.143	0.090	0.139	0.090	0.135		0.090	0.132	0.090	0.121
0.090	0.095	0.164	0.095	0.150	0.095	0.145	0.095	0.141		0.095	0.137	0.095	0.125
0.095	0.101	0.173	0.101	0.158	0.101	0.152	0.101	0.147		0.101	0.142	0.101	0.128
0.100	0.107	0.183	0.107	0.165	0.107	0.159	0.107	0.153		0.107	0.148	0.106	0.131
0.105	0.113	0.193	0.112	0.173	0.112	0.165	0.112	0.158		0.112	0.153	0.112	0.134
0.110	0.118	0.203	0.118	0.181	0.118	0.172	0.118	0.164		0.118	0.158	0.117	0.137
0.115	0.124	0.213	0.124	0.188	0.123	0.178	0.123	0.170		0.123	0.163	0.123	0.139
0.120	0.130	0.223	0.129	0.196	0.129	0.185	0.129	0.176		0.129	0.168	0.129	0.142
0.125	0.136	0.233	0.135	0.203	0.135	0.192	0.135	0.181		0.134	0.173	0.134	0.145
0.130	0.142	0.244	0.141	0.211	0.141	0.198	0.140	0.187		0.140	0.177	0.140	0.147
0.135	0.148	0.254	0.147	0.219	0.146	0.205	0.146	0.193		0.146	0.182	0.145	0.150
0.140	0.154	0.265	0.153	0.227	0.152	0.211	0.152	0.198		0.152	0.187	0.151	0.152
0.145	0.161	0.275	0.158	0.234	0.158	0.218	0.158	0.204		0.157	0.192	0.156	0.155
0.150	0.167	0.286	0.164	0.242	0.164	0.225	0.163	0.210		0.163	0.197	0.162	0.157
0.155	0.173	0.297	0.170	0.250	0.170	0.231	0.169	0.215		0.168	0.202	0.168	0.159
0.160	0.179	0.308	0.176	0.258	0.175	0.238	0.175	0.221		0.174	0.207	0.173	0.162
0.165	0.186	0.319	0.182	0.266	0.181	0.245	0.181	0.227		0.180	0.211	0.179	0.164
0.170	0.192	0.330	0.188	0.274	0.187	0.251	0.186	0.232		0.186	0.216	0.184	0.166
0.175	0.199	0.341	0.194	0.282	0.193	0.258	0.192	0.238		0.191	0.221	0.190	0.168
0.180	0.205	0.352	0.200	0.290	0.199	0.265	0.198	0.244		0.197	0.225	0.195	0.170
0.185	0.212	0.364	0.206	0.298	0.205	0.272	0.204	0.249		0.203	0.230	0.201	0.172
0.190	0.219	0.375	0.213	0.306	0.211	0.278	0.210	0.255		0.208	0.235	0.207	0.174
0.195	0.226	0.387	0.219	0.314	0.217	0.285	0.215	0.260		0.214	0.239	0.212	0.176
0.200	0.233	0.399	0.225	0.322	0.223	0.292	0.221	0.266		0.220	0.244	0.218	0.178
0.205	0.240	0.411	0.231	0.330	0.229	0.299	0.227	0.272		0.226	0.249	0.223	0.180
0.210	0.247	0.423	0.238	0.339	0.235	0.305	0.233	0.277		0.232	0.253	0.229	0.182
0.215	0.254	0.435	0.244	0.347	0.241	0.312	0.239	0.283		0.238	0.258	0.234	0.183
0.220	0.261	0.448	0.250	0.355	0.247	0.319	0.245	0.289		0.243	0.263	0.240	0.185
0.225	0.269	0.460	0.257	0.364	0.253	0.326	0.251	0.294		0.249	0.267	0.246	0.187
0.230	0.276	0.473	0.263	0.372	0.259	0.333	0.257	0.300		0.255	0.272	0.251	0.189
0.235	0.284	0.486	0.270	0.381	0.266	0.340	0.263	0.305		0.261	0.276	0.257	0.190
0.240	0.291	0.499	0.276	0.389	0.272	0.347	0.269	0.311		0.266	0.281	0.263	0.192
0.245	0.299	0.513	0.283	0.398	0.278	0.354	0.275	0.317		0.272	0.286	0.268	0.194
0.250	0.307	0.526	0.289	0.406	0.284	0.361	0.281	0.322		0.278	0.290	0.274	0.195

TABLE 2_Concrete class C90_Steel class S500_a/d=0.10 (cont.)
Doubly reinforced sections under bending moment

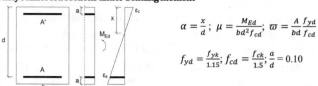

$$\alpha = \frac{x}{d}; \quad \mu = \frac{M_{Ed}}{bd^2 f_{cd}}; \quad \varpi = \frac{A\,f_{yd}}{bd\,f_{cd}}$$

$$f_{yd} = \frac{f_{yk}}{1.15}; \quad f_{cd} = \frac{f_{ck}}{1.5}; \quad \frac{a}{d} = 0.10$$

μ	$A'/A=0.0$		$A'/A=0.2$		$A'/A=0.3$		$A'/A=0.4$		$A'/A=0.5$		$A'/A=1.0$	
	ϖ	α	ϖ	α	ϖ	α	μ	ϖ	α	ϖ	α	ϖ
0.255	0.315	0.540	0.296	0.415	0.290	0.368	0.287	0.328	0.284	0.295	0.279	0.197
0.260	0.335	0.554	0.303	0.424	0.297	0.375	0.293	0.334	0.290	0.299	0.285	0.199
0.265	0.364	0.568	0.309	0.433	0.303	0.382	0.299	0.340	0.296	0.304	0.291	0.200
0.270	0.396	0.582	0.316	0.441	0.309	0.389	0.305	0.345	0.302	0.308	0.296	0.202
0.275	0.432	0.597	0.323	0.450	0.316	0.396	0.311	0.351	0.308	0.313	0.302	0.203
0.280	0.471	0.612	0.330	0.459	0.322	0.403	0.317	0.357	0.314	0.318	0.307	0.205
0.285	0.515	0.627	0.336	0.468	0.329	0.410	0.323	0.362	0.319	0.322	0.313	0.206
0.290	0.565	0.643	0.343	0.477	0.335	0.418	0.330	0.368	0.326	0.327	0.318	0.208
0.295	0.620	0.659	0.350	0.486	0.342	0.425	0.335	0.374	0.331	0.331	0.324	0.209
0.300	0.684	0.675	0.357	0.496	0.348	0.432	0.342	0.379	0.337	0.336	0.330	0.211
0.310	0.839	0.708	0.372	0.514	0.361	0.447	0.354	0.391	0.349	0.345	0.341	0.213
0.320	1.052	0.744	0.386	0.533	0.374	0.461	0.367	0.402	0.361	0.354	0.352	0.216
0.330	1.356	0.781	0.407	0.549	0.388	0.476	0.379	0.414	0.373	0.363	0.363	0.219
0.340	1.822	0.820	0.440	0.560	0.401	0.491	0.392	0.426	0.386	0.372	0.375	0.221
0.350	2.632	0.862	0.474	0.571	0.415	0.506	0.404	0.437	0.398	0.381	0.386	0.224
0.360			0.509	0.581	0.429	0.522	0.417	0.449	0.410	0.390	0.397	0.226
0.370			0.545	0.590	0.442	0.537	0.430	0.461	0.422	0.399	0.408	0.229
0.380			0.582	0.599	0.463	0.549	0.443	0.473	0.434	0.408	0.419	0.231
0.390			0.620	0.608	0.489	0.556	0.456	0.485	0.446	0.417	0.431	0.234
0.400			0.659	0.616	0.517	0.563	0.469	0.497	0.458	0.426	0.442	0.236
0.410			0.700	0.624	0.544	0.569	0.482	0.509	0.471	0.435	0.453	0.238
0.420			0.742	0.632	0.573	0.575	0.495	0.520	0.483	0.444	0.464	0.240
0.430			0.785	0.639	0.600	0.581	0.508	0.533	0.496	0.454	0.476	0.242
0.440			0.828	0.646	0.630	0.587	0.523	0.545	0.508	0.463	0.487	0.245
0.450			0.872	0.653	0.659	0.593	0.544	0.550	0.520	0.472	0.498	0.247
0.460			0.917	0.659	0.689	0.598	0.567	0.554	0.533	0.481	0.509	0.249
0.470			0.962	0.665	0.718	0.603	0.590	0.559	0.545	0.490	0.520	0.251
0.480			1.008	0.670	0.748	0.607	0.613	0.563	0.558	0.499	0.531	0.253
0.490			1.055	0.676	0.779	0.612	0.636	0.567	0.571	0.508	0.543	0.255
0.500			1.102	0.681	0.811	0.617	0.659	0.571	0.584	0.518	0.554	0.257
0.510			1.149	0.686	0.843	0.621	0.684	0.575	0.596	0.527	0.565	0.258
0.520			1.197	0.690	0.875	0.625	0.707	0.578	0.609	0.536	0.576	0.260
0.530			1.245	0.695	0.905	0.629	0.732	0.582	0.622	0.545	0.588	0.262
0.540			1.295	0.699	0.940	0.633	0.756	0.585	0.642	0.548	0.599	0.264
0.550			1.343	0.703	0.972	0.637	0.778	0.588	0.662	0.551	0.610	0.266
0.560			1.393	0.707	1.004	0.640	0.804	0.591	0.681	0.554	0.621	0.268
0.570			1.442	0.710	1.036	0.644	0.827	0.594	0.701	0.556	0.633	0.269
0.580			1.493	0.714	1.070	0.647	0.853	0.597	0.722	0.559	0.644	0.271
0.590			1.542	0.717	1.104	0.650	0.877	0.600	0.742	0.562	0.655	0.273
0.600			1.592	0.721	1.138	0.653	0.902	0.603	0.762	0.564	0.666	0.274
0.610			1.644	0.724	1.173	0.656	0.926	0.605	0.780	0.566	0.677	0.276
0.620			1.694	0.727	1.206	0.659	0.951	0.608	0.802	0.569	0.689	0.278
0.630			1.746	0.729	1.240	0.662	0.976	0.610	0.822	0.571	0.700	0.279
0.640			1.798	0.732	1.272	0.664	1.002	0.613	0.843	0.573	0.711	0.281
0.650			1.850	0.735	1.306	0.667	1.026	0.615	0.861	0.575	0.722	0.282
0.660			1.901	0.737	1.340	0.669	1.054	0.617	0.881	0.577	0.733	0.284
0.670			1.951	0.740	1.376	0.672	1.079	0.619	0.904	0.579	0.745	0.285
0.680			2.003	0.742	1.410	0.674	1.105	0.622	0.921	0.581	0.756	0.287
0.690			2.055	0.744	1.444	0.676	1.129	0.624	0.943	0.583	0.767	0.288
0.700			2.108	0.746	1.480	0.678	1.157	0.626	0.965	0.585	0.778	0.290

5.3   Bending and Axial Force of Rectangular Sections

Purpose of the Tables in This Section
This section contains tables to be used in the design of reinforced rectangular cross–sections under bending moment and axial force. TABLE 3 is used when a symmetric reinforcement is intended for the cross–section, as it is the common case of building columns. This table gives similar results as design Chart 1 in Chap. 6, although the table provides for more precision. TABLE 4 is applied when the bending moment effect prevails over the axial force, as in the case of prestressed concrete beams.

How to Use the Tables in This Section
The example regarding these tables is n° 3 in Chap. 4. When using TABLE 4 note that, in the case of compression force, if the area $A = \frac{1}{f_{yd}}\left(\varpi_{1,s}bdf_{cd} + N_{Ed}\right)$ becomes negative it means that this table can not be used, that is, there is no solution with simply reinforced section. Note that TABLE 4 has other useful informations besides the neutral axis position α, the normalized lever arm of internal forces under bending ζ, the strains in concrete ε_{c2} and steel ε_{s1} and the steel stress σ_{sd}.

Cautions When Using the Tables in This Section
It is noted that both bending moment and axial load are evaluated at the centroid of the section. The tables are valid with the definitions presented in the figures attatched to each table and the corresponding evaluation of non–dimensional values.

5.3.1   *Symmetric Reinforcement—TABLE 3*

Tab. 5.11

TABLE 3_Concrete classes C12-C50_Steel class S500_a/h=0.10

Bending and axial force with symmetric reinforcement

$$\alpha = \frac{x}{h}; \quad \mu = \frac{M_{Ed}}{bh^2 f_{cd}}; \nu = \frac{N_{Ed}}{bh f_{cd}}; \varpi = \frac{A_s}{bh}\frac{f_{yd}}{f_{cd}}$$

$$A_s = A + A'; A = A'$$

$$f_{yd} = \frac{f_{yk}}{1.15}; f_{cd} = \frac{f_{ck}}{1.5}; \frac{a}{h} = 0.10$$

	$\nu = 0.0$		$\nu = 0.1$		$\nu = 0.2$		$\nu = 0.3$		$\nu = 0.4$		$\nu = 0.5$	
μ	ϖ	α	ϖ	α	ϖ	α	ϖ	α	ϖ	α	ϖ	α
0.000	0.000	-	0.000	-	0.000	-	0.000	-	0.000	-	0.000	-
0.005	0.010	0.027	0.000	-	0.000	-	0.000	-	0.000	-	0.000	-
0.010	0.021	0.039	0.000	-	0.000	-	0.000	-	0.000	-	0.000	-
0.015	0.032	0.048	0.000	-	0.000	-	0.000	-	0.000	-	0.000	-
0.020	0.043	0.056	0.000	-	0.000	-	0.000	-	0.000	-	0.000	-
0.025	0.054	0.062	0.000	-	0.000	-	0.000	-	0.000	-	0.000	-
0.030	0.066	0.069	0.000	-	0.000	-	0.000	-	0.000	-	0.000	-
0.035	0.078	0.074	0.000	-	0.000	-	0.000	-	0.000	-	0.000	-
0.040	0.090	0.079	0.000	-	0.000	-	0.000	-	0.000	-	0.000	-
0.045	0.102	0.083	0.000	-	0.000	-	0.000	-	0.000	-	0.000	-
0.050	0.114	0.087	0.013	0.129	0.000	-	0.000	-	0.000	-	0.000	-
0.055	0.126	0.091	0.025	0.133	0.000	-	0.000	-	0.000	-	0.000	-
0.060	0.138	0.094	0.038	0.137	0.000	-	0.000	-	0.000	-	0.000	-
0.065	0.151	0.097	0.050	0.140	0.000	-	0.000	-	0.000	-	0.000	-
0.070	0.163	0.100	0.063	0.144	0.000	-	0.000	-	0.000	-	0.000	-
0.075	0.176	0.103	0.076	0.146	0.000	-	0.000	-	0.000	-	0.000	-
0.080	0.188	0.106	0.089	0.149	0.001	0.247	0.000	-	0.000	-	0.000	-
0.085	0.201	0.108	0.101	0.152	0.014	0.247	0.000	-	0.000	-	0.000	-
0.090	0.213	0.111	0.114	0.154	0.027	0.248	0.000	-	0.000	-	0.000	-
0.095	0.225	0.113	0.127	0.156	0.039	0.248	0.000	-	0.000	-	0.000	-
0.100	0.238	0.115	0.139	0.158	0.052	0.248	0.000	-	0.000	-	0.000	-
0.105	0.250	0.118	0.152	0.160	0.064	0.249	0.003	0.370	0.000	-	0.000	-
0.110	0.263	0.120	0.164	0.162	0.077	0.249	0.016	0.370	0.000	-	0.000	-
0.115	0.275	0.122	0.177	0.164	0.089	0.249	0.028	0.370	0.000	-	0.000	-
0.120	0.288	0.123	0.190	0.166	0.102	0.249	0.041	0.370	0.006	0.494	0.000	-
0.125	0.300	0.125	0.202	0.167	0.115	0.250	0.053	0.370	0.018	0.494	0.010	0.616
0.130	0.313	0.127	0.215	0.169	0.127	0.250	0.066	0.370	0.031	0.494	0.024	0.614
0.135	0.325	0.129	0.227	0.170	0.140	0.250	0.078	0.370	0.043	0.494	0.038	0.612
0.140	0.338	0.130	0.240	0.172	0.152	0.250	0.091	0.370	0.056	0.494	0.052	0.610
0.145	0.350	0.132	0.253	0.173	0.165	0.250	0.103	0.370	0.068	0.494	0.066	0.608
0.150	0.363	0.134	0.266	0.175	0.177	0.251	0.116	0.370	0.081	0.494	0.080	0.607
0.155	0.376	0.135	0.278	0.176	0.190	0.251	0.128	0.370	0.093	0.494	0.093	0.605
0.160	0.388	0.136	0.291	0.177	0.202	0.251	0.141	0.370	0.106	0.494	0.107	0.604
0.165	0.401	0.138	0.303	0.178	0.215	0.251	0.153	0.370	0.118	0.494	0.121	0.602
0.170	0.413	0.139	0.316	0.179	0.227	0.251	0.166	0.370	0.131	0.494	0.134	0.601
0.175	0.426	0.141	0.328	0.181	0.240	0.252	0.178	0.370	0.143	0.494	0.148	0.600
0.180	0.438	0.142	0.341	0.182	0.252	0.252	0.191	0.370	0.156	0.494	0.161	0.599
0.185	0.451	0.143	0.354	0.183	0.265	0.252	0.203	0.370	0.168	0.494	0.175	0.598
0.190	0.463	0.144	0.366	0.184	0.278	0.252	0.216	0.370	0.181	0.494	0.188	0.597
0.195	0.476	0.146	0.379	0.185	0.290	0.252	0.228	0.370	0.193	0.494	0.201	0.596
0.200	0.488	0.147	0.391	0.186	0.303	0.252	0.241	0.370	0.206	0.494	0.215	0.595
0.205	0.501	0.148	0.404	0.187	0.315	0.252	0.253	0.370	0.218	0.494	0.228	0.594
0.210	0.514	0.149	0.417	0.188	0.328	0.253	0.266	0.370	0.231	0.494	0.241	0.593
0.215	0.526	0.150	0.429	0.188	0.340	0.253	0.278	0.370	0.243	0.494	0.254	0.592
0.220	0.539	0.151	0.442	0.189	0.352	0.253	0.291	0.370	0.256	0.494	0.267	0.591
0.225	0.551	0.152	0.454	0.190	0.365	0.253	0.303	0.370	0.268	0.494	0.280	0.591
0.230	0.564	0.153	0.467	0.191	0.378	0.253	0.316	0.370	0.281	0.494	0.293	0.590
0.235	0.576	0.154	0.480	0.192	0.390	0.253	0.328	0.370	0.293	0.494	0.307	0.589
0.240	0.589	0.155	0.493	0.192	0.403	0.253	0.341	0.370	0.306	0.494	0.320	0.588
0.245	0.602	0.156	0.505	0.193	0.415	0.253	0.353	0.370	0.318	0.494	0.333	0.588

TABLE 3_Concrete classes C12-C50_Steel class S500_a/h=0.10 (cont.)
Bending and axial force with symmetric reinforcement

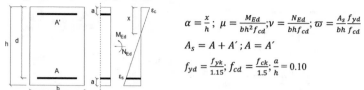

$$\alpha = \frac{x}{h}; \quad \mu = \frac{M_{Ed}}{bh^2 f_{cd}}; \nu = \frac{N_{Ed}}{bh f_{cd}}; \varpi = \frac{A_s \, f_{yd}}{bh \, f_{cd}}$$

$$A_s = A + A'; A = A'$$

$$f_{yd} = \frac{f_{yk}}{1.15}; f_{cd} = \frac{f_{ck}}{1.5}; \frac{a}{h} = 0.10$$

μ	$\nu=0.6$ ϖ	α	$\nu=0.8$ ϖ	α	$\nu=1.0$ ϖ	α	$\nu=1.2$ ϖ	α	$\nu=1.4$ ϖ	α	$\nu=1.6$ ϖ	α
0.000	0.000	-	0.000	-	0.000	-	0.220	-	0.437	-	0.650	-
0.005	0.000	-	0.000	-	0.014	2.690	0.221	7.540	0.438	13.600	0.000	-
0.010	0.000	-	0.000	-	0.027	2.090	0.226	4.340	0.441	7.220	0.658	10.300
0.015	0.000	-	0.000	-	0.041	1.800	0.239	3.030	0.445	5.100	0.661	7.130
0.020	0.000	-	0.000	-	0.054	1.640	0.252	2.470	0.451	3.970	0.664	5.530
0.025	0.000	-	0.000	-	0.068	1.520	0.265	2.160	0.463	3.180	0.668	4.570
0.030	0.000	-	0.000	-	0.081	1.440	0.278	1.950	0.476	2.710	0.676	3.820
0.035	0.000	-	0.000	-	0.095	1.380	0.291	1.790	0.489	2.400	0.688	3.250
0.040	0.000	-	0.000	-	0.109	1.320	0.304	1.680	0.502	2.190	0.701	2.860
0.045	0.000	-	0.000	-	0.122	1.280	0.318	1.590	0.515	2.020	0.714	2.580
0.050	0.000	-	0.000	-	0.135	1.240	0.331	1.520	0.528	1.900	0.727	2.370
0.055	0.000	-	0.000	-	0.149	1.210	0.344	1.460	0.541	1.790	0.740	2.200
0.060	0.000	-	0.000	-	0.162	1.180	0.357	1.410	0.554	1.710	0.753	2.070
0.065	0.000	-	0.000	-	0.175	1.160	0.370	1.370	0.567	1.640	0.765	1.960
0.070	0.000	-	0.000	-	0.188	1.140	0.383	1.330	0.580	1.580	0.778	1.870
0.075	0.000	-	0.011	0.981	0.201	1.120	0.397	1.300	0.593	1.520	0.791	1.790
0.080	0.000	-	0.025	0.971	0.215	1.100	0.410	1.270	0.606	1.480	0.804	1.720
0.085	0.000	-	0.039	0.961	0.228	1.090	0.423	1.240	0.619	1.440	0.817	1.660
0.090	0.000	-	0.054	0.952	0.241	1.070	0.436	1.220	0.632	1.400	0.830	1.610
0.095	0.000	-	0.069	0.943	0.254	1.060	0.449	1.200	0.645	1.370	0.843	1.560
0.100	0.000	-	0.083	0.934	0.268	1.050	0.462	1.180	0.658	1.340	0.856	1.520
0.105	0.000	-	0.098	0.925	0.281	1.030	0.475	1.160	0.671	1.310	0.869	1.490
0.110	0.000	-	0.113	0.917	0.294	1.020	0.488	1.150	0.684	1.290	0.881	1.450
0.115	0.000	-	0.128	0.908	0.307	1.010	0.501	1.130	0.697	1.270	0.894	1.420
0.120	0.016	0.735	0.143	0.900	0.320	1.010	0.514	1.120	0.710	1.250	0.907	1.390
0.125	0.032	0.729	0.158	0.892	0.333	0.997	0.527	1.100	0.723	1.230	0.920	1.370
0.130	0.048	0.723	0.173	0.885	0.347	0.990	0.540	1.090	0.736	1.210	0.933	1.340
0.135	0.064	0.718	0.187	0.877	0.361	0.983	0.553	1.080	0.749	1.190	0.946	1.320
0.140	0.079	0.713	0.203	0.870	0.374	0.975	0.566	1.070	0.762	1.180	0.959	1.300
0.145	0.095	0.708	0.218	0.863	0.388	0.968	0.579	1.060	0.775	1.160	0.972	1.280
0.150	0.110	0.704	0.232	0.856	0.402	0.961	0.592	1.050	0.788	1.150	0.985	1.260
0.155	0.125	0.700	0.248	0.850	0.416	0.955	0.605	1.040	0.801	1.140	0.998	1.250
0.160	0.141	0.695	0.262	0.844	0.430	0.948	0.619	1.030	0.814	1.130	1.010	1.230
0.165	0.155	0.692	0.277	0.838	0.444	0.942	0.632	1.020	0.827	1.110	1.023	1.220
0.170	0.170	0.688	0.292	0.832	0.458	0.935	0.644	1.020	0.840	1.100	1.036	1.200
0.175	0.185	0.685	0.308	0.826	0.472	0.929	0.657	1.010	0.853	1.090	1.049	1.190
0.180	0.200	0.682	0.322	0.820	0.486	0.923	0.670	1.000	0.866	1.080	1.062	1.180
0.185	0.214	0.678	0.337	0.815	0.500	0.917	0.684	0.995	0.879	1.080	1.075	1.160
0.190	0.229	0.676	0.352	0.810	0.514	0.911	0.697	0.989	0.891	1.070	1.088	1.150
0.195	0.243	0.673	0.367	0.805	0.529	0.906	0.711	0.984	0.904	1.060	1.101	1.140
0.200	0.258	0.670	0.382	0.800	0.542	0.900	0.724	0.978	0.917	1.050	1.113	1.130
0.205	0.272	0.667	0.397	0.795	0.557	0.895	0.738	0.972	0.930	1.040	1.126	1.120
0.210	0.286	0.665	0.411	0.791	0.571	0.889	0.751	0.967	0.943	1.040	1.139	1.110
0.215	0.300	0.663	0.426	0.786	0.585	0.884	0.765	0.962	0.956	1.030	1.152	1.100
0.220	0.314	0.660	0.440	0.782	0.599	0.879	0.778	0.956	0.969	1.020	1.165	1.100
0.225	0.328	0.658	0.455	0.778	0.614	0.874	0.792	0.951	0.982	1.020	1.178	1.090
0.230	0.342	0.656	0.469	0.774	0.628	0.869	0.805	0.946	0.995	1.010	1.191	1.080
0.235	0.356	0.654	0.484	0.770	0.642	0.865	0.819	0.941	1.008	1.000	1.204	1.070
0.240	0.370	0.652	0.498	0.767	0.656	0.860	0.833	0.936	1.021	0.999	1.216	1.070
0.245	0.384	0.651	0.513	0.763	0.670	0.856	0.847	0.931	1.034	0.994	1.229	1.060

TABLE 3_Concrete classes C12-C50_Steel class S500_a/h=0.10 (cont.)
Bending and axial force with symmetric reinforcement

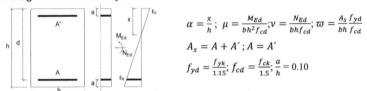

$$\alpha = \frac{x}{h}; \quad \mu = \frac{M_{Ed}}{bh^2 f_{cd}}; \nu = \frac{N_{Ed}}{bh f_{cd}}; \varpi = \frac{A_s f_{yd}}{bh f_{cd}}$$

$$A_s = A + A'; A = A'$$

$$f_{yd} = \frac{f_{yk}}{1.15}; f_{cd} = \frac{f_{ck}}{1.5}; \frac{a}{h} = 0.10$$

	$\nu=0.0$		$\nu=0.1$		$\nu=0.2$		$\nu=0.3$		$\nu=0.4$		$\nu=0.5$	
μ	ϖ	α	ϖ	α	ϖ	α	ϖ	α	ϖ	α	ϖ	α
0.250	0.614	0.157	0.517	0.194	0.428	0.254	0.366	0.370	0.331	0.494	0.346	0.587
0.255	0.627	0.158	0.530	0.195	0.440	0.254	0.378	0.370	0.343	0.494	0.359	0.587
0.260	0.639	0.159	0.542	0.195	0.453	0.254	0.391	0.370	0.356	0.494	0.372	0.586
0.265	0.652	0.160	0.555	0.196	0.465	0.254	0.403	0.370	0.368	0.494	0.385	0.586
0.270	0.664	0.161	0.568	0.197	0.478	0.254	0.416	0.370	0.381	0.494	0.397	0.585
0.275	0.677	0.162	0.581	0.197	0.491	0.254	0.428	0.370	0.393	0.494	0.410	0.584
0.280	0.689	0.162	0.593	0.198	0.503	0.254	0.441	0.370	0.406	0.494	0.424	0.584
0.285	0.702	0.163	0.606	0.199	0.516	0.254	0.453	0.370	0.418	0.494	0.436	0.584
0.290	0.715	0.164	0.618	0.199	0.528	0.254	0.466	0.370	0.431	0.494	0.449	0.583
0.295	0.727	0.165	0.631	0.200	0.541	0.255	0.478	0.370	0.443	0.494	0.462	0.583
0.300	0.740	0.166	0.643	0.200	0.553	0.255	0.491	0.370	0.456	0.494	0.475	0.582
0.305	0.752	0.166	0.656	0.201	0.565	0.255	0.503	0.370	0.468	0.494	0.488	0.582
0.310	0.765	0.167	0.668	0.202	0.578	0.255	0.516	0.370	0.481	0.494	0.501	0.581
0.315	0.777	0.168	0.681	0.202	0.590	0.255	0.528	0.370	0.493	0.494	0.513	0.581
0.320	0.790	0.168	0.693	0.203	0.603	0.255	0.541	0.370	0.506	0.494	0.526	0.581
0.325	0.802	0.169	0.706	0.203	0.616	0.255	0.553	0.370	0.518	0.494	0.539	0.580
0.330	0.815	0.170	0.719	0.204	0.628	0.255	0.566	0.370	0.531	0.494	0.552	0.580
0.335	0.827	0.171	0.731	0.204	0.641	0.255	0.578	0.370	0.543	0.494	0.565	0.579
0.340	0.840	0.171	0.744	0.205	0.653	0.255	0.591	0.370	0.556	0.494	0.578	0.579
0.345	0.853	0.172	0.756	0.205	0.666	0.255	0.603	0.370	0.568	0.494	0.590	0.579
0.350	0.865	0.173	0.769	0.206	0.678	0.255	0.616	0.370	0.581	0.494	0.603	0.578
0.355	0.878	0.173	0.782	0.206	0.691	0.256	0.628	0.370	0.593	0.494	0.616	0.578
0.360	0.890	0.174	0.794	0.207	0.703	0.256	0.641	0.370	0.606	0.494	0.629	0.578
0.365	0.903	0.174	0.807	0.207	0.716	0.256	0.653	0.370	0.618	0.494	0.641	0.578
0.370	0.915	0.175	0.819	0.207	0.728	0.256	0.666	0.370	0.631	0.494	0.654	0.577
0.375	0.928	0.176	0.832	0.208	0.741	0.256	0.678	0.370	0.643	0.494	0.667	0.577
0.380	0.941	0.176	0.844	0.208	0.753	0.256	0.691	0.370	0.656	0.494	0.680	0.577
0.385	0.953	0.177	0.857	0.209	0.766	0.256	0.703	0.370	0.668	0.494	0.692	0.576
0.390	0.966	0.177	0.869	0.209	0.779	0.256	0.716	0.370	0.681	0.494	0.705	0.576
0.395	0.978	0.178	0.882	0.210	0.791	0.256	0.728	0.370	0.693	0.494	0.718	0.576
0.400	0.991	0.179	0.895	0.210	0.803	0.256	0.741	0.370	0.706	0.494	0.730	0.576
0.410	1.016	0.180	0.920	0.211	0.828	0.256	0.766	0.370	0.731	0.494	0.756	0.575
0.420	1.041	0.181	0.945	0.212	0.854	0.256	0.791	0.370	0.756	0.494	0.781	0.575
0.430	1.066	0.182	0.970	0.212	0.879	0.257	0.816	0.370	0.781	0.494	0.807	0.574
0.440	1.091	0.183	0.995	0.213	0.904	0.257	0.841	0.370	0.806	0.494	0.832	0.574
0.450	1.116	0.184	1.020	0.214	0.929	0.257	0.866	0.370	0.831	0.494	0.857	0.574
0.460	1.141	0.185	1.046	0.214	0.954	0.257	0.891	0.370	0.856	0.494	0.883	0.573
0.470	1.167	0.186	1.070	0.215	0.979	0.257	0.916	0.370	0.881	0.494	0.908	0.573
0.480	1.192	0.186	1.096	0.216	1.003	0.257	0.941	0.370	0.906	0.494	0.933	0.572
0.490	1.217	0.187	1.120	0.216	1.029	0.257	0.966	0.370	0.931	0.494	0.959	0.572
0.500	1.242	0.188	1.146	0.217	1.053	0.257	0.991	0.370	0.956	0.494	0.984	0.572
0.510	1.267	0.189	1.171	0.218	1.079	0.257	1.016	0.370	0.981	0.494	1.009	0.571
0.520	1.292	0.190	1.196	0.218	1.103	0.257	1.041	0.370	1.006	0.494	1.034	0.571
0.530	1.317	0.191	1.221	0.219	1.129	0.258	1.066	0.370	1.031	0.494	1.059	0.571
0.540	1.342	0.191	1.246	0.219	1.153	0.258	1.091	0.370	1.056	0.494	1.085	0.571
0.550	1.367	0.192	1.271	0.220	1.179	0.258	1.116	0.370	1.081	0.494	1.110	0.570
0.560	1.392	0.193	1.296	0.220	1.204	0.258	1.141	0.370	1.106	0.494	1.135	0.570
0.570	1.417	0.194	1.321	0.221	1.229	0.258	1.166	0.370	1.131	0.494	1.161	0.570
0.580	1.443	0.194	1.346	0.221	1.254	0.258	1.191	0.370	1.156	0.494	1.186	0.570
0.590	1.467	0.195	1.371	0.222	1.279	0.258	1.216	0.370	1.181	0.494	1.211	0.569

TABLE 3_Concrete classes C12-C50_Steel class S500_a/h=0.10 (cont.)
Bending and axial force with symmetric reinforcement

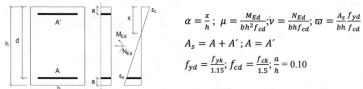

$$\alpha = \frac{x}{h}; \quad \mu = \frac{M_{Ed}}{bh^2 f_{cd}}; \quad \nu = \frac{N_{Ed}}{bh f_{cd}}; \quad \varpi = \frac{A_s f_{yd}}{bh f_{cd}}$$

$$A_s = A + A'; A = A'$$

$$f_{yd} = \frac{f_{yk}}{1.15}; f_{cd} = \frac{f_{ck}}{1.5}; \frac{a}{h} = 0.10$$

μ	$\nu=0.6$ ϖ	α	$\nu=0.8$ ϖ	α	$\nu=1.0$ ϖ	α	$\nu=1.2$ ϖ	α	$\nu=1.4$ ϖ	α	$\nu=1.6$ ϖ	α
0.250	0.398	0.649	0.527	0.760	0.685	0.851	0.860	0.927	1.047	0.989	1.242	1.050
0.255	0.412	0.647	0.541	0.756	0.698	0.847	0.874	0.922	1.060	0.984	1.255	1.050
0.260	0.425	0.645	0.556	0.753	0.713	0.843	0.888	0.917	1.074	0.980	1.268	1.040
0.265	0.439	0.644	0.570	0.750	0.727	0.839	0.902	0.913	1.087	0.975	1.281	1.030
0.270	0.453	0.642	0.585	0.747	0.741	0.835	0.915	0.909	1.100	0.971	1.294	1.030
0.275	0.466	0.641	0.599	0.744	0.755	0.831	0.929	0.904	1.114	0.966	1.306	1.020
0.280	0.480	0.639	0.612	0.741	0.769	0.827	0.943	0.900	1.127	0.962	1.319	1.020
0.285	0.493	0.638	0.627	0.738	0.784	0.824	0.956	0.896	1.140	0.957	1.332	1.010
0.290	0.506	0.637	0.641	0.735	0.797	0.820	0.970	0.892	1.154	0.953	1.345	1.010
0.295	0.520	0.635	0.655	0.733	0.812	0.817	0.984	0.888	1.167	0.949	1.358	1.000
0.300	0.534	0.634	0.669	0.730	0.825	0.813	0.998	0.884	1.181	0.945	1.371	0.997
0.305	0.547	0.633	0.683	0.727	0.840	0.810	1.011	0.880	1.194	0.941	1.384	0.993
0.310	0.560	0.632	0.697	0.725	0.854	0.806	1.026	0.876	1.208	0.937	1.397	0.989
0.315	0.574	0.631	0.711	0.723	0.868	0.803	1.039	0.873	1.221	0.933	1.410	0.985
0.320	0.587	0.629	0.725	0.720	0.882	0.800	1.053	0.869	1.234	0.929	1.423	0.981
0.325	0.600	0.628	0.739	0.718	0.896	0.797	1.067	0.866	1.248	0.925	1.437	0.977
0.330	0.614	0.627	0.752	0.716	0.910	0.794	1.080	0.862	1.262	0.921	1.450	0.973
0.335	0.627	0.626	0.766	0.714	0.924	0.791	1.094	0.859	1.275	0.918	1.463	0.969
0.340	0.640	0.625	0.780	0.712	0.937	0.788	1.108	0.855	1.288	0.914	1.476	0.965
0.345	0.653	0.624	0.794	0.710	0.951	0.786	1.122	0.852	1.302	0.910	1.489	0.962
0.350	0.667	0.623	0.807	0.708	0.965	0.783	1.135	0.849	1.315	0.907	1.503	0.958
0.355	0.680	0.623	0.821	0.706	0.979	0.780	1.149	0.846	1.329	0.904	1.516	0.954
0.360	0.693	0.622	0.835	0.704	0.993	0.778	1.163	0.843	1.342	0.900	1.529	0.951
0.365	0.706	0.621	0.849	0.702	1.007	0.775	1.177	0.840	1.356	0.897	1.543	0.947
0.370	0.719	0.620	0.863	0.700	1.020	0.773	1.190	0.837	1.370	0.893	1.556	0.944
0.375	0.732	0.619	0.876	0.699	1.034	0.770	1.204	0.834	1.383	0.890	1.569	0.940
0.380	0.746	0.618	0.890	0.697	1.048	0.768	1.218	0.831	1.397	0.887	1.582	0.937
0.385	0.759	0.618	0.903	0.695	1.061	0.765	1.232	0.828	1.410	0.884	1.596	0.934
0.390	0.772	0.617	0.917	0.693	1.075	0.763	1.245	0.825	1.424	0.881	1.609	0.930
0.395	0.785	0.616	0.930	0.692	1.089	0.761	1.259	0.823	1.437	0.878	1.622	0.927
0.400	0.798	0.615	0.944	0.690	1.103	0.759	1.272	0.820	1.451	0.875	1.636	0.924
0.410	0.824	0.614	0.971	0.687	1.130	0.754	1.300	0.815	1.478	0.869	1.662	0.918
0.420	0.850	0.613	0.998	0.684	1.158	0.750	1.327	0.810	1.505	0.863	1.689	0.912
0.430	0.876	0.611	1.025	0.682	1.184	0.746	1.355	0.805	1.532	0.858	1.716	0.906
0.440	0.902	0.610	1.051	0.679	1.211	0.743	1.381	0.800	1.559	0.853	1.743	0.900
0.450	0.928	0.609	1.078	0.677	1.239	0.739	1.409	0.796	1.586	0.848	1.769	0.894
0.460	0.954	0.608	1.104	0.674	1.266	0.735	1.436	0.791	1.613	0.843	1.796	0.889
0.470	0.980	0.607	1.131	0.672	1.293	0.732	1.463	0.787	1.640	0.838	1.823	0.884
0.480	1.006	0.606	1.158	0.669	1.319	0.729	1.490	0.783	1.667	0.833	1.850	0.879
0.490	1.031	0.605	1.185	0.667	1.347	0.726	1.517	0.779	1.694	0.829	1.876	0.874
0.500	1.057	0.604	1.211	0.665	1.373	0.723	1.544	0.775	1.721	0.824	1.903	0.869
0.510	1.083	0.603	1.237	0.663	1.400	0.720	1.571	0.772	1.748	0.820	1.930	0.864
0.520	1.109	0.602	1.264	0.661	1.427	0.717	1.598	0.768	1.774	0.816	1.956	0.860
0.530	1.134	0.601	1.290	0.659	1.454	0.714	1.625	0.765	1.801	0.812	1.983	0.855
0.540	1.160	0.600	1.316	0.658	1.480	0.711	1.651	0.761	1.828	0.808	2.010	0.851
0.550	1.185	0.600	1.343	0.656	1.507	0.709	1.678	0.758	1.854	0.804	2.036	0.847
0.560	1.211	0.599	1.369	0.654	1.533	0.706	1.705	0.755	1.882	0.800	2.063	0.842
0.570	1.237	0.598	1.395	0.653	1.560	0.704	1.731	0.752	1.908	0.797	2.090	0.838
0.580	1.262	0.597	1.421	0.651	1.586	0.702	1.758	0.749	1.935	0.793	2.116	0.835
0.590	1.288	0.597	1.447	0.649	1.613	0.699	1.785	0.746	1.961	0.790	2.143	0.831

Tab. 5.12

TABLE 3_Concrete classes C12-C50_Steel class S500_a/h=0.15

Bending and axial force with symmetric reinforcement

$$\alpha = \frac{x}{h}; \quad \mu = \frac{M_{Ed}}{bh^2 f_{cd}}; \quad v = \frac{N_{Ed}}{bh f_{cd}}; \quad \varpi = \frac{A_s}{bh}\frac{f_{yd}}{f_{cd}}$$

$$A_s = A + A'; \quad A = A'$$

$$f_{yd} = \frac{f_{yk}}{1.15}; \quad f_{cd} = \frac{f_{ck}}{1.5}; \quad \frac{a}{h} = 0.15$$

	$v=0.0$		$v=0.1$		$v=0.2$		$v=0.3$		$v=0.4$		$v=0.5$	
μ	ϖ	α	ϖ	α	ϖ	α	ϖ	α	ϖ	α	ϖ	α
0.000	0.000	-	0.000	-	0.000	-	0.000	-	0.000	-	0.000	-
0.005	0.010	0.026	0.000	-	0.000	-	0.000	-	0.000	-	0.000	-
0.010	0.021	0.038	0.000	-	0.000	-	0.000	-	0.000	-	0.000	-
0.015	0.031	0.049	0.000	-	0.000	-	0.000	-	0.000	-	0.000	-
0.020	0.042	0.060	0.000	-	0.000	-	0.000	-	0.000	-	0.000	-
0.025	0.053	0.070	0.000	-	0.000	-	0.000	-	0.000	-	0.000	-
0.030	0.064	0.081	0.000	-	0.000	-	0.000	-	0.000	-	0.000	-
0.035	0.076	0.093	0.000	-	0.000	-	0.000	-	0.000	-	0.000	-
0.040	0.089	0.100	0.000	-	0.000	-	0.000	-	0.000	-	0.000	-
0.045	0.103	0.106	0.001	0.124	0.000	-	0.000	-	0.000	-	0.000	-
0.050	0.116	0.112	0.014	0.134	0.000	-	0.000	-	0.000	-	0.000	-
0.055	0.130	0.117	0.027	0.142	0.000	-	0.000	-	0.000	-	0.000	-
0.060	0.144	0.122	0.041	0.149	0.000	-	0.000	-	0.000	-	0.000	-
0.065	0.157	0.126	0.055	0.156	0.000	-	0.000	-	0.000	-	0.000	-
0.070	0.171	0.131	0.069	0.161	0.000	-	0.000	-	0.000	-	0.000	-
0.075	0.185	0.135	0.083	0.167	0.000	-	0.000	-	0.000	-	0.000	-
0.080	0.199	0.139	0.097	0.171	0.002	0.247	0.000	-	0.000	-	0.000	-
0.085	0.213	0.143	0.112	0.176	0.016	0.251	0.000	-	0.000	-	0.000	-
0.090	0.227	0.146	0.126	0.180	0.031	0.254	0.000	-	0.000	-	0.000	-
0.095	0.241	0.150	0.140	0.184	0.046	0.256	0.000	-	0.000	-	0.000	-
0.100	0.255	0.153	0.155	0.188	0.060	0.259	0.000	-	0.000	-	0.000	-
0.105	0.269	0.156	0.169	0.191	0.075	0.262	0.004	0.371	0.000	-	0.000	-
0.110	0.284	0.159	0.183	0.195	0.090	0.264	0.018	0.371	0.000	-	0.000	-
0.115	0.298	0.162	0.198	0.198	0.104	0.266	0.032	0.371	0.000	-	0.000	-
0.120	0.312	0.165	0.212	0.201	0.119	0.268	0.047	0.372	0.006	0.494	0.000	-
0.125	0.326	0.168	0.226	0.204	0.134	0.270	0.061	0.372	0.021	0.494	0.012	0.615
0.130	0.340	0.170	0.241	0.207	0.148	0.272	0.076	0.372	0.035	0.494	0.029	0.611
0.135	0.355	0.173	0.256	0.209	0.163	0.274	0.090	0.373	0.049	0.494	0.047	0.607
0.140	0.369	0.175	0.270	0.212	0.178	0.276	0.105	0.373	0.064	0.494	0.063	0.604
0.145	0.383	0.178	0.284	0.214	0.192	0.278	0.119	0.373	0.078	0.494	0.080	0.601
0.150	0.397	0.180	0.299	0.217	0.207	0.279	0.133	0.374	0.092	0.494	0.097	0.598
0.155	0.412	0.182	0.313	0.219	0.222	0.281	0.148	0.374	0.106	0.494	0.113	0.596
0.160	0.426	0.184	0.327	0.221	0.236	0.283	0.162	0.374	0.121	0.494	0.129	0.593
0.165	0.440	0.186	0.342	0.223	0.251	0.284	0.176	0.374	0.135	0.494	0.146	0.591
0.170	0.455	0.189	0.356	0.225	0.265	0.286	0.191	0.375	0.149	0.494	0.162	0.589
0.175	0.469	0.191	0.371	0.227	0.280	0.287	0.205	0.375	0.164	0.494	0.178	0.587
0.180	0.483	0.192	0.386	0.229	0.295	0.288	0.220	0.375	0.178	0.494	0.193	0.585
0.185	0.498	0.194	0.400	0.231	0.309	0.290	0.234	0.375	0.192	0.494	0.209	0.584
0.190	0.512	0.196	0.415	0.233	0.324	0.291	0.248	0.376	0.206	0.494	0.225	0.582
0.195	0.526	0.198	0.429	0.235	0.338	0.292	0.263	0.376	0.221	0.494	0.240	0.580
0.200	0.541	0.200	0.443	0.236	0.352	0.293	0.277	0.376	0.235	0.494	0.256	0.579
0.205	0.555	0.202	0.458	0.238	0.367	0.295	0.291	0.376	0.249	0.494	0.271	0.577
0.210	0.569	0.203	0.472	0.240	0.381	0.296	0.306	0.377	0.264	0.494	0.287	0.576
0.215	0.584	0.205	0.486	0.241	0.396	0.297	0.320	0.377	0.278	0.494	0.302	0.575
0.220	0.598	0.207	0.501	0.243	0.411	0.298	0.335	0.377	0.292	0.494	0.317	0.574
0.225	0.612	0.208	0.515	0.244	0.425	0.299	0.349	0.377	0.306	0.494	0.333	0.573
0.230	0.627	0.210	0.530	0.246	0.440	0.300	0.363	0.377	0.321	0.494	0.348	0.571
0.235	0.641	0.211	0.544	0.247	0.454	0.301	0.378	0.378	0.335	0.494	0.363	0.570
0.240	0.656	0.213	0.559	0.249	0.469	0.302	0.392	0.378	0.349	0.494	0.378	0.569
0.245	0.670	0.214	0.574	0.250	0.483	0.303	0.406	0.378	0.364	0.494	0.393	0.568

TABLE 3_Concrete classes C12-C50_Steel class S500_a/h=0.15 (cont.)
Bending and axial force with symmetric reinforcement

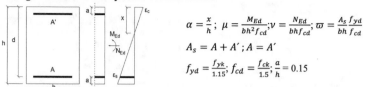

$$\alpha = \frac{x}{h}; \quad \mu = \frac{M_{Ed}}{bh^2 f_{cd}}; \nu = \frac{N_{Ed}}{bh f_{cd}}; \varpi = \frac{A_s\, f_{yd}}{bh\, f_{cd}}$$

$$A_s = A + A'\,; A = A'$$

$$f_{yd} = \frac{f_{yk}}{1.15}; \, f_{cd} = \frac{f_{ck}}{1.5}, \frac{a}{h} = 0.15$$

μ	ν=0.6 ϖ	ν=0.6 α	ν=0.8 ϖ	ν=0.8 α	ν=1.0 ϖ	ν=1.0 α	ν=1.2 ϖ	ν=1.2 α	ν=1.4 ϖ	ν=1.4 α	ν=1.6 ϖ	ν=1.6 α
0.000	0.000	-	0.000	-	0.000	-	0.219	-	0.435	-	0.650	-
0.005	0.000	-	0.000	-	0.014	2.690	0.222	6.180	0.438	10.900	0.656	15.477
0.010	0.000	-	0.000	-	0.028	2.050	0.229	3.690	0.443	5.830	0.659	8.146
0.015	0.000	-	0.000	-	0.042	1.770	0.243	2.690	0.448	4.200	0.664	5.690
0.020	0.000	-	0.000	-	0.056	1.610	0.257	2.240	0.457	3.270	0.668	4.467
0.025	0.000	-	0.000	-	0.070	1.500	0.271	1.970	0.471	2.700	0.673	3.730
0.030	0.000	-	0.000	-	0.084	1.410	0.285	1.790	0.486	2.340	0.686	3.113
0.035	0.000	-	0.000	-	0.099	1.350	0.299	1.670	0.499	2.110	0.700	2.700
0.040	0.000	-	0.000	-	0.113	1.290	0.313	1.570	0.514	1.940	0.714	2.415
0.045	0.000	-	0.000	-	0.127	1.250	0.328	1.490	0.528	1.810	0.728	2.207
0.050	0.000	-	0.000	-	0.141	1.220	0.342	1.430	0.542	1.710	0.742	2.046
0.055	0.000	-	0.000	-	0.155	1.180	0.356	1.380	0.556	1.620	0.757	1.920
0.060	0.000	-	0.000	-	0.169	1.160	0.370	1.330	0.571	1.550	0.771	1.816
0.065	0.000	-	0.000	-	0.183	1.130	0.384	1.300	0.585	1.490	0.785	1.730
0.070	0.000	-	0.000	-	0.197	1.110	0.398	1.260	0.599	1.440	0.799	1.658
0.075	0.000	-	0.012	0.979	0.211	1.090	0.412	1.230	0.613	1.400	0.813	1.595
0.080	0.000	-	0.027	0.968	0.226	1.080	0.426	1.210	0.627	1.360	0.828	1.541
0.085	0.000	-	0.043	0.957	0.240	1.060	0.440	1.180	0.641	1.330	0.842	1.493
0.090	0.000	-	0.058	0.946	0.254	1.040	0.455	1.160	0.655	1.300	0.856	1.451
0.095	0.000	-	0.074	0.936	0.268	1.030	0.469	1.140	0.669	1.270	0.870	1.414
0.100	0.000	-	0.091	0.925	0.282	1.020	0.483	1.120	0.684	1.240	0.884	1.380
0.105	0.000	-	0.106	0.915	0.296	1.010	0.497	1.110	0.698	1.220	0.899	1.349
0.110	0.000	-	0.123	0.905	0.311	0.998	0.512	1.090	0.712	1.200	0.913	1.321
0.115	0.000	-	0.140	0.895	0.325	0.989	0.526	1.080	0.726	1.180	0.927	1.296
0.120	0.019	0.732	0.157	0.885	0.340	0.980	0.540	1.060	0.740	1.160	0.941	1.273
0.125	0.038	0.724	0.174	0.876	0.355	0.972	0.554	1.050	0.755	1.150	0.955	1.251
0.130	0.057	0.717	0.190	0.867	0.370	0.963	0.568	1.040	0.769	1.130	0.970	1.231
0.135	0.075	0.710	0.207	0.858	0.385	0.955	0.582	1.030	0.783	1.120	0.984	1.213
0.140	0.094	0.703	0.224	0.850	0.401	0.947	0.597	1.020	0.797	1.100	0.998	1.196
0.145	0.112	0.697	0.241	0.842	0.416	0.939	0.611	1.010	0.812	1.090	1.012	1.180
0.150	0.130	0.691	0.258	0.834	0.432	0.931	0.625	1.000	0.826	1.080	1.026	1.164
0.155	0.148	0.685	0.275	0.826	0.448	0.923	0.640	0.995	0.840	1.070	1.040	1.150
0.160	0.165	0.680	0.293	0.819	0.464	0.916	0.654	0.988	0.854	1.060	1.055	1.137
0.165	0.183	0.675	0.310	0.812	0.479	0.909	0.669	0.981	0.868	1.050	1.069	1.125
0.170	0.200	0.671	0.327	0.805	0.495	0.901	0.684	0.974	0.882	1.040	1.083	1.113
0.175	0.217	0.667	0.344	0.798	0.511	0.894	0.699	0.967	0.897	1.030	1.097	1.102
0.180	0.234	0.662	0.361	0.792	0.528	0.887	0.713	0.960	0.911	1.020	1.111	1.091
0.185	0.251	0.659	0.378	0.786	0.543	0.881	0.729	0.954	0.925	1.010	1.126	1.081
0.190	0.268	0.655	0.395	0.780	0.560	0.874	0.744	0.947	0.939	1.010	1.140	1.071
0.195	0.284	0.652	0.412	0.774	0.575	0.868	0.759	0.941	0.953	0.999	1.154	1.062
0.200	0.301	0.648	0.429	0.769	0.592	0.862	0.775	0.934	0.968	0.993	1.168	1.054
0.205	0.318	0.645	0.445	0.763	0.608	0.856	0.790	0.928	0.983	0.987	1.182	1.045
0.210	0.334	0.642	0.462	0.758	0.624	0.850	0.805	0.922	0.997	0.981	1.197	1.037
0.215	0.351	0.639	0.479	0.753	0.640	0.844	0.821	0.916	1.012	0.975	1.211	1.030
0.220	0.367	0.637	0.496	0.749	0.657	0.839	0.836	0.911	1.027	0.969	1.225	1.023
0.225	0.383	0.634	0.513	0.744	0.673	0.833	0.852	0.905	1.042	0.964	1.239	1.016
0.230	0.399	0.632	0.529	0.740	0.689	0.828	0.867	0.899	1.057	0.958	1.253	1.009
0.235	0.415	0.629	0.546	0.735	0.705	0.823	0.883	0.894	1.072	0.953	1.268	1.003
0.240	0.431	0.627	0.562	0.732	0.722	0.818	0.898	0.889	1.087	0.947	1.282	0.997
0.245	0.447	0.625	0.579	0.728	0.738	0.813	0.914	0.883	1.102	0.942	1.297	0.992

TABLE 3_Concrete classes C12-C50_Steel class S500_a/h=0.15 (cont.)
Bending and axial force with symmetric reinforcement

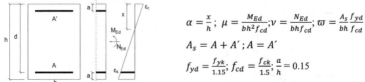

$$\alpha = \frac{x}{h}; \ \mu = \frac{M_{Ed}}{bh^2 f_{cd}}; v = \frac{N_{Ed}}{bhf_{cd}}; \ \varpi = \frac{A_s\, f_{yd}}{bh\, f_{cd}}$$

$$A_s = A + A'\ ; A = A'$$

$$f_{yd} = \frac{f_{yk}}{1.15}; \ f_{cd} = \frac{f_{ck}}{1.5}; \ \frac{a}{h} = 0.15$$

	$v=0.0$		$v=0.1$		$v=0.2$		$v=0.3$		$v=0.4$		$v=0.5$	
μ	ϖ	α	ϖ	α	ϖ	α	ϖ	α	ϖ	α	ϖ	α
0.250	0.684	0.216	0.588	0.251	0.498	0.304	0.421	0.378	0.378	0.494	0.408	0.568
0.255	0.699	0.217	0.602	0.252	0.512	0.305	0.435	0.378	0.392	0.494	0.424	0.567
0.260	0.713	0.218	0.617	0.254	0.527	0.306	0.450	0.378	0.406	0.494	0.439	0.566
0.265	0.727	0.220	0.631	0.255	0.541	0.307	0.464	0.379	0.421	0.494	0.454	0.565
0.270	0.742	0.221	0.646	0.256	0.556	0.307	0.478	0.379	0.435	0.494	0.468	0.564
0.275	0.756	0.222	0.660	0.257	0.570	0.308	0.493	0.379	0.449	0.494	0.483	0.563
0.280	0.771	0.224	0.675	0.259	0.585	0.309	0.507	0.379	0.464	0.494	0.498	0.563
0.285	0.785	0.225	0.689	0.260	0.599	0.310	0.521	0.379	0.478	0.494	0.513	0.562
0.290	0.799	0.226	0.704	0.261	0.614	0.311	0.536	0.379	0.492	0.494	0.528	0.561
0.295	0.814	0.227	0.718	0.262	0.628	0.311	0.550	0.380	0.506	0.494	0.543	0.561
0.300	0.828	0.229	0.733	0.263	0.642	0.312	0.564	0.380	0.521	0.494	0.558	0.560
0.305	0.843	0.230	0.747	0.264	0.657	0.313	0.578	0.380	0.535	0.494	0.573	0.560
0.310	0.857	0.231	0.761	0.265	0.672	0.314	0.593	0.380	0.549	0.494	0.587	0.559
0.315	0.871	0.232	0.776	0.266	0.686	0.314	0.607	0.380	0.564	0.494	0.602	0.558
0.320	0.886	0.233	0.790	0.267	0.700	0.315	0.621	0.380	0.578	0.494	0.617	0.558
0.325	0.900	0.234	0.805	0.268	0.715	0.316	0.636	0.380	0.592	0.494	0.632	0.557
0.330	0.915	0.235	0.819	0.269	0.730	0.316	0.650	0.381	0.606	0.494	0.647	0.557
0.335	0.929	0.236	0.834	0.270	0.744	0.317	0.665	0.381	0.621	0.494	0.661	0.556
0.340	0.943	0.237	0.848	0.271	0.758	0.318	0.679	0.381	0.635	0.494	0.676	0.556
0.345	0.958	0.238	0.862	0.272	0.773	0.318	0.693	0.381	0.649	0.494	0.691	0.555
0.350	0.972	0.239	0.877	0.273	0.787	0.319	0.708	0.381	0.664	0.494	0.705	0.555
0.355	0.987	0.240	0.891	0.274	0.802	0.319	0.722	0.381	0.678	0.494	0.720	0.554
0.360	1.001	0.241	0.906	0.274	0.816	0.320	0.736	0.381	0.692	0.494	0.735	0.554
0.365	1.015	0.242	0.920	0.275	0.831	0.321	0.751	0.381	0.706	0.494	0.749	0.554
0.370	1.030	0.243	0.935	0.276	0.845	0.321	0.765	0.381	0.721	0.494	0.764	0.553
0.375	1.044	0.244	0.949	0.277	0.860	0.322	0.779	0.382	0.735	0.494	0.779	0.553
0.380	1.059	0.245	0.964	0.278	0.874	0.322	0.794	0.382	0.749	0.494	0.793	0.552
0.385	1.073	0.246	0.978	0.279	0.888	0.323	0.808	0.382	0.764	0.494	0.808	0.552
0.390	1.087	0.247	0.993	0.279	0.903	0.323	0.822	0.382	0.778	0.494	0.823	0.552
0.395	1.102	0.248	1.007	0.280	0.917	0.324	0.837	0.382	0.792	0.494	0.837	0.551
0.400	1.116	0.249	1.021	0.281	0.932	0.324	0.851	0.382	0.806	0.494	0.852	0.551
0.410	1.145	0.251	1.050	0.282	0.960	0.325	0.879	0.382	0.835	0.494	0.881	0.550
0.420	1.174	0.252	1.079	0.284	0.990	0.326	0.908	0.383	0.864	0.494	0.910	0.550
0.430	1.203	0.254	1.108	0.285	1.018	0.327	0.937	0.383	0.892	0.494	0.939	0.549
0.440	1.231	0.255	1.137	0.287	1.047	0.328	0.965	0.383	0.921	0.494	0.968	0.549
0.450	1.260	0.257	1.166	0.288	1.076	0.329	0.994	0.383	0.949	0.494	0.997	0.548
0.460	1.289	0.258	1.195	0.289	1.104	0.330	1.023	0.383	0.978	0.494	1.026	0.548
0.470	1.318	0.260	1.224	0.291	1.133	0.331	1.051	0.383	1.006	0.494	1.055	0.547
0.480	1.346	0.261	1.252	0.292	1.162	0.332	1.080	0.384	1.035	0.494	1.084	0.547
0.490	1.375	0.263	1.281	0.293	1.191	0.333	1.108	0.384	1.064	0.494	1.113	0.546
0.500	1.404	0.264	1.310	0.294	1.220	0.333	1.137	0.384	1.092	0.494	1.142	0.546
0.510	1.433	0.265	1.339	0.295	1.249	0.334	1.166	0.384	1.121	0.494	1.171	0.545
0.520	1.462	0.267	1.367	0.296	1.277	0.335	1.194	0.384	1.149	0.494	1.200	0.545
0.530	1.490	0.268	1.396	0.298	1.306	0.336	1.223	0.384	1.178	0.494	1.229	0.544
0.540	1.519	0.269	1.425	0.299	1.335	0.336	1.252	0.385	1.206	0.494	1.258	0.544
0.550	1.548	0.271	1.454	0.300	1.364	0.337	1.281	0.385	1.235	0.494	1.287	0.544
0.560	1.577	0.272	1.483	0.301	1.393	0.338	1.309	0.385	1.264	0.494	1.316	0.543
0.570	1.606	0.273	1.511	0.302	1.422	0.338	1.338	0.385	1.292	0.494	1.345	0.543
0.580	1.634	0.274	1.540	0.303	1.450	0.339	1.366	0.385	1.321	0.494	1.374	0.543
0.590	1.663	0.275	1.569	0.304	1.479	0.340	1.395	0.385	1.349	0.494	1.402	0.542

TABLE 3_Concrete classes C12-C50_Steel class S500_a/h=0.15 (cont.)
Bending and axial force with symmetric reinforcement

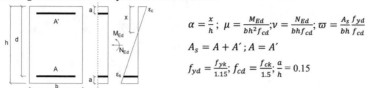

$$\alpha = \frac{x}{h}\,; \quad \mu = \frac{M_{Ed}}{bh^2 f_{cd}}\,; \nu = \frac{N_{Ed}}{bh f_{cd}}\,; \varpi = \frac{A_s\,f_{yd}}{bh\,f_{cd}}$$

$$A_s = A + A'\,; A = A'$$

$$f_{yd} = \frac{f_{yk}}{1.15}\,; f_{cd} = \frac{f_{ck}}{1.5}\,, \frac{a}{h} = 0.15$$

μ	$\nu=0.6$ ϖ	α	$\nu=0.8$ ϖ	α	$\nu=1.0$ ϖ	α	$\nu=1.2$ ϖ	α	$\nu=1.4$ ϖ	α	$\nu=1.6$ ϖ	α
0.250	0.463	0.623	0.595	0.724	0.754	0.808	0.930	0.878	1.117	0.937	1.311	0.986
0.255	0.479	0.621	0.612	0.720	0.770	0.804	0.946	0.873	1.132	0.932	1.326	0.981
0.260	0.495	0.619	0.628	0.717	0.786	0.799	0.961	0.868	1.147	0.927	1.341	0.976
0.265	0.510	0.617	0.645	0.713	0.802	0.795	0.977	0.864	1.162	0.922	1.355	0.971
0.270	0.526	0.616	0.660	0.710	0.819	0.791	0.993	0.859	1.178	0.917	1.370	0.966
0.275	0.541	0.614	0.677	0.707	0.835	0.787	1.008	0.854	1.193	0.912	1.385	0.961
0.280	0.557	0.612	0.693	0.704	0.851	0.783	1.024	0.850	1.208	0.907	1.400	0.957
0.285	0.572	0.611	0.709	0.701	0.867	0.779	1.040	0.846	1.223	0.903	1.415	0.952
0.290	0.588	0.609	0.726	0.698	0.883	0.775	1.056	0.841	1.239	0.898	1.430	0.947
0.295	0.604	0.608	0.741	0.695	0.899	0.771	1.071	0.837	1.254	0.894	1.445	0.943
0.300	0.619	0.606	0.757	0.692	0.915	0.768	1.087	0.833	1.270	0.889	1.460	0.938
0.305	0.634	0.605	0.773	0.690	0.931	0.764	1.103	0.829	1.285	0.885	1.475	0.934
0.310	0.649	0.604	0.789	0.687	0.947	0.761	1.119	0.825	1.300	0.881	1.490	0.930
0.315	0.665	0.602	0.805	0.685	0.963	0.758	1.134	0.821	1.316	0.877	1.505	0.925
0.320	0.680	0.601	0.821	0.682	0.979	0.754	1.150	0.817	1.331	0.873	1.520	0.921
0.325	0.696	0.6	0.837	0.680	0.995	0.751	1.166	0.814	1.347	0.869	1.535	0.917
0.330	0.711	0.599	0.852	0.678	1.011	0.748	1.181	0.810	1.362	0.865	1.550	0.913
0.335	0.726	0.598	0.868	0.675	1.027	0.745	1.197	0.807	1.378	0.861	1.565	0.909
0.340	0.741	0.597	0.884	0.673	1.042	0.742	1.213	0.803	1.393	0.857	1.580	0.905
0.345	0.756	0.596	0.900	0.671	1.058	0.739	1.229	0.800	1.409	0.853	1.595	0.901
0.350	0.771	0.594	0.916	0.669	1.074	0.737	1.245	0.797	1.424	0.850	1.611	0.897
0.355	0.786	0.593	0.931	0.667	1.090	0.734	1.260	0.793	1.440	0.846	1.626	0.894
0.360	0.801	0.593	0.947	0.665	1.105	0.731	1.276	0.790	1.455	0.843	1.641	0.890
0.365	0.817	0.592	0.962	0.663	1.121	0.729	1.292	0.787	1.470	0.840	1.656	0.886
0.370	0.832	0.591	0.978	0.661	1.137	0.726	1.307	0.784	1.486	0.836	1.671	0.883
0.375	0.847	0.59	0.993	0.660	1.152	0.724	1.323	0.781	1.502	0.833	1.687	0.879
0.380	0.862	0.589	1.009	0.658	1.168	0.721	1.338	0.778	1.517	0.830	1.702	0.876
0.385	0.876	0.588	1.024	0.656	1.184	0.719	1.354	0.775	1.533	0.826	1.717	0.872
0.390	0.892	0.587	1.040	0.655	1.200	0.717	1.370	0.773	1.548	0.823	1.732	0.869
0.395	0.907	0.586	1.055	0.653	1.215	0.714	1.385	0.770	1.563	0.820	1.748	0.866
0.400	0.922	0.586	1.070	0.651	1.231	0.712	1.401	0.767	1.579	0.817	1.763	0.862
0.410	0.951	0.584	1.101	0.648	1.262	0.708	1.432	0.762	1.610	0.811	1.793	0.856
0.420	0.981	0.583	1.132	0.646	1.293	0.704	1.463	0.757	1.641	0.806	1.824	0.850
0.430	1.011	0.581	1.163	0.643	1.324	0.700	1.494	0.752	1.671	0.800	1.855	0.844
0.440	1.041	0.58	1.193	0.640	1.355	0.696	1.525	0.748	1.702	0.795	1.885	0.838
0.450	1.070	0.579	1.223	0.638	1.385	0.693	1.556	0.743	1.733	0.790	1.916	0.833
0.460	1.100	0.578	1.254	0.635	1.416	0.689	1.587	0.739	1.764	0.785	1.946	0.827
0.470	1.129	0.576	1.284	0.633	1.447	0.686	1.618	0.735	1.794	0.780	1.977	0.822
0.480	1.159	0.575	1.314	0.631	1.478	0.683	1.649	0.731	1.825	0.776	2.007	0.817
0.490	1.188	0.574	1.344	0.629	1.509	0.680	1.679	0.727	1.856	0.771	2.038	0.812
0.500	1.218	0.573	1.375	0.627	1.539	0.677	1.710	0.724	1.887	0.767	2.068	0.807
0.510	1.247	0.572	1.404	0.625	1.569	0.674	1.740	0.720	1.917	0.763	2.099	0.803
0.520	1.277	0.571	1.435	0.623	1.600	0.671	1.771	0.717	1.948	0.759	2.129	0.798
0.530	1.306	0.571	1.465	0.621	1.630	0.669	1.801	0.713	1.978	0.755	2.160	0.794
0.540	1.335	0.57	1.495	0.619	1.660	0.666	1.832	0.710	2.009	0.751	2.190	0.790
0.550	1.364	0.569	1.524	0.618	1.691	0.664	1.863	0.707	2.039	0.748	2.221	0.786
0.560	1.393	0.568	1.554	0.616	1.721	0.661	1.893	0.704	2.070	0.744	2.251	0.782
0.570	1.423	0.567	1.584	0.614	1.751	0.659	1.924	0.701	2.100	0.741	2.281	0.778
0.580	1.452	0.567	1.614	0.613	1.781	0.657	1.953	0.698	2.131	0.737	2.312	0.774
0.590	1.481	0.566	1.644	0.611	1.811	0.655	1.984	0.695	2.161	0.734	2.342	0.770

Tab. 5.13

TABLE 3_Concrete classes C12-C50_Steel class S400_a/h=0.10

Bending and axial force with symmetric reinforcement

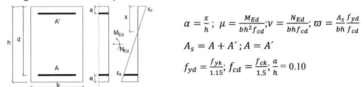

$$\alpha = \frac{x}{h}; \quad \mu = \frac{M_{Ed}}{bh^2 f_{cd}}; \nu = \frac{N_{Ed}}{bh f_{cd}}; \varpi = \frac{A_s f_{yd}}{bh f_{cd}}$$

$$A_s = A + A'; A = A'$$

$$f_{yd} = \frac{f_{yk}}{1.15}; f_{cd} = \frac{f_{ck}}{1.5}; \frac{a}{h} = 0.10$$

μ	$\nu=0.0$ ϖ	α	$\nu=0.1$ ϖ	α	$\nu=0.2$ ϖ	α	$\nu=0.3$ ϖ	α	$\nu=0.4$ ϖ	α	$\nu=0.5$ ϖ	α
0.000	0.000	-	0.000	-	0.000	-	0.000	-	0.000	-	0.000	-
0.005	0.010	0.029	0.000	-	0.000	-	0.000	-	0.000	-	0.000	-
0.010	0.021	0.042	0.000	-	0.000	-	0.000	-	0.000	-	0.000	-
0.015	0.032	0.051	0.000	-	0.000	-	0.000	-	0.000	-	0.000	-
0.020	0.043	0.059	0.000	-	0.000	-	0.000	-	0.000	-	0.000	-
0.025	0.054	0.066	0.000	-	0.000	-	0.000	-	0.000	-	0.000	-
0.030	0.066	0.071	0.000	-	0.000	-	0.000	-	0.000	-	0.000	-
0.035	0.078	0.077	0.000	-	0.000	-	0.000	-	0.000	-	0.000	-
0.040	0.090	0.081	0.000	-	0.000	-	0.000	-	0.000	-	0.000	-
0.045	0.102	0.086	0.000	-	0.000	-	0.000	-	0.000	-	0.000	-
0.050	0.114	0.089	0.013	0.128	0.000	-	0.000	-	0.000	-	0.000	-
0.055	0.126	0.093	0.025	0.132	0.000	-	0.000	-	0.000	-	0.000	-
0.060	0.139	0.096	0.038	0.135	0.000	-	0.000	-	0.000	-	0.000	-
0.065	0.151	0.099	0.051	0.138	0.000	-	0.000	-	0.000	-	0.000	-
0.070	0.163	0.101	0.063	0.140	0.000	-	0.000	-	0.000	-	0.000	-
0.075	0.176	0.104	0.076	0.142	0.000	-	0.000	-	0.000	-	0.000	-
0.080	0.188	0.106	0.088	0.144	0.002	0.247	0.000	-	0.000	-	0.000	-
0.085	0.201	0.108	0.101	0.146	0.014	0.247	0.000	-	0.000	-	0.000	-
0.090	0.213	0.109	0.114	0.148	0.027	0.247	0.000	-	0.000	-	0.000	-
0.095	0.226	0.111	0.126	0.150	0.039	0.247	0.000	-	0.000	-	0.000	-
0.100	0.238	0.113	0.139	0.151	0.052	0.247	0.000	-	0.000	-	0.000	-
0.105	0.250	0.115	0.151	0.152	0.064	0.247	0.003	0.370	0.000	-	0.000	-
0.110	0.263	0.116	0.164	0.154	0.077	0.247	0.016	0.370	0.000	-	0.000	-
0.115	0.275	0.118	0.176	0.155	0.089	0.247	0.028	0.370	0.000	-	0.000	-
0.120	0.288	0.119	0.189	0.156	0.102	0.247	0.041	0.370	0.006	0.494	0.000	-
0.125	0.300	0.121	0.202	0.157	0.114	0.247	0.053	0.370	0.018	0.494	0.009	0.617
0.130	0.313	0.122	0.214	0.158	0.127	0.247	0.066	0.370	0.031	0.494	0.022	0.617
0.135	0.325	0.124	0.227	0.159	0.139	0.247	0.078	0.370	0.043	0.494	0.035	0.616
0.140	0.338	0.125	0.239	0.160	0.152	0.247	0.091	0.370	0.056	0.494	0.048	0.616
0.145	0.350	0.126	0.252	0.161	0.164	0.247	0.103	0.370	0.068	0.494	0.061	0.615
0.150	0.363	0.127	0.264	0.162	0.177	0.247	0.116	0.370	0.081	0.494	0.073	0.615
0.155	0.375	0.128	0.277	0.162	0.189	0.247	0.128	0.370	0.093	0.494	0.086	0.614
0.160	0.388	0.130	0.289	0.163	0.202	0.247	0.141	0.370	0.106	0.494	0.099	0.614
0.165	0.401	0.131	0.302	0.164	0.214	0.247	0.153	0.370	0.118	0.494	0.112	0.613
0.170	0.413	0.132	0.314	0.164	0.227	0.247	0.166	0.370	0.131	0.494	0.125	0.613
0.175	0.426	0.133	0.327	0.165	0.239	0.247	0.178	0.370	0.143	0.494	0.137	0.613
0.180	0.438	0.134	0.340	0.166	0.252	0.247	0.191	0.370	0.156	0.494	0.150	0.613
0.185	0.451	0.135	0.352	0.166	0.264	0.247	0.203	0.370	0.168	0.494	0.163	0.612
0.190	0.463	0.135	0.365	0.167	0.277	0.247	0.216	0.370	0.181	0.494	0.176	0.612
0.195	0.476	0.136	0.377	0.168	0.289	0.247	0.228	0.370	0.193	0.494	0.189	0.612
0.200	0.488	0.137	0.390	0.168	0.302	0.247	0.241	0.370	0.206	0.494	0.201	0.611
0.205	0.501	0.138	0.402	0.169	0.314	0.247	0.253	0.370	0.218	0.494	0.214	0.611
0.210	0.513	0.139	0.415	0.169	0.327	0.247	0.266	0.370	0.231	0.494	0.226	0.611
0.215	0.526	0.140	0.427	0.170	0.339	0.247	0.278	0.370	0.243	0.494	0.239	0.611
0.220	0.538	0.140	0.440	0.170	0.352	0.247	0.291	0.370	0.256	0.494	0.252	0.611
0.225	0.551	0.141	0.453	0.170	0.364	0.247	0.303	0.370	0.268	0.494	0.265	0.610
0.230	0.563	0.142	0.465	0.171	0.377	0.247	0.316	0.370	0.281	0.494	0.277	0.610
0.235	0.576	0.142	0.478	0.171	0.389	0.247	0.328	0.370	0.293	0.494	0.290	0.610
0.240	0.588	0.143	0.490	0.172	0.402	0.247	0.341	0.370	0.306	0.494	0.302	0.610
0.245	0.601	0.144	0.502	0.172	0.414	0.247	0.353	0.370	0.318	0.494	0.315	0.610

TABLE 3_Concrete classes C12-C50_Steel class S400_a/h=0.10 (cont.)
Bending and axial force with symmetric reinforcement

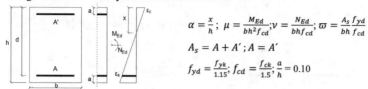

$$\alpha = \frac{x}{h}; \quad \mu = \frac{M_{Ed}}{bh^2 f_{cd}}; \nu = \frac{N_{Ed}}{bh f_{cd}}; \varpi = \frac{A_s}{bh} \frac{f_{yd}}{f_{cd}}$$

$$A_s = A + A'; A = A'$$

$$f_{yd} = \frac{f_{yk}}{1.15}; f_{cd} = \frac{f_{ck}}{1.5}; \frac{a}{h} = 0.10$$

μ	$\nu=0.6$ ϖ	α	$\nu=0.8$ ϖ	α	$\nu=1.0$ ϖ	α	$\nu=1.2$ ϖ	α	$\nu=1.4$ ϖ	α	$\nu=1.6$ ϖ	α
0.000	0.000	-	0.000	-	0.000	-	0.200	-	0.400	-	0.600	-
0.005	0.000	-	0.000	-	0.014	2.590	0.214	3.150	0.413	3.440	0.613	3.570
0.010	0.000	-	0.000	-	0.028	2.000	0.227	2.520	0.426	2.860	0.626	3.080
0.015	0.000	-	0.000	-	0.041	1.740	0.240	2.170	0.439	2.490	0.639	2.740
0.020	0.000	-	0.000	-	0.055	1.590	0.253	1.950	0.452	2.250	0.652	2.490
0.025	0.000	-	0.000	-	0.068	1.480	0.266	1.800	0.465	2.070	0.664	2.300
0.030	0.000	-	0.000	-	0.082	1.400	0.279	1.680	0.478	1.930	0.677	2.150
0.035	0.000	-	0.000	-	0.095	1.340	0.293	1.590	0.491	1.820	0.690	2.020
0.040	0.000	-	0.000	-	0.109	1.290	0.306	1.520	0.504	1.730	0.703	1.920
0.045	0.000	-	0.000	-	0.123	1.250	0.319	1.460	0.517	1.650	0.716	1.830
0.050	0.000	-	0.000	-	0.136	1.220	0.332	1.410	0.530	1.590	0.729	1.760
0.055	0.000	-	0.000	-	0.149	1.190	0.345	1.360	0.543	1.540	0.742	1.700
0.060	0.000	-	0.000	-	0.162	1.160	0.358	1.330	0.556	1.490	0.755	1.640
0.065	0.000	-	0.000	-	0.176	1.140	0.371	1.290	0.569	1.450	0.768	1.590
0.070	0.000	-	0.000	-	0.189	1.120	0.385	1.270	0.582	1.410	0.780	1.550
0.075	0.000	-	0.011	0.980	0.202	1.100	0.398	1.240	0.595	1.380	0.793	1.510
0.080	0.000	-	0.025	0.970	0.215	1.090	0.411	1.220	0.608	1.350	0.806	1.470
0.085	0.000	-	0.040	0.961	0.229	1.070	0.424	1.200	0.621	1.320	0.819	1.440
0.090	0.000	-	0.054	0.951	0.242	1.060	0.437	1.180	0.634	1.290	0.832	1.410
0.095	0.000	-	0.069	0.942	0.255	1.050	0.450	1.160	0.647	1.270	0.845	1.380
0.100	0.000	-	0.083	0.933	0.268	1.040	0.463	1.140	0.660	1.250	0.858	1.360
0.105	0.000	-	0.098	0.924	0.281	1.030	0.476	1.130	0.673	1.230	0.871	1.340
0.110	0.000	-	0.113	0.916	0.295	1.020	0.489	1.110	0.686	1.210	0.884	1.310
0.115	0.000	-	0.128	0.908	0.307	1.010	0.502	1.100	0.699	1.200	0.896	1.290
0.120	0.015	0.736	0.143	0.900	0.321	0.998	0.515	1.090	0.712	1.180	0.909	1.280
0.125	0.031	0.731	0.158	0.893	0.334	0.991	0.528	1.080	0.725	1.170	0.922	1.260
0.130	0.046	0.726	0.173	0.885	0.348	0.984	0.541	1.070	0.738	1.160	0.935	1.240
0.135	0.061	0.722	0.187	0.878	0.361	0.977	0.555	1.060	0.751	1.140	0.948	1.230
0.140	0.076	0.718	0.202	0.872	0.376	0.970	0.567	1.050	0.763	1.130	0.961	1.210
0.145	0.091	0.714	0.217	0.865	0.389	0.963	0.581	1.040	0.777	1.120	0.974	1.200
0.150	0.106	0.711	0.231	0.859	0.403	0.957	0.594	1.030	0.789	1.110	0.987	1.190
0.155	0.120	0.708	0.246	0.853	0.417	0.950	0.607	1.020	0.802	1.100	1.000	1.180
0.160	0.135	0.705	0.261	0.847	0.431	0.944	0.619	1.020	0.815	1.090	1.012	1.160
0.165	0.149	0.701	0.276	0.842	0.445	0.938	0.632	1.010	0.828	1.080	1.025	1.150
0.170	0.163	0.699	0.291	0.836	0.459	0.932	0.645	1.000	0.841	1.070	1.038	1.140
0.175	0.178	0.696	0.305	0.831	0.473	0.927	0.658	0.997	0.854	1.060	1.051	1.130
0.180	0.192	0.694	0.320	0.826	0.487	0.921	0.672	0.991	0.867	1.060	1.064	1.120
0.185	0.206	0.691	0.335	0.821	0.501	0.916	0.685	0.985	0.880	1.050	1.077	1.110
0.190	0.220	0.689	0.349	0.817	0.515	0.910	0.698	0.980	0.893	1.040	1.090	1.110
0.195	0.234	0.687	0.364	0.812	0.529	0.905	0.712	0.975	0.906	1.040	1.102	1.100
0.200	0.247	0.685	0.378	0.808	0.543	0.900	0.725	0.970	0.919	1.030	1.115	1.090
0.205	0.261	0.683	0.392	0.804	0.557	0.895	0.739	0.965	0.931	1.020	1.128	1.080
0.210	0.275	0.681	0.406	0.800	0.571	0.890	0.752	0.960	0.944	1.020	1.141	1.080
0.215	0.289	0.679	0.421	0.796	0.585	0.886	0.766	0.955	0.957	1.010	1.154	1.070
0.220	0.302	0.678	0.435	0.793	0.599	0.881	0.779	0.950	0.970	1.010	1.167	1.060
0.225	0.316	0.676	0.449	0.789	0.613	0.877	0.793	0.945	0.983	1.000	1.179	1.060
0.230	0.329	0.675	0.464	0.786	0.627	0.873	0.806	0.941	1.009	0.991	1.192	1.050
0.235	0.343	0.673	0.478	0.782	0.641	0.868	0.820	0.936	1.009	0.991	1.205	1.040
0.240	0.357	0.672	0.492	0.779	0.655	0.864	0.834	0.932	1.022	0.987	1.218	1.040
0.245	0.370	0.670	0.506	0.776	0.668	0.861	0.847	0.928	1.036	0.982	1.231	1.030

TABLE 3_Concrete classes C12-C50_Steel class S400_a/h=0.10 (cont.)
Bending and axial force with symmetric reinforcement

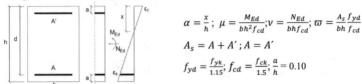

$$\alpha = \frac{x}{h}; \quad \mu = \frac{M_{Ed}}{bh^2 f_{cd}}; \nu = \frac{N_{Ed}}{bh f_{cd}}; \varpi = \frac{A_s f_{yd}}{bh f_{cd}}$$

$$A_s = A + A'; A = A'$$

$$f_{yd} = \frac{f_{yk}}{1.15}; f_{cd} = \frac{f_{ck}}{1.5}; \frac{a}{h} = 0.10$$

| | ν=0.0 | | ν=0.1 | | ν=0.2 | | ν=0.3 | | ν=0.4 | | ν=0.5 | |
μ	ϖ	α	ϖ	α	ϖ	α	ϖ	α	ϖ	α	ϖ	α
0.250	0.613	0.144	0.515	0.173	0.427	0.247	0.366	0.370	0.331	0.494	0.328	0.609
0.255	0.626	0.145	0.528	0.173	0.439	0.247	0.378	0.370	0.343	0.494	0.340	0.609
0.260	0.638	0.146	0.540	0.173	0.452	0.247	0.391	0.370	0.356	0.494	0.353	0.609
0.265	0.651	0.146	0.553	0.174	0.464	0.247	0.403	0.370	0.368	0.494	0.366	0.609
0.270	0.663	0.147	0.565	0.174	0.477	0.247	0.416	0.370	0.381	0.494	0.378	0.609
0.275	0.676	0.147	0.578	0.174	0.489	0.247	0.428	0.370	0.393	0.494	0.391	0.609
0.280	0.689	0.148	0.590	0.175	0.502	0.247	0.441	0.370	0.406	0.494	0.403	0.609
0.285	0.701	0.148	0.603	0.175	0.514	0.247	0.453	0.370	0.418	0.494	0.416	0.608
0.290	0.714	0.149	0.615	0.175	0.527	0.247	0.466	0.370	0.431	0.494	0.429	0.608
0.295	0.726	0.150	0.628	0.176	0.539	0.247	0.478	0.370	0.443	0.494	0.441	0.608
0.300	0.739	0.150	0.641	0.176	0.552	0.247	0.491	0.370	0.456	0.494	0.454	0.608
0.305	0.751	0.151	0.653	0.176	0.564	0.247	0.503	0.370	0.468	0.494	0.466	0.608
0.310	0.764	0.151	0.665	0.176	0.577	0.247	0.516	0.370	0.481	0.494	0.479	0.608
0.315	0.776	0.151	0.678	0.177	0.589	0.247	0.528	0.370	0.493	0.494	0.491	0.608
0.320	0.789	0.152	0.691	0.177	0.602	0.247	0.541	0.370	0.506	0.494	0.504	0.608
0.325	0.801	0.152	0.703	0.177	0.614	0.247	0.553	0.370	0.518	0.494	0.517	0.608
0.330	0.814	0.153	0.716	0.177	0.627	0.247	0.566	0.370	0.531	0.494	0.529	0.608
0.335	0.826	0.153	0.728	0.178	0.639	0.247	0.578	0.370	0.543	0.494	0.542	0.607
0.340	0.839	0.154	0.741	0.178	0.652	0.247	0.591	0.370	0.556	0.494	0.555	0.607
0.345	0.851	0.154	0.753	0.178	0.664	0.247	0.603	0.370	0.568	0.494	0.567	0.607
0.350	0.864	0.155	0.766	0.178	0.677	0.247	0.616	0.370	0.581	0.494	0.579	0.607
0.355	0.876	0.155	0.778	0.179	0.689	0.247	0.628	0.370	0.593	0.494	0.592	0.607
0.360	0.889	0.155	0.791	0.179	0.702	0.247	0.641	0.370	0.606	0.494	0.605	0.607
0.365	0.901	0.156	0.803	0.179	0.714	0.247	0.653	0.370	0.618	0.494	0.617	0.607
0.370	0.914	0.156	0.816	0.179	0.727	0.247	0.666	0.370	0.631	0.494	0.630	0.607
0.375	0.927	0.156	0.828	0.179	0.739	0.247	0.678	0.370	0.643	0.494	0.642	0.607
0.380	0.939	0.157	0.841	0.180	0.752	0.247	0.691	0.370	0.656	0.494	0.655	0.607
0.385	0.952	0.157	0.853	0.180	0.764	0.247	0.703	0.370	0.668	0.494	0.667	0.607
0.390	0.964	0.158	0.866	0.180	0.777	0.247	0.716	0.370	0.681	0.494	0.680	0.607
0.395	0.977	0.158	0.879	0.180	0.789	0.247	0.728	0.370	0.693	0.494	0.692	0.606
0.400	0.989	0.158	0.891	0.180	0.802	0.247	0.741	0.370	0.706	0.494	0.705	0.606
0.410	1.014	0.159	0.916	0.181	0.827	0.247	0.766	0.370	0.731	0.494	0.730	0.606
0.420	1.039	0.160	0.941	0.181	0.852	0.247	0.791	0.370	0.756	0.494	0.755	0.606
0.430	1.064	0.160	0.966	0.181	0.877	0.247	0.816	0.370	0.781	0.494	0.780	0.606
0.440	1.089	0.161	0.991	0.182	0.902	0.247	0.841	0.370	0.806	0.494	0.805	0.606
0.450	1.114	0.161	1.016	0.182	0.927	0.247	0.866	0.370	0.831	0.494	0.831	0.606
0.460	1.139	0.162	1.041	0.182	0.952	0.247	0.891	0.370	0.856	0.494	0.855	0.606
0.470	1.164	0.162	1.066	0.183	0.977	0.247	0.916	0.370	0.881	0.494	0.881	0.606
0.480	1.189	0.163	1.091	0.183	1.001	0.247	0.941	0.370	0.906	0.494	0.906	0.606
0.490	1.215	0.164	1.116	0.183	1.026	0.247	0.966	0.370	0.931	0.494	0.931	0.605
0.500	1.239	0.164	1.141	0.183	1.051	0.247	0.991	0.370	0.956	0.494	0.956	0.605
0.510	1.264	0.165	1.166	0.184	1.076	0.247	1.016	0.370	0.981	0.494	0.981	0.605
0.520	1.290	0.165	1.191	0.184	1.101	0.247	1.041	0.370	1.006	0.494	1.006	0.605
0.530	1.314	0.165	1.216	0.184	1.126	0.247	1.066	0.370	1.031	0.494	1.031	0.605
0.540	1.340	0.166	1.241	0.184	1.151	0.247	1.091	0.370	1.056	0.494	1.056	0.605
0.550	1.365	0.166	1.266	0.185	1.176	0.247	1.116	0.370	1.081	0.494	1.081	0.605
0.560	1.390	0.167	1.291	0.185	1.201	0.247	1.141	0.370	1.106	0.494	1.106	0.605
0.570	1.415	0.167	1.317	0.185	1.226	0.247	1.166	0.370	1.131	0.494	1.131	0.605
0.580	1.440	0.168	1.342	0.185	1.251	0.247	1.191	0.370	1.156	0.494	1.156	0.605
0.590	1.465	0.168	1.367	0.185	1.276	0.247	1.216	0.370	1.181	0.494	1.181	0.605

TABLE 3_Concrete classes C12-C50_Steel class S400_a/h=0.10 (cont.)
Bending and axial force with symmetric reinforcement

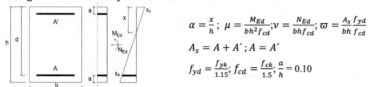

$$\alpha = \frac{x}{h}; \quad \mu = \frac{M_{Ed}}{bh^2 f_{cd}}; \nu = \frac{N_{Ed}}{bh f_{cd}}; \varpi = \frac{A_s f_{yd}}{bh f_{cd}}$$

$$A_s = A + A'; A = A'$$

$$f_{yd} = \frac{f_{yk}}{1.15}; f_{cd} = \frac{f_{ck}}{1.5}; \frac{a}{h} = 0.10$$

μ	ν=0.6 ϖ	α	ν=0.8 ϖ	α	ν=1.0 ϖ	α	ν=1.2 ϖ	α	ν=1.4 ϖ	α	ν=1.6 ϖ	α
0.250	0.383	0.669	0.520	0.773	0.682	0.857	0.861	0.923	1.049	0.978	1.244	1.030
0.255	0.397	0.668	0.534	0.770	0.696	0.853	0.875	0.919	1.062	0.974	1.256	1.020
0.260	0.410	0.667	0.548	0.767	0.710	0.849	0.888	0.915	1.076	0.969	1.269	1.020
0.265	0.423	0.665	0.562	0.765	0.724	0.846	0.902	0.911	1.089	0.965	1.282	1.010
0.270	0.437	0.664	0.576	0.762	0.738	0.842	0.915	0.908	1.102	0.961	1.295	1.010
0.275	0.450	0.663	0.590	0.760	0.752	0.839	0.929	0.904	1.115	0.957	1.308	1.000
0.280	0.463	0.662	0.604	0.757	0.766	0.836	0.943	0.900	1.128	0.954	1.321	0.999
0.285	0.476	0.661	0.617	0.755	0.780	0.832	0.956	0.896	1.142	0.950	1.334	0.995
0.290	0.490	0.660	0.631	0.752	0.794	0.829	0.970	0.893	1.155	0.946	1.347	0.991
0.295	0.503	0.659	0.645	0.750	0.807	0.826	0.983	0.889	1.169	0.942	1.360	0.987
0.300	0.516	0.658	0.658	0.748	0.821	0.823	0.997	0.886	1.182	0.939	1.373	0.984
0.305	0.529	0.657	0.673	0.746	0.835	0.820	1.010	0.883	1.195	0.935	1.386	0.980
0.310	0.542	0.656	0.686	0.744	0.849	0.818	1.024	0.879	1.209	0.932	1.399	0.976
0.315	0.555	0.656	0.700	0.742	0.862	0.815	1.038	0.876	1.222	0.928	1.412	0.973
0.320	0.568	0.655	0.713	0.740	0.876	0.812	1.052	0.873	1.236	0.925	1.425	0.969
0.325	0.581	0.654	0.727	0.738	0.890	0.809	1.065	0.870	1.249	0.922	1.439	0.966
0.330	0.595	0.653	0.740	0.736	0.903	0.807	1.079	0.867	1.262	0.918	1.452	0.963
0.335	0.608	0.652	0.754	0.734	0.917	0.804	1.092	0.864	1.276	0.915	1.465	0.959
0.340	0.621	0.652	0.768	0.732	0.931	0.802	1.106	0.861	1.289	0.912	1.478	0.956
0.345	0.634	0.651	0.781	0.731	0.944	0.800	1.119	0.858	1.302	0.909	1.491	0.953
0.350	0.647	0.650	0.795	0.729	0.958	0.797	1.133	0.856	1.316	0.906	1.504	0.950
0.355	0.659	0.650	0.808	0.727	0.972	0.795	1.147	0.853	1.329	0.903	1.518	0.946
0.360	0.673	0.649	0.822	0.726	0.985	0.793	1.160	0.850	1.343	0.900	1.531	0.943
0.365	0.686	0.648	0.835	0.724	0.999	0.790	1.173	0.848	1.356	0.897	1.544	0.940
0.370	0.698	0.648	0.849	0.723	1.012	0.788	1.187	0.845	1.369	0.894	1.557	0.937
0.375	0.711	0.647	0.862	0.721	1.026	0.786	1.200	0.843	1.383	0.892	1.571	0.934
0.380	0.724	0.647	0.875	0.720	1.039	0.784	1.214	0.840	1.396	0.889	1.584	0.932
0.385	0.737	0.646	0.889	0.718	1.053	0.782	1.227	0.838	1.410	0.886	1.597	0.929
0.390	0.750	0.646	0.902	0.717	1.066	0.780	1.241	0.835	1.423	0.884	1.610	0.926
0.395	0.763	0.645	0.915	0.716	1.080	0.778	1.254	0.833	1.436	0.881	1.623	0.923
0.400	0.776	0.644	0.928	0.715	1.093	0.776	1.268	0.831	1.450	0.878	1.637	0.920
0.410	0.802	0.643	0.955	0.712	1.120	0.773	1.295	0.826	1.476	0.873	1.663	0.915
0.420	0.827	0.642	0.981	0.710	1.147	0.769	1.322	0.822	1.503	0.869	1.690	0.910
0.430	0.853	0.642	1.008	0.707	1.174	0.766	1.348	0.818	1.530	0.864	1.716	0.905
0.440	0.879	0.641	1.034	0.705	1.200	0.763	1.375	0.814	1.556	0.859	1.743	0.900
0.450	0.904	0.640	1.061	0.703	1.227	0.759	1.402	0.810	1.583	0.855	1.769	0.895
0.460	0.930	0.639	1.087	0.701	1.254	0.756	1.429	0.806	1.610	0.851	1.796	0.891
0.470	0.956	0.638	1.113	0.699	1.281	0.754	1.455	0.803	1.636	0.847	1.822	0.886
0.480	0.981	0.637	1.139	0.697	1.307	0.751	1.482	0.799	1.663	0.843	1.848	0.882
0.490	1.007	0.637	1.165	0.695	1.334	0.748	1.509	0.796	1.690	0.839	1.875	0.878
0.500	1.032	0.636	1.192	0.693	1.360	0.746	1.535	0.793	1.716	0.835	1.901	0.874
0.510	1.058	0.635	1.218	0.692	1.386	0.743	1.562	0.789	1.742	0.832	1.928	0.870
0.520	1.083	0.635	1.244	0.690	1.413	0.741	1.588	0.786	1.769	0.828	1.954	0.866
0.530	1.108	0.634	1.270	0.689	1.439	0.738	1.615	0.783	1.795	0.825	1.980	0.862
0.540	1.134	0.634	1.296	0.687	1.466	0.736	1.641	0.781	1.822	0.821	2.007	0.858
0.550	1.159	0.633	1.322	0.686	1.492	0.734	1.667	0.778	1.848	0.818	2.033	0.855
0.560	1.185	0.632	1.348	0.684	1.518	0.732	1.694	0.775	1.875	0.815	2.060	0.851
0.570	1.210	0.632	1.374	0.683	1.544	0.730	1.720	0.773	1.901	0.812	2.086	0.848
0.580	1.236	0.631	1.400	0.682	1.570	0.728	1.747	0.770	1.927	0.809	2.112	0.844
0.590	1.261	0.631	1.425	0.680	1.597	0.726	1.773	0.768	1.954	0.806	2.138	0.841

Tab. 5.14

TABLE 3_Concrete classes C12-C50_Steel class S400_a/h=0.15

Bending and axial force with symmetric reinforcement

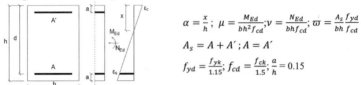

$$\alpha = \frac{x}{h}; \quad \mu = \frac{M_{Ed}}{bh^2 f_{cd}}; \nu = \frac{N_{Ed}}{bh f_{cd}}; \quad \varpi = \frac{A_s \, f_{yd}}{bh \, f_{cd}}$$

$$A_s = A + A'; A = A'$$

$$f_{yd} = \frac{f_{yk}}{1.15}; f_{cd} = \frac{f_{ck}}{1.5}, \frac{a}{h} = 0.15$$

	$\nu=0.0$		$\nu=0.1$		$\nu=0.2$		$\nu=0.3$		$\nu=0.4$		$\nu=0.5$	
μ	ϖ	α	ϖ	α	ϖ	α	ϖ	α	ϖ	α	ϖ	α
0.000	0.000	-	0.000	-	0.000	-	0.000	-	0.000	-	0.000	-
0.005	0.010	0.028	0.000	-	0.000	-	0.000	-	0.000	-	0.000	-
0.010	0.021	0.041	0.000	-	0.000	-	0.000	-	0.000	-	0.000	-
0.015	0.031	0.052	0.000	-	0.000	-	0.000	-	0.000	-	0.000	-
0.020	0.042	0.063	0.000	-	0.000	-	0.000	-	0.000	-	0.000	-
0.025	0.053	0.074	0.000	-	0.000	-	0.000	-	0.000	-	0.000	-
0.030	0.064	0.085	0.000	-	0.000	-	0.000	-	0.000	-	0.000	-
0.035	0.076	0.096	0.000	-	0.000	-	0.000	-	0.000	-	0.000	-
0.040	0.089	0.104	0.000	-	0.000	-	0.000	-	0.000	-	0.000	-
0.045	0.102	0.110	0.000	-	0.000	-	0.000	-	0.000	-	0.000	-
0.050	0.116	0.115	0.014	0.134	0.000	-	0.000	-	0.000	-	0.000	-
0.055	0.129	0.120	0.027	0.142	0.000	-	0.000	-	0.000	-	0.000	-
0.060	0.143	0.125	0.041	0.149	0.000	-	0.000	-	0.000	-	0.000	-
0.065	0.157	0.129	0.055	0.155	0.000	-	0.000	-	0.000	-	0.000	-
0.070	0.171	0.133	0.069	0.161	0.000	-	0.000	-	0.000	-	0.000	-
0.075	0.185	0.137	0.083	0.165	0.000	-	0.000	-	0.000	-	0.000	-
0.080	0.199	0.140	0.097	0.170	0.001	0.247	0.000	-	0.000	-	0.000	-
0.085	0.213	0.144	0.112	0.174	0.016	0.249	0.000	-	0.000	-	0.000	-
0.090	0.227	0.147	0.126	0.177	0.031	0.251	0.000	-	0.000	-	0.000	-
0.095	0.241	0.150	0.140	0.181	0.045	0.252	0.000	-	0.000	-	0.000	-
0.100	0.255	0.152	0.155	0.184	0.060	0.254	0.000	-	0.000	-	0.000	-
0.105	0.269	0.155	0.169	0.187	0.074	0.255	0.004	0.370	0.000	-	0.000	-
0.110	0.284	0.158	0.183	0.189	0.089	0.256	0.018	0.370	0.000	-	0.000	-
0.115	0.298	0.160	0.197	0.192	0.103	0.257	0.032	0.370	0.000	-	0.000	-
0.120	0.312	0.163	0.212	0.194	0.118	0.258	0.047	0.370	0.006	0.494	0.000	-
0.125	0.326	0.165	0.226	0.197	0.132	0.259	0.061	0.370	0.021	0.494	0.011	0.616
0.130	0.340	0.167	0.241	0.199	0.146	0.260	0.075	0.370	0.035	0.494	0.027	0.614
0.135	0.355	0.169	0.255	0.201	0.161	0.261	0.089	0.370	0.049	0.494	0.043	0.612
0.140	0.369	0.171	0.269	0.203	0.175	0.262	0.104	0.370	0.064	0.494	0.059	0.610
0.145	0.383	0.173	0.284	0.205	0.190	0.263	0.118	0.370	0.078	0.494	0.074	0.608
0.150	0.397	0.175	0.298	0.206	0.204	0.264	0.132	0.370	0.092	0.494	0.090	0.607
0.155	0.412	0.177	0.313	0.208	0.219	0.264	0.147	0.370	0.106	0.494	0.105	0.605
0.160	0.426	0.178	0.327	0.210	0.233	0.265	0.161	0.370	0.121	0.494	0.121	0.604
0.165	0.440	0.180	0.341	0.211	0.247	0.266	0.175	0.370	0.135	0.494	0.136	0.603
0.170	0.454	0.182	0.355	0.213	0.261	0.266	0.189	0.370	0.149	0.494	0.151	0.602
0.175	0.469	0.183	0.370	0.214	0.276	0.267	0.204	0.370	0.164	0.494	0.166	0.601
0.180	0.483	0.185	0.384	0.216	0.290	0.268	0.218	0.370	0.178	0.494	0.182	0.600
0.185	0.498	0.186	0.399	0.217	0.305	0.268	0.232	0.370	0.192	0.494	0.196	0.599
0.190	0.512	0.188	0.413	0.218	0.319	0.269	0.247	0.370	0.206	0.494	0.211	0.598
0.195	0.526	0.189	0.427	0.219	0.334	0.269	0.261	0.370	0.221	0.494	0.227	0.597
0.200	0.540	0.191	0.442	0.221	0.348	0.270	0.275	0.370	0.235	0.494	0.241	0.596
0.205	0.555	0.192	0.456	0.222	0.362	0.270	0.289	0.370	0.249	0.494	0.256	0.596
0.210	0.569	0.193	0.470	0.223	0.376	0.271	0.304	0.370	0.264	0.494	0.271	0.595
0.215	0.583	0.194	0.485	0.224	0.391	0.271	0.318	0.370	0.278	0.494	0.286	0.594
0.220	0.598	0.196	0.499	0.225	0.406	0.272	0.332	0.370	0.292	0.494	0.301	0.593
0.225	0.612	0.197	0.514	0.226	0.420	0.272	0.347	0.370	0.306	0.494	0.316	0.593
0.230	0.626	0.198	0.528	0.227	0.434	0.273	0.361	0.370	0.321	0.494	0.330	0.592
0.235	0.640	0.199	0.542	0.228	0.448	0.273	0.375	0.370	0.335	0.494	0.345	0.592
0.240	0.655	0.200	0.557	0.229	0.463	0.273	0.389	0.370	0.349	0.494	0.360	0.591
0.245	0.669	0.201	0.571	0.230	0.477	0.274	0.404	0.370	0.364	0.494	0.374	0.591

TABLE 3_Concrete classes C12-C50_Steel class S400_a/h=0.15 (cont.)
Bending and axial force with symmetric reinforcement

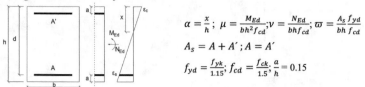

$$\alpha = \frac{x}{h}\,; \quad \mu = \frac{M_{Ed}}{bh^2 f_{cd}}; \nu = \frac{N_{Ed}}{bh f_{cd}}; \varpi = \frac{A_s\, f_{yd}}{bh\, f_{cd}}$$

$$A_s = A + A'\,; A = A'$$

$$f_{yd} = \frac{f_{yk}}{1.15}; f_{cd} = \frac{f_{ck}}{1.5}; \frac{a}{h} = 0.15$$

	$\nu=0.6$		$\nu=0.8$		$\nu=1.0$		$\nu=1.2$		$\nu=1.4$		$\nu=1.6$	
μ	ϖ	α	ϖ	α	ϖ	α	ϖ	α	ϖ	α	ϖ	α
0.000	0.000	-	0.000	-	0.000	-	0.200	-	0.400	-	0.600	-
0.005	0.000	-	0.000	-	0.014	2.590	0.214	2.950	0.414	3.160	0.614	3.260
0.010	0.000	-	0.000	-	0.028	1.990	0.228	2.350	0.428	2.620	0.628	2.790
0.015	0.000	-	0.000	-	0.042	1.720	0.242	2.040	0.443	2.290	0.643	2.480
0.020	0.000	-	0.000	-	0.056	1.560	0.257	1.830	0.457	2.060	0.657	2.250
0.025	0.000	-	0.000	-	0.071	1.450	0.271	1.690	0.471	1.900	0.671	2.080
0.030	0.000	-	0.000	-	0.084	1.380	0.285	1.580	0.485	1.780	0.685	1.940
0.035	0.000	-	0.000	-	0.098	1.320	0.299	1.500	0.499	1.680	0.699	1.840
0.040	0.000	-	0.000	-	0.112	1.270	0.313	1.440	0.513	1.600	0.714	1.750
0.045	0.000	-	0.000	-	0.126	1.230	0.327	1.380	0.528	1.530	0.728	1.670
0.050	0.000	-	0.000	-	0.141	1.190	0.341	1.340	0.542	1.470	0.742	1.610
0.055	0.000	-	0.000	-	0.155	1.160	0.355	1.300	0.556	1.430	0.756	1.550
0.060	0.000	-	0.000	-	0.169	1.140	0.370	1.260	0.570	1.380	0.770	1.500
0.065	0.000	-	0.000	-	0.183	1.110	0.384	1.230	0.584	1.350	0.785	1.460
0.070	0.000	-	0.000	-	0.197	1.090	0.398	1.200	0.598	1.310	0.799	1.420
0.075	0.000	-	0.012	0.979	0.211	1.080	0.412	1.180	0.612	1.280	0.813	1.390
0.080	0.000	-	0.027	0.967	0.226	1.060	0.426	1.160	0.627	1.260	0.827	1.350
0.085	0.000	-	0.043	0.956	0.239	1.050	0.441	1.140	0.641	1.230	0.841	1.330
0.090	0.000	-	0.059	0.945	0.254	1.030	0.455	1.120	0.655	1.210	0.856	1.300
0.095	0.000	-	0.074	0.934	0.268	1.020	0.469	1.100	0.669	1.190	0.870	1.280
0.100	0.000	-	0.091	0.923	0.282	1.010	0.483	1.090	0.684	1.170	0.884	1.260
0.105	0.000	-	0.107	0.913	0.296	0.998	0.497	1.080	0.698	1.160	0.898	1.240
0.110	0.000	-	0.124	0.903	0.311	0.989	0.511	1.060	0.712	1.140	0.912	1.220
0.115	0.000	-	0.140	0.893	0.326	0.980	0.526	1.050	0.726	1.130	0.927	1.200
0.120	0.019	0.733	0.157	0.884	0.341	0.972	0.540	1.040	0.740	1.110	0.941	1.180
0.125	0.037	0.726	0.174	0.875	0.356	0.963	0.554	1.030	0.754	1.100	0.955	1.170
0.130	0.055	0.720	0.191	0.866	0.372	0.955	0.568	1.020	0.769	1.090	0.969	1.150
0.135	0.073	0.714	0.207	0.858	0.387	0.947	0.582	1.010	0.783	1.080	0.983	1.140
0.140	0.091	0.708	0.224	0.850	0.402	0.939	0.596	1.000	0.797	1.070	0.997	1.130
0.145	0.108	0.703	0.241	0.842	0.418	0.932	0.611	0.995	0.811	1.060	1.012	1.120
0.150	0.125	0.698	0.258	0.835	0.434	0.924	0.625	0.988	0.825	1.050	1.026	1.110
0.155	0.142	0.693	0.275	0.828	0.449	0.917	0.640	0.981	0.840	1.040	1.040	1.100
0.160	0.159	0.689	0.292	0.821	0.465	0.910	0.655	0.974	0.854	1.030	1.054	1.090
0.165	0.176	0.685	0.308	0.815	0.480	0.903	0.670	0.967	0.868	1.020	1.068	1.080
0.170	0.193	0.681	0.326	0.808	0.497	0.896	0.685	0.961	0.882	1.010	1.083	1.070
0.175	0.210	0.678	0.342	0.802	0.512	0.890	0.700	0.955	0.896	1.010	1.097	1.060
0.180	0.226	0.675	0.359	0.797	0.528	0.884	0.715	0.948	0.911	0.999	1.111	1.050
0.185	0.242	0.672	0.375	0.791	0.544	0.878	0.730	0.942	0.925	0.993	1.125	1.040
0.190	0.258	0.669	0.392	0.786	0.560	0.872	0.746	0.936	0.940	0.987	1.139	1.030
0.195	0.275	0.666	0.409	0.781	0.576	0.866	0.761	0.931	0.954	0.981	1.154	1.030
0.200	0.291	0.663	0.425	0.776	0.592	0.861	0.776	0.925	0.969	0.976	1.168	1.020
0.205	0.307	0.661	0.442	0.771	0.608	0.855	0.792	0.919	0.984	0.970	1.182	1.010
0.210	0.323	0.659	0.458	0.767	0.624	0.850	0.807	0.914	0.999	0.965	1.196	1.010
0.215	0.339	0.656	0.474	0.763	0.640	0.845	0.823	0.908	1.014	0.959	1.210	1.000
0.220	0.354	0.654	0.491	0.759	0.656	0.840	0.838	0.903	1.029	0.954	1.225	0.996
0.225	0.370	0.652	0.507	0.755	0.672	0.835	0.853	0.898	1.044	0.949	1.239	0.991
0.230	0.386	0.650	0.524	0.751	0.688	0.831	0.869	0.893	1.059	0.944	1.254	0.986
0.235	0.401	0.648	0.540	0.747	0.704	0.826	0.884	0.888	1.074	0.939	1.269	0.982
0.240	0.417	0.647	0.556	0.744	0.720	0.822	0.900	0.884	1.089	0.935	1.283	0.977
0.245	0.432	0.645	0.572	0.740	0.736	0.817	0.915	0.879	1.104	0.930	1.298	0.972

TABLE 3_Concrete classes C12-C50_Steel class S400_a/h=0.15 (cont.)
Bending and axial force with symmetric reinforcement

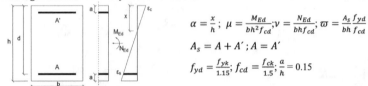

$$\alpha = \frac{x}{h}; \quad \mu = \frac{M_{Ed}}{bh^2 f_{cd}}; \nu = \frac{N_{Ed}}{bh f_{cd}}; \quad \varpi = \frac{A_s f_{yd}}{bh f_{cd}}$$

$$A_s = A + A'; A = A'$$

$$f_{yd} = \frac{f_{yk}}{1.15}; f_{cd} = \frac{f_{ck}}{1.5}; \frac{a}{h} = 0.15$$

	$\nu=0.0$		$\nu=0.1$		$\nu=0.2$		$\nu=0.3$		$\nu=0.4$		$\nu=0.5$	
μ	ϖ	α	ϖ	α	ϖ	α	ϖ	α	ϖ	α	ϖ	α
0.250	0.684	0.202	0.585	0.231	0.492	0.274	0.418	0.370	0.378	0.494	0.389	0.590
0.255	0.698	0.203	0.600	0.231	0.506	0.274	0.432	0.370	0.392	0.494	0.404	0.590
0.260	0.712	0.204	0.614	0.232	0.520	0.275	0.447	0.370	0.406	0.494	0.418	0.589
0.265	0.727	0.205	0.628	0.233	0.535	0.275	0.461	0.370	0.421	0.494	0.433	0.589
0.270	0.741	0.206	0.643	0.234	0.549	0.275	0.475	0.370	0.435	0.494	0.448	0.588
0.275	0.755	0.207	0.658	0.235	0.563	0.276	0.489	0.370	0.449	0.494	0.462	0.588
0.280	0.769	0.208	0.672	0.235	0.578	0.276	0.504	0.370	0.464	0.494	0.477	0.588
0.285	0.784	0.209	0.686	0.236	0.592	0.276	0.518	0.370	0.478	0.494	0.492	0.587
0.290	0.798	0.210	0.700	0.237	0.607	0.277	0.532	0.370	0.492	0.494	0.506	0.587
0.295	0.813	0.211	0.715	0.238	0.621	0.277	0.547	0.370	0.506	0.494	0.521	0.587
0.300	0.827	0.212	0.729	0.238	0.635	0.277	0.561	0.370	0.521	0.494	0.535	0.586
0.305	0.841	0.212	0.744	0.239	0.649	0.277	0.575	0.370	0.535	0.494	0.550	0.586
0.310	0.856	0.213	0.758	0.240	0.664	0.278	0.589	0.370	0.549	0.494	0.564	0.586
0.315	0.870	0.214	0.772	0.240	0.678	0.278	0.604	0.370	0.564	0.494	0.579	0.585
0.320	0.884	0.215	0.787	0.241	0.692	0.278	0.618	0.370	0.578	0.494	0.593	0.585
0.325	0.899	0.215	0.801	0.241	0.707	0.278	0.632	0.370	0.592	0.494	0.608	0.585
0.330	0.913	0.216	0.815	0.242	0.721	0.279	0.647	0.370	0.606	0.494	0.622	0.585
0.335	0.927	0.217	0.830	0.243	0.735	0.279	0.661	0.370	0.621	0.494	0.637	0.584
0.340	0.942	0.218	0.844	0.243	0.750	0.279	0.675	0.370	0.635	0.494	0.652	0.584
0.345	0.956	0.218	0.859	0.244	0.764	0.279	0.689	0.370	0.649	0.494	0.666	0.584
0.350	0.970	0.219	0.873	0.244	0.778	0.280	0.704	0.370	0.664	0.494	0.681	0.584
0.355	0.985	0.220	0.887	0.245	0.792	0.280	0.718	0.370	0.678	0.494	0.695	0.583
0.360	0.999	0.220	0.902	0.245	0.807	0.280	0.732	0.370	0.692	0.494	0.710	0.583
0.365	1.013	0.221	0.916	0.246	0.821	0.280	0.747	0.370	0.706	0.494	0.724	0.583
0.370	1.028	0.222	0.930	0.246	0.835	0.280	0.761	0.370	0.721	0.494	0.739	0.583
0.375	1.042	0.222	0.944	0.247	0.850	0.280	0.775	0.370	0.735	0.494	0.753	0.582
0.380	1.056	0.223	0.959	0.247	0.864	0.281	0.789	0.370	0.749	0.494	0.767	0.582
0.385	1.071	0.224	0.973	0.248	0.878	0.281	0.804	0.370	0.764	0.494	0.781	0.582
0.390	1.085	0.224	0.988	0.248	0.893	0.281	0.818	0.370	0.778	0.494	0.796	0.582
0.395	1.099	0.225	1.002	0.249	0.907	0.281	0.832	0.370	0.792	0.494	0.811	0.582
0.400	1.113	0.225	1.016	0.249	0.922	0.281	0.847	0.370	0.806	0.494	0.825	0.582
0.410	1.142	0.226	1.045	0.250	0.950	0.282	0.875	0.370	0.835	0.494	0.854	0.581
0.420	1.171	0.227	1.074	0.251	0.979	0.282	0.904	0.370	0.864	0.494	0.883	0.581
0.430	1.200	0.229	1.102	0.251	1.007	0.282	0.932	0.370	0.892	0.494	0.911	0.581
0.440	1.228	0.230	1.131	0.252	1.036	0.283	0.961	0.370	0.921	0.494	0.940	0.580
0.450	1.257	0.231	1.160	0.253	1.065	0.283	0.989	0.370	0.949	0.494	0.969	0.580
0.460	1.285	0.231	1.188	0.254	1.093	0.283	1.018	0.370	0.978	0.494	0.998	0.580
0.470	1.314	0.232	1.217	0.254	1.122	0.283	1.047	0.370	1.006	0.494	1.027	0.579
0.480	1.343	0.233	1.246	0.255	1.151	0.284	1.075	0.370	1.035	0.494	1.056	0.579
0.490	1.372	0.234	1.274	0.256	1.179	0.284	1.104	0.370	1.064	0.494	1.084	0.579
0.500	1.400	0.235	1.303	0.256	1.208	0.284	1.132	0.370	1.092	0.494	1.113	0.579
0.510	1.429	0.236	1.332	0.257	1.236	0.284	1.161	0.370	1.121	0.494	1.142	0.579
0.520	1.458	0.237	1.360	0.258	1.265	0.285	1.189	0.370	1.149	0.494	1.171	0.578
0.530	1.486	0.237	1.389	0.258	1.294	0.285	1.218	0.370	1.178	0.494	1.199	0.578
0.540	1.515	0.238	1.418	0.259	1.322	0.285	1.247	0.370	1.206	0.494	1.228	0.578
0.550	1.543	0.239	1.446	0.259	1.351	0.285	1.275	0.370	1.235	0.494	1.257	0.578
0.560	1.572	0.240	1.475	0.260	1.380	0.285	1.304	0.370	1.264	0.494	1.286	0.578
0.570	1.601	0.240	1.504	0.260	1.408	0.286	1.332	0.370	1.292	0.494	1.314	0.577
0.580	1.629	0.241	1.532	0.261	1.437	0.286	1.361	0.370	1.321	0.494	1.343	0.577
0.590	1.658	0.242	1.561	0.261	1.465	0.286	1.389	0.370	1.349	0.494	1.372	0.577

TABLE 3_Concrete classes C12-C50_Steel class S400_a/h=0.15 (cont.)
Bending and axial force with symmetric reinforcement

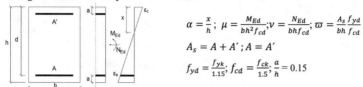

$$\alpha = \frac{x}{h}; \quad \mu = \frac{M_{Ed}}{bh^2 f_{cd}}; \nu = \frac{N_{Ed}}{bh f_{cd}}; \quad \varpi = \frac{A_s \, f_{yd}}{bh \, f_{cd}}$$

$$A_S = A + A'; A = A'$$

$$f_{yd} = \frac{f_{yk}}{1.15}; f_{cd} = \frac{f_{ck}}{1.5}; \frac{a}{h} = 0.15$$

	$\nu=0.6$		$\nu=0.8$		$\nu=1.0$		$\nu=1.2$		$\nu=1.4$		$\nu=1.6$	
μ	ϖ	α	ϖ	α	ϖ	α	ϖ	α	ϖ	α	ϖ	α
0.250	0.448	0.643	0.588	0.737	0.752	0.813	0.931	0.875	1.119	0.925	1.313	0.967
0.255	0.463	0.642	0.604	0.734	0.768	0.809	0.946	0.870	1.134	0.921	1.328	0.963
0.260	0.479	0.640	0.620	0.731	0.784	0.805	0.962	0.866	1.149	0.916	1.342	0.958
0.265	0.494	0.639	0.636	0.728	0.800	0.802	0.978	0.862	1.164	0.912	1.357	0.954
0.270	0.509	0.638	0.652	0.725	0.815	0.798	0.993	0.858	1.179	0.908	1.372	0.950
0.275	0.524	0.636	0.668	0.722	0.831	0.794	1.008	0.854	1.195	0.904	1.387	0.946
0.280	0.539	0.635	0.684	0.720	0.847	0.791	1.024	0.850	1.210	0.899	1.402	0.941
0.285	0.555	0.634	0.699	0.717	0.863	0.788	1.039	0.846	1.225	0.895	1.417	0.937
0.290	0.570	0.633	0.715	0.715	0.879	0.784	1.055	0.842	1.240	0.892	1.432	0.933
0.295	0.585	0.631	0.731	0.712	0.894	0.781	1.071	0.839	1.256	0.888	1.447	0.929
0.300	0.600	0.630	0.746	0.710	0.910	0.778	1.086	0.835	1.271	0.884	1.462	0.925
0.305	0.615	0.629	0.762	0.708	0.926	0.775	1.102	0.832	1.286	0.880	1.477	0.922
0.310	0.630	0.628	0.778	0.706	0.941	0.772	1.117	0.828	1.302	0.876	1.492	0.918
0.315	0.645	0.627	0.793	0.704	0.957	0.769	1.133	0.825	1.317	0.873	1.507	0.914
0.320	0.660	0.626	0.809	0.702	0.973	0.766	1.149	0.822	1.332	0.869	1.522	0.911
0.325	0.676	0.625	0.824	0.700	0.988	0.764	1.164	0.819	1.348	0.866	1.537	0.907
0.330	0.690	0.625	0.840	0.698	1.004	0.761	1.180	0.816	1.363	0.863	1.552	0.904
0.335	0.705	0.624	0.855	0.696	1.019	0.759	1.195	0.813	1.378	0.859	1.567	0.900
0.340	0.720	0.623	0.870	0.694	1.035	0.756	1.210	0.810	1.394	0.856	1.582	0.897
0.345	0.735	0.622	0.886	0.692	1.051	0.754	1.226	0.807	1.409	0.853	1.597	0.893
0.350	0.750	0.621	0.901	0.690	1.066	0.751	1.241	0.804	1.424	0.850	1.612	0.890
0.355	0.765	0.620	0.917	0.689	1.082	0.749	1.257	0.801	1.439	0.847	1.627	0.887
0.360	0.780	0.620	0.932	0.687	1.097	0.747	1.272	0.798	1.455	0.844	1.643	0.884
0.365	0.795	0.619	0.947	0.686	1.112	0.744	1.288	0.796	1.470	0.841	1.658	0.881
0.370	0.809	0.618	0.962	0.684	1.128	0.742	1.303	0.793	1.485	0.838	1.673	0.878
0.375	0.824	0.618	0.978	0.683	1.143	0.740	1.318	0.791	1.501	0.835	1.688	0.875
0.380	0.839	0.617	0.993	0.681	1.159	0.738	1.334	0.788	1.516	0.833	1.703	0.872
0.385	0.854	0.616	1.008	0.680	1.174	0.736	1.349	0.786	1.531	0.830	1.718	0.869
0.390	0.869	0.616	1.024	0.678	1.190	0.734	1.365	0.783	1.546	0.827	1.733	0.866
0.395	0.883	0.615	1.039	0.677	1.205	0.732	1.380	0.781	1.562	0.825	1.748	0.863
0.400	0.898	0.614	1.054	0.676	1.220	0.730	1.395	0.779	1.577	0.822	1.764	0.860
0.410	0.927	0.613	1.084	0.673	1.251	0.727	1.426	0.775	1.608	0.817	1.794	0.855
0.420	0.957	0.612	1.114	0.671	1.281	0.723	1.457	0.770	1.638	0.812	1.824	0.850
0.430	0.986	0.611	1.145	0.668	1.312	0.720	1.487	0.766	1.669	0.808	1.854	0.845
0.440	1.016	0.610	1.174	0.666	1.342	0.717	1.518	0.762	1.699	0.803	1.885	0.840
0.450	1.045	0.609	1.205	0.664	1.373	0.714	1.548	0.758	1.729	0.799	1.915	0.835
0.460	1.074	0.608	1.234	0.662	1.403	0.711	1.579	0.755	1.760	0.795	1.945	0.831
0.470	1.104	0.607	1.264	0.660	1.433	0.708	1.609	0.751	1.790	0.791	1.975	0.826
0.480	1.133	0.607	1.294	0.658	1.464	0.705	1.639	0.748	1.820	0.787	2.005	0.822
0.490	1.162	0.606	1.324	0.656	1.493	0.703	1.669	0.745	1.850	0.783	2.036	0.818
0.500	1.191	0.605	1.354	0.655	1.524	0.700	1.700	0.742	1.881	0.779	2.066	0.814
0.510	1.221	0.604	1.384	0.653	1.554	0.698	1.730	0.739	1.911	0.776	2.096	0.810
0.520	1.249	0.604	1.413	0.651	1.584	0.695	1.760	0.736	1.941	0.772	2.126	0.806
0.530	1.279	0.603	1.443	0.650	1.614	0.693	1.790	0.733	1.971	0.769	2.156	0.803
0.540	1.308	0.602	1.473	0.648	1.644	0.691	1.821	0.730	2.001	0.766	2.186	0.799
0.550	1.337	0.602	1.502	0.647	1.674	0.689	1.850	0.727	2.031	0.763	2.216	0.796
0.560	1.366	0.601	1.532	0.645	1.704	0.687	1.880	0.725	2.062	0.760	2.246	0.792
0.570	1.395	0.600	1.562	0.644	1.733	0.685	1.911	0.722	2.092	0.757	2.276	0.789
0.580	1.424	0.600	1.591	0.643	1.763	0.683	1.940	0.720	2.122	0.754	2.306	0.786
0.590	1.453	0.599	1.620	0.642	1.793	0.681	1.970	0.718	2.152	0.751	2.336	0.783

Tab. 5.15

TABLE 3_Concrete class C55_Steel class S500_a/h=0.10

Bending and axial force with symmetric reinforcement

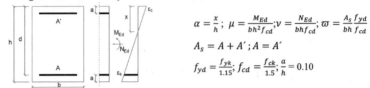

$$\alpha = \frac{x}{h}\ ;\ \ \mu = \frac{M_{Ed}}{bh^2 f_{cd}}; \nu = \frac{N_{Ed}}{bh f_{cd}};\ \varpi = \frac{A_s\ f_{yd}}{bh\ f_{cd}}$$

$$A_s = A + A'\ ;\ A = A'$$

$$f_{yd} = \frac{f_{yk}}{1.15};\ f_{cd} = \frac{f_{ck}}{1.5};\ \frac{a}{h} = 0.10$$

	$\nu=0.0$		$\nu=0.1$		$\nu=0.2$		$\nu=0.3$		$\nu=0.4$		$\nu=0.5$	
μ	ϖ	α	ϖ	α	ϖ	α	ϖ	α	ϖ	α	ϖ	α
0.000	0.000	-	0.000	-	0.000	-	0.000	-	0.000	-	0.000	-
0.005	0.010	0.031	0.000	-	0.000	-	0.000	-	0.000	-	0.000	-
0.010	0.021	0.043	0.000	-	0.000	-	0.000	-	0.000	-	0.000	-
0.015	0.032	0.053	0.000	-	0.000	-	0.000	-	0.000	-	0.000	-
0.020	0.043	0.060	0.000	-	0.000	-	0.000	-	0.000	-	0.000	-
0.025	0.055	0.067	0.000	-	0.000	-	0.000	-	0.000	-	0.000	-
0.030	0.067	0.073	0.000	-	0.000	-	0.000	-	0.000	-	0.000	-
0.035	0.078	0.078	0.000	-	0.000	-	0.000	-	0.000	-	0.000	-
0.040	0.090	0.083	0.000	-	0.000	-	0.000	-	0.000	-	0.000	-
0.045	0.102	0.087	0.001	0.135	0.000	-	0.000	-	0.000	-	0.000	-
0.050	0.114	0.091	0.014	0.140	0.000	-	0.000	-	0.000	-	0.000	-
0.055	0.127	0.094	0.026	0.145	0.000	-	0.000	-	0.000	-	0.000	-
0.060	0.139	0.098	0.038	0.149	0.000	-	0.000	-	0.000	-	0.000	-
0.065	0.151	0.101	0.051	0.152	0.000	-	0.000	-	0.000	-	0.000	-
0.070	0.164	0.104	0.064	0.156	0.000	-	0.000	-	0.000	-	0.000	-
0.075	0.176	0.107	0.076	0.159	0.000	-	0.000	-	0.000	-	0.000	-
0.080	0.188	0.110	0.089	0.162	0.003	0.270	0.000	-	0.000	-	0.000	-
0.085	0.201	0.113	0.102	0.165	0.015	0.271	0.000	-	0.000	-	0.000	-
0.090	0.213	0.116	0.115	0.168	0.028	0.271	0.000	-	0.000	-	0.000	-
0.095	0.226	0.118	0.127	0.170	0.041	0.272	0.000	-	0.000	-	0.000	-
0.100	0.238	0.121	0.140	0.172	0.054	0.273	0.000	-	0.000	-	0.000	-
0.105	0.251	0.123	0.153	0.175	0.066	0.274	0.006	0.404	0.000	-	0.000	-
0.110	0.263	0.126	0.165	0.177	0.079	0.275	0.019	0.404	0.000	-	0.000	-
0.115	0.276	0.128	0.178	0.179	0.091	0.275	0.031	0.404	0.000	-	0.000	-
0.120	0.288	0.130	0.190	0.181	0.104	0.276	0.044	0.404	0.012	0.539	0.007	0.672
0.125	0.301	0.132	0.203	0.183	0.117	0.277	0.056	0.404	0.024	0.538	0.023	0.666
0.130	0.313	0.134	0.215	0.185	0.130	0.277	0.069	0.404	0.037	0.538	0.039	0.661
0.135	0.326	0.136	0.229	0.187	0.142	0.278	0.081	0.404	0.050	0.538	0.055	0.657
0.140	0.338	0.138	0.241	0.188	0.155	0.279	0.094	0.404	0.063	0.538	0.070	0.652
0.145	0.351	0.140	0.254	0.190	0.167	0.279	0.106	0.404	0.075	0.537	0.086	0.648
0.150	0.363	0.142	0.266	0.192	0.180	0.280	0.119	0.404	0.088	0.537	0.101	0.644
0.155	0.376	0.144	0.279	0.193	0.193	0.280	0.131	0.404	0.101	0.537	0.116	0.641
0.160	0.388	0.145	0.292	0.195	0.205	0.281	0.144	0.404	0.114	0.537	0.131	0.637
0.165	0.401	0.147	0.304	0.196	0.218	0.281	0.156	0.404	0.126	0.536	0.146	0.634
0.170	0.414	0.149	0.317	0.198	0.230	0.282	0.169	0.404	0.139	0.536	0.161	0.631
0.175	0.426	0.150	0.330	0.199	0.243	0.282	0.181	0.404	0.152	0.536	0.176	0.629
0.180	0.439	0.152	0.342	0.201	0.256	0.283	0.194	0.404	0.164	0.536	0.190	0.626
0.185	0.451	0.153	0.354	0.202	0.268	0.283	0.206	0.404	0.177	0.536	0.204	0.623
0.190	0.464	0.155	0.367	0.203	0.281	0.284	0.219	0.404	0.190	0.535	0.219	0.621
0.195	0.476	0.156	0.380	0.204	0.294	0.284	0.231	0.404	0.202	0.535	0.233	0.619
0.200	0.489	0.158	0.392	0.206	0.306	0.285	0.244	0.404	0.215	0.535	0.247	0.617
0.205	0.501	0.159	0.405	0.207	0.319	0.285	0.256	0.404	0.227	0.535	0.261	0.614
0.210	0.514	0.160	0.418	0.208	0.331	0.286	0.269	0.404	0.240	0.535	0.275	0.613
0.215	0.527	0.162	0.430	0.209	0.344	0.286	0.281	0.404	0.253	0.535	0.289	0.611
0.220	0.539	0.163	0.443	0.210	0.357	0.287	0.294	0.404	0.265	0.535	0.303	0.609
0.225	0.552	0.164	0.456	0.211	0.369	0.287	0.306	0.404	0.278	0.534	0.317	0.607
0.230	0.564	0.166	0.468	0.212	0.382	0.288	0.319	0.404	0.291	0.534	0.331	0.606
0.235	0.577	0.167	0.481	0.213	0.394	0.288	0.331	0.404	0.303	0.534	0.344	0.604
0.240	0.589	0.168	0.494	0.214	0.407	0.288	0.344	0.404	0.316	0.534	0.358	0.603
0.245	0.602	0.169	0.506	0.215	0.420	0.289	0.356	0.404	0.328	0.534	0.372	0.601

TABLE 3_Concrete class C55_Steel class S500_a/h=0.10 (cont.)
Bending and axial force with symmetric reinforcement

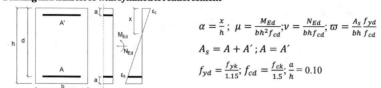

$$\alpha = \frac{x}{h}; \ \mu = \frac{M_{Ed}}{bh^2 f_{cd}}; \nu = \frac{N_{Ed}}{bh f_{cd}}; \ \varpi = \frac{A_s f_{yd}}{bh \, f_{cd}}$$

$$A_s = A + A' ; A = A'$$

$$f_{yd} = \frac{f_{yk}}{1.15}; \ f_{cd} = \frac{f_{ck}}{1.5}; \frac{a}{h} = 0.10$$

μ	ν=0.6 ϖ	α	ν=0.8 ϖ	α	ν=1.0 ϖ	α	ν=1.2 ϖ	α	ν=1.4 ϖ	α	ν=1.6 ϖ	α
0.000	0.000	-	0.000	-	0.000	-	0.201	-	0.400	-	0.600	-
0.005	0.000	-	0.000	-	0.016	4.000	0.214	6.970	0.413	10.000	0.613	13.000
0.010	0.000	-	0.000	-	0.031	2.830	0.227	4.420	0.427	6.110	0.626	7.840
0.015	0.000	-	0.000	-	0.046	2.360	0.242	3.410	0.440	4.600	0.639	5.800
0.020	0.000	-	0.000	-	0.062	2.080	0.256	2.880	0.454	3.760	0.652	4.690
0.025	0.000	-	0.000	-	0.076	1.890	0.270	2.530	0.468	3.240	0.666	4.000
0.030	0.000	-	0.000	-	0.091	1.760	0.285	2.290	0.481	2.890	0.679	3.520
0.035	0.000	-	0.000	-	0.106	1.650	0.299	2.110	0.495	2.620	0.693	3.160
0.040	0.000	-	0.000	-	0.121	1.570	0.313	1.970	0.509	2.410	0.706	2.890
0.045	0.000	-	0.000	-	0.135	1.500	0.327	1.860	0.522	2.260	0.719	2.680
0.050	0.000	-	0.000	-	0.149	1.440	0.341	1.760	0.536	2.120	0.733	2.510
0.055	0.000	-	0.000	-	0.164	1.390	0.355	1.690	0.550	2.020	0.746	2.360
0.060	0.000	-	0.000	-	0.179	1.350	0.369	1.620	0.563	1.920	0.760	2.240
0.065	0.000	-	0.008	1.100	0.193	1.310	0.383	1.560	0.577	1.840	0.773	2.130
0.070	0.000	-	0.023	1.080	0.207	1.280	0.397	1.510	0.591	1.770	0.787	2.040
0.075	0.000	-	0.037	1.060	0.221	1.250	0.411	1.470	0.605	1.710	0.800	1.960
0.080	0.000	-	0.052	1.050	0.236	1.230	0.425	1.430	0.618	1.660	0.814	1.890
0.085	0.000	-	0.066	1.030	0.250	1.200	0.439	1.390	0.632	1.610	0.827	1.830
0.090	0.000	-	0.080	1.020	0.264	1.180	0.453	1.360	0.646	1.560	0.841	1.770
0.095	0.000	-	0.094	1.010	0.278	1.160	0.467	1.330	0.660	1.520	0.854	1.720
0.100	0.000	-	0.109	0.995	0.292	1.140	0.481	1.310	0.673	1.490	0.868	1.680
0.105	0.000	-	0.123	0.985	0.306	1.120	0.495	1.280	0.687	1.450	0.881	1.630
0.110	0.000	-	0.139	0.975	0.320	1.110	0.508	1.260	0.700	1.420	0.895	1.600
0.115	0.018	0.799	0.154	0.965	0.334	1.090	0.522	1.240	0.714	1.400	0.908	1.560
0.120	0.035	0.790	0.169	0.955	0.348	1.080	0.536	1.220	0.728	1.370	0.922	1.530
0.125	0.052	0.781	0.184	0.946	0.361	1.070	0.550	1.200	0.742	1.340	0.935	1.500
0.130	0.069	0.773	0.199	0.937	0.376	1.060	0.563	1.180	0.755	1.320	0.949	1.470
0.135	0.086	0.765	0.215	0.928	0.389	1.040	0.577	1.170	0.769	1.300	0.962	1.440
0.140	0.102	0.758	0.230	0.919	0.403	1.030	0.591	1.150	0.782	1.280	0.976	1.420
0.145	0.119	0.751	0.245	0.910	0.417	1.020	0.605	1.140	0.796	1.260	0.989	1.400
0.150	0.135	0.745	0.261	0.902	0.431	1.010	0.618	1.130	0.809	1.250	1.002	1.370
0.155	0.151	0.739	0.276	0.894	0.444	1.000	0.632	1.110	0.823	1.230	1.016	1.350
0.160	0.167	0.733	0.291	0.886	0.458	0.996	0.646	1.100	0.836	1.220	1.029	1.330
0.165	0.183	0.727	0.307	0.879	0.473	0.989	0.659	1.090	0.850	1.200	1.043	1.320
0.170	0.199	0.722	0.322	0.871	0.487	0.981	0.673	1.080	0.863	1.190	1.056	1.300
0.175	0.215	0.717	0.338	0.864	0.501	0.974	0.687	1.070	0.877	1.170	1.070	1.280
0.180	0.230	0.712	0.353	0.857	0.516	0.966	0.700	1.060	0.890	1.160	1.083	1.270
0.185	0.245	0.708	0.369	0.850	0.530	0.959	0.713	1.050	0.904	1.150	1.096	1.250
0.190	0.261	0.703	0.384	0.844	0.545	0.952	0.727	1.040	0.917	1.140	1.110	1.240
0.195	0.276	0.699	0.399	0.838	0.559	0.945	0.741	1.030	0.931	1.130	1.123	1.230
0.200	0.291	0.695	0.415	0.832	0.574	0.939	0.754	1.030	0.944	1.120	1.136	1.210
0.205	0.306	0.691	0.429	0.826	0.588	0.932	0.768	1.020	0.958	1.110	1.150	1.200
0.210	0.321	0.688	0.445	0.820	0.603	0.926	0.781	1.010	0.971	1.100	1.163	1.190
0.215	0.336	0.684	0.460	0.815	0.617	0.919	0.794	1.000	0.985	1.090	1.176	1.180
0.220	0.351	0.681	0.475	0.809	0.632	0.913	0.809	0.997	0.998	1.080	1.190	1.170
0.225	0.365	0.678	0.491	0.804	0.647	0.907	0.822	0.991	1.011	1.070	1.203	1.160
0.230	0.380	0.675	0.505	0.799	0.661	0.901	0.836	0.985	1.024	1.060	1.216	1.150
0.235	0.394	0.672	0.521	0.794	0.675	0.896	0.850	0.979	1.038	1.060	1.230	1.140
0.240	0.409	0.669	0.535	0.790	0.690	0.890	0.864	0.973	1.051	1.050	1.243	1.130
0.245	0.423	0.666	0.550	0.785	0.705	0.885	0.878	0.967	1.065	1.040	1.256	1.120

TABLE 3_Concrete class C55_Steel class S500_a/h=0.10 (cont.)

Bending and axial force with symmetric reinforcement

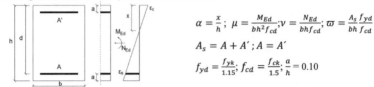

$$\alpha = \frac{x}{h}\ ;\ \mu = \frac{M_{Ed}}{bh^2 f_{cd}};\nu = \frac{N_{Ed}}{bh f_{cd}};\ \varpi = \frac{A_s}{bh}\frac{f_{yd}}{f_{cd}}$$

$$A_s = A + A'\ ;\ A = A'$$

$$f_{yd} = \frac{f_{yk}}{1.15};\ f_{cd} = \frac{f_{ck}}{1.5};\ \frac{a}{h} = 0.10$$

	$\nu = 0.0$		$\nu = 0.1$		$\nu = 0.2$		$\nu = 0.3$		$\nu = 0.4$		$\nu = 0.5$	
μ	ϖ	α	ϖ	α	ϖ	α	ϖ	α	ϖ	α	ϖ	α
0.250	0.615	0.170	0.519	0.216	0.432	0.289	0.369	0.404	0.341	0.534	0.385	0.600
0.255	0.627	0.171	0.531	0.217	0.445	0.290	0.381	0.404	0.353	0.534	0.399	0.598
0.260	0.640	0.173	0.544	0.218	0.457	0.290	0.394	0.404	0.366	0.534	0.412	0.597
0.265	0.652	0.174	0.557	0.219	0.470	0.290	0.406	0.404	0.379	0.534	0.426	0.596
0.270	0.665	0.175	0.569	0.220	0.482	0.291	0.419	0.404	0.391	0.534	0.439	0.595
0.275	0.678	0.176	0.582	0.221	0.495	0.291	0.431	0.404	0.404	0.534	0.453	0.593
0.280	0.690	0.177	0.595	0.222	0.508	0.291	0.444	0.404	0.416	0.534	0.466	0.592
0.285	0.702	0.178	0.607	0.223	0.520	0.292	0.456	0.404	0.429	0.533	0.480	0.591
0.290	0.715	0.179	0.620	0.223	0.533	0.292	0.469	0.404	0.441	0.533	0.493	0.590
0.295	0.728	0.180	0.633	0.224	0.545	0.292	0.481	0.404	0.454	0.533	0.506	0.589
0.300	0.740	0.181	0.645	0.225	0.558	0.293	0.494	0.404	0.467	0.533	0.520	0.588
0.305	0.753	0.182	0.000	0.000	0.571	0.293	0.506	0.404	0.479	0.533	0.533	0.587
0.310	0.765	0.183	0.670	0.227	0.583	0.293	0.519	0.404	0.492	0.533	0.546	0.586
0.315	0.778	0.184	0.683	0.227	0.596	0.294	0.531	0.404	0.504	0.533	0.560	0.586
0.320	0.791	0.185	0.696	0.228	0.608	0.294	0.544	0.404	0.517	0.533	0.572	0.585
0.325	0.803	0.185	0.708	0.229	0.621	0.294	0.556	0.404	0.529	0.533	0.585	0.584
0.330	0.816	0.186	0.721	0.230	0.633	0.294	0.569	0.404	0.542	0.533	0.599	0.583
0.335	0.828	0.187	0.733	0.230	0.646	0.295	0.581	0.404	0.554	0.533	0.612	0.582
0.340	0.841	0.188	0.746	0.231	0.659	0.295	0.594	0.404	0.567	0.533	0.625	0.582
0.345	0.853	0.189	0.758	0.232	0.671	0.295	0.606	0.404	0.580	0.533	0.638	0.581
0.350	0.866	0.190	0.772	0.232	0.684	0.296	0.619	0.404	0.592	0.533	0.651	0.580
0.355	0.879	0.191	0.784	0.233	0.696	0.296	0.631	0.404	0.605	0.533	0.664	0.579
0.360	0.891	0.191	0.796	0.234	0.709	0.296	0.644	0.404	0.617	0.533	0.677	0.579
0.365	0.904	0.192	0.809	0.234	0.721	0.296	0.656	0.404	0.630	0.532	0.691	0.578
0.370	0.916	0.193	0.822	0.235	0.734	0.297	0.669	0.404	0.642	0.532	0.703	0.577
0.375	0.929	0.194	0.834	0.236	0.747	0.297	0.681	0.404	0.655	0.532	0.717	0.577
0.380	0.941	0.195	0.847	0.236	0.759	0.297	0.694	0.404	0.667	0.532	0.729	0.576
0.385	0.954	0.195	0.860	0.237	0.772	0.297	0.706	0.404	0.680	0.532	0.742	0.576
0.390	0.967	0.196	0.872	0.237	0.784	0.298	0.719	0.404	0.692	0.532	0.755	0.575
0.395	0.979	0.197	0.885	0.238	0.797	0.298	0.731	0.404	0.705	0.532	0.768	0.575
0.400	0.992	0.198	0.897	0.239	0.809	0.298	0.744	0.404	0.717	0.532	0.781	0.574
0.410	1.017	0.199	0.922	0.240	0.835	0.299	0.769	0.404	0.743	0.532	0.807	0.573
0.420	1.042	0.200	0.947	0.241	0.860	0.299	0.794	0.404	0.768	0.532	0.833	0.572
0.430	1.067	0.202	0.973	0.242	0.885	0.299	0.819	0.404	0.793	0.532	0.859	0.571
0.440	1.092	0.203	0.998	0.243	0.910	0.300	0.844	0.404	0.818	0.532	0.884	0.570
0.450	1.117	0.204	1.023	0.244	0.935	0.300	0.869	0.404	0.843	0.532	0.911	0.569
0.460	1.143	0.206	1.049	0.245	0.960	0.301	0.894	0.404	0.868	0.532	0.936	0.568
0.470	1.168	0.207	1.073	0.246	0.985	0.301	0.919	0.404	0.893	0.532	0.962	0.567
0.480	1.193	0.208	1.099	0.247	1.010	0.302	0.944	0.404	0.918	0.532	0.988	0.567
0.490	1.218	0.209	1.124	0.248	1.035	0.302	0.969	0.404	0.943	0.532	1.013	0.566
0.500	1.243	0.211	1.149	0.249	1.060	0.302	0.994	0.404	0.968	0.532	1.039	0.565
0.510	1.268	0.212	1.174	0.250	1.085	0.303	1.019	0.404	0.993	0.532	1.064	0.565
0.520	1.293	0.213	1.199	0.251	1.110	0.303	1.044	0.404	1.018	0.531	1.090	0.564
0.530	1.319	0.214	1.224	0.251	1.135	0.303	1.069	0.404	1.043	0.531	1.115	0.563
0.540	1.344	0.215	1.249	0.252	1.161	0.304	1.094	0.404	1.068	0.531	1.141	0.563
0.550	1.369	0.216	1.274	0.253	1.186	0.304	1.119	0.404	1.093	0.531	1.166	0.562
0.560	1.394	0.217	1.299	0.254	1.211	0.304	1.144	0.404	1.118	0.531	1.192	0.561
0.570	1.419	0.218	1.325	0.255	1.236	0.305	1.169	0.404	1.143	0.531	1.218	0.561
0.580	1.444	0.219	1.350	0.255	1.261	0.305	1.194	0.404	1.168	0.531	1.243	0.560
0.590	1.469	0.220	1.375	0.256	1.286	0.305	1.219	0.404	1.193	0.531	1.268	0.560

TABLE 3_Concrete class C55_Steel class S500_a/h=0.10 (cont.)
Bending and axial force with symmetric reinforcement

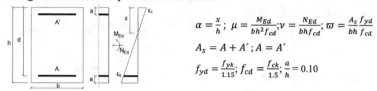

$$\alpha = \frac{x}{h}; \ \mu = \frac{M_{Ed}}{bh^2 f_{cd}}; v = \frac{N_{Ed}}{bhf_{cd}}; \varpi = \frac{A_s \, f_{yd}}{bh \, f_{cd}}$$

$$A_s = A + A'; A = A'$$

$$f_{yd} = \frac{f_{yk}}{1.15}; f_{cd} = \frac{f_{ck}}{1.5}; \frac{a}{h} = 0.10$$

μ	v=0.6 ϖ	α	v=0.8 ϖ	α	v=1.0 ϖ	α	v=1.2 ϖ	α	v=1.4 ϖ	α	v=1.6 ϖ	α
0.250	0.437	0.664	0.565	0.781	0.719	0.879	0.892	0.961	1.078	1.030	1.270	1.110
0.255	0.452	0.661	0.580	0.776	0.734	0.874	0.906	0.956	1.092	1.030	1.283	1.100
0.260	0.466	0.659	0.595	0.772	0.748	0.869	0.920	0.950	1.105	1.020	1.296	1.100
0.265	0.480	0.656	0.609	0.768	0.763	0.864	0.934	0.945	1.118	1.010	1.309	1.090
0.270	0.494	0.654	0.624	0.764	0.778	0.859	0.948	0.940	1.132	1.010	1.323	1.080
0.275	0.508	0.652	0.638	0.761	0.792	0.855	0.963	0.935	1.145	1.000	1.336	1.070
0.280	0.522	0.650	0.653	0.757	0.806	0.850	0.977	0.930	1.158	0.998	1.349	1.070
0.285	0.536	0.648	0.667	0.753	0.821	0.846	0.990	0.925	1.172	0.992	1.363	1.060
0.290	0.550	0.646	0.682	0.750	0.835	0.841	1.004	0.920	1.185	0.987	1.376	1.050
0.295	0.564	0.644	0.696	0.747	0.850	0.837	1.019	0.915	1.199	0.982	1.389	1.050
0.300	0.578	0.642	0.711	0.743	0.864	0.833	1.033	0.910	1.213	0.977	1.402	1.040
0.305	0.592	0.640	0.725	0.740	0.878	0.829	1.047	0.906	1.227	0.972	1.415	1.030
0.310	0.605	0.639	0.740	0.737	0.893	0.825	1.061	0.901	1.240	0.968	1.429	1.030
0.315	0.619	0.637	0.754	0.734	0.907	0.821	1.075	0.897	1.254	0.963	1.442	1.020
0.320	0.633	0.635	0.768	0.731	0.922	0.817	1.089	0.892	1.268	0.958	1.455	1.020
0.325	0.647	0.634	0.782	0.728	0.936	0.813	1.104	0.888	1.281	0.954	1.468	1.010
0.330	0.660	0.632	0.797	0.726	0.950	0.810	1.117	0.884	1.295	0.949	1.482	1.010
0.335	0.673	0.631	0.811	0.723	0.964	0.806	1.131	0.880	1.309	0.945	1.495	1.000
0.340	0.687	0.629	0.825	0.720	0.979	0.803	1.146	0.876	1.323	0.941	1.508	0.998
0.345	0.701	0.628	0.839	0.718	0.993	0.799	1.160	0.872	1.337	0.936	1.522	0.994
0.350	0.715	0.627	0.853	0.715	1.007	0.796	1.174	0.868	1.350	0.932	1.535	0.989
0.355	0.728	0.625	0.867	0.713	1.021	0.793	1.188	0.864	1.365	0.928	1.549	0.985
0.360	0.741	0.624	0.881	0.710	1.035	0.789	1.202	0.861	1.378	0.924	1.562	0.981
0.365	0.755	0.623	0.895	0.708	1.049	0.786	1.216	0.857	1.392	0.920	1.576	0.976
0.370	0.768	0.622	0.909	0.706	1.063	0.783	1.230	0.853	1.406	0.916	1.589	0.972
0.375	0.781	0.620	0.923	0.704	1.077	0.780	1.244	0.850	1.419	0.912	1.603	0.968
0.380	0.795	0.619	0.937	0.702	1.091	0.778	1.258	0.846	1.433	0.908	1.616	0.964
0.385	0.809	0.618	0.950	0.700	1.106	0.775	1.272	0.843	1.447	0.905	1.630	0.960
0.390	0.822	0.617	0.964	0.697	1.119	0.772	1.286	0.840	1.461	0.901	1.643	0.956
0.395	0.835	0.616	0.978	0.695	1.134	0.769	1.300	0.836	1.475	0.897	1.657	0.952
0.400	0.848	0.615	0.992	0.694	1.148	0.767	1.314	0.833	1.489	0.894	1.671	0.949
0.410	0.875	0.613	1.019	0.690	1.175	0.761	1.341	0.827	1.516	0.887	1.698	0.941
0.420	0.902	0.611	1.047	0.686	1.203	0.756	1.369	0.821	1.544	0.880	1.725	0.934
0.430	0.928	0.609	1.074	0.683	1.231	0.752	1.397	0.815	1.572	0.874	1.752	0.927
0.440	0.954	0.607	1.101	0.680	1.259	0.747	1.425	0.810	1.599	0.867	1.779	0.920
0.450	0.980	0.606	1.128	0.676	1.286	0.743	1.453	0.804	1.626	0.861	1.807	0.913
0.460	1.007	0.604	1.156	0.673	1.314	0.739	1.480	0.799	1.654	0.855	1.834	0.907
0.470	1.033	0.603	1.183	0.670	1.341	0.734	1.508	0.794	1.681	0.850	1.861	0.901
0.480	1.059	0.601	1.210	0.668	1.368	0.731	1.535	0.789	1.709	0.844	1.888	0.895
0.490	1.085	0.600	1.236	0.665	1.396	0.727	1.563	0.785	1.736	0.839	1.916	0.889
0.500	1.112	0.598	1.263	0.662	1.423	0.723	1.590	0.780	1.764	0.834	1.943	0.883
0.510	1.138	0.597	1.290	0.660	1.450	0.720	1.617	0.776	1.791	0.829	1.970	0.878
0.520	1.164	0.596	1.317	0.658	1.477	0.716	1.645	0.772	1.818	0.824	1.997	0.872
0.530	1.189	0.595	1.343	0.655	1.504	0.713	1.672	0.768	1.845	0.819	2.024	0.867
0.540	1.215	0.593	1.370	0.653	1.531	0.710	1.699	0.764	1.873	0.814	2.051	0.862
0.550	1.242	0.592	1.397	0.651	1.558	0.707	1.726	0.760	1.900	0.810	2.078	0.857
0.560	1.267	0.591	1.423	0.649	1.585	0.704	1.753	0.756	1.927	0.806	2.105	0.852
0.570	1.293	0.590	1.449	0.647	1.612	0.701	1.781	0.753	1.954	0.801	2.132	0.847
0.580	1.319	0.589	1.476	0.645	1.639	0.698	1.807	0.749	1.981	0.797	2.159	0.843
0.590	1.345	0.588	1.502	0.643	1.665	0.696	1.834	0.746	2.008	0.793	2.187	0.838

Tab. 5.16

TABLE 3_Concrete class C60_Steel class S500_a/h=0.10

Bending and axial force with symmetric reinforcement

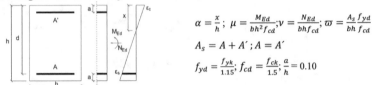

$$\alpha = \frac{x}{h}; \quad \mu = \frac{M_{Ed}}{bh^2 f_{cd}}; \nu = \frac{N_{Ed}}{bh f_{cd}}; \quad \varpi = \frac{A_s}{bh}\frac{f_{yd}}{f_{cd}}$$

$$A_s = A + A'; A = A'$$

$$f_{yd} = \frac{f_{yk}}{1.15}; f_{cd} = \frac{f_{ck}}{1.5}; \frac{a}{h} = 0.10$$

μ	$\nu=0.0$ ϖ	α	$\nu=0.1$ ϖ	α	$\nu=0.2$ ϖ	α	$\nu=0.3$ ϖ	α	$\nu=0.4$ ϖ	α	$\nu=0.5$ ϖ	α
0.000	0.000	0.000	0.000	-	0.000	-	0.000	-	0.000	-	0.000	-
0.005	0.010	0.033	0.000	-	0.000	-	0.000	-	0.000	-	0.000	-
0.010	0.021	0.045	0.000	-	0.000	-	0.000	-	0.000	-	0.000	-
0.015	0.032	0.055	0.000	-	0.000	-	0.000	-	0.000	-	0.000	-
0.020	0.044	0.062	0.000	-	0.000	-	0.000	-	0.000	-	0.000	-
0.025	0.055	0.069	0.000	-	0.000	-	0.000	-	0.000	-	0.000	-
0.030	0.067	0.074	0.000	-	0.000	-	0.000	-	0.000	-	0.000	-
0.035	0.079	0.080	0.000	-	0.000	-	0.000	-	0.000	-	0.000	-
0.040	0.091	0.084	0.000	-	0.000	-	0.000	-	0.000	-	0.000	-
0.045	0.103	0.088	0.002	0.145	0.000	-	0.000	-	0.000	-	0.000	-
0.050	0.115	0.092	0.014	0.149	0.000	-	0.000	-	0.000	-	0.000	-
0.055	0.127	0.096	0.027	0.154	0.000	-	0.000	-	0.000	-	0.000	-
0.060	0.139	0.100	0.039	0.158	0.000	-	0.000	-	0.000	-	0.000	-
0.065	0.152	0.104	0.051	0.162	0.000	-	0.000	-	0.000	-	0.000	-
0.070	0.164	0.107	0.064	0.166	0.000	-	0.000	-	0.000	-	0.000	-
0.075	0.176	0.111	0.077	0.169	0.000	-	0.000	-	0.000	-	0.000	-
0.080	0.189	0.114	0.089	0.172	0.004	0.288	0.000	-	0.000	-	0.000	-
0.085	0.201	0.117	0.102	0.175	0.017	0.289	0.000	-	0.000	-	0.000	-
0.090	0.214	0.120	0.115	0.178	0.030	0.290	0.000	-	0.000	-	0.000	-
0.095	0.226	0.123	0.127	0.181	0.043	0.292	0.000	-	0.000	-	0.000	-
0.100	0.239	0.125	0.140	0.184	0.055	0.293	0.000	-	0.000	-	0.000	-
0.105	0.251	0.128	0.153	0.186	0.068	0.294	0.010	0.432	0.000	-	0.000	-
0.110	0.263	0.130	0.166	0.189	0.081	0.295	0.022	0.432	0.000	-	0.000	-
0.115	0.276	0.133	0.178	0.191	0.094	0.296	0.035	0.432	0.005	0.575	0.002	0.718
0.120	0.288	0.135	0.191	0.193	0.106	0.297	0.047	0.432	0.020	0.572	0.020	0.710
0.125	0.301	0.138	0.204	0.195	0.119	0.298	0.060	0.432	0.034	0.570	0.038	0.703
0.130	0.314	0.140	0.216	0.197	0.132	0.298	0.072	0.432	0.048	0.568	0.054	0.696
0.135	0.326	0.142	0.229	0.199	0.144	0.299	0.085	0.432	0.062	0.566	0.071	0.689
0.140	0.339	0.144	0.242	0.201	0.157	0.300	0.097	0.432	0.076	0.564	0.087	0.683
0.145	0.351	0.146	0.254	0.203	0.170	0.301	0.110	0.432	0.090	0.563	0.104	0.677
0.150	0.364	0.148	0.266	0.205	0.182	0.302	0.122	0.432	0.104	0.561	0.120	0.672
0.155	0.376	0.150	0.280	0.207	0.195	0.303	0.135	0.432	0.118	0.560	0.136	0.667
0.160	0.389	0.152	0.292	0.208	0.208	0.304	0.147	0.432	0.131	0.558	0.152	0.663
0.165	0.401	0.154	0.305	0.210	0.220	0.304	0.160	0.432	0.145	0.557	0.167	0.658
0.170	0.414	0.156	0.318	0.212	0.233	0.305	0.172	0.432	0.158	0.556	0.183	0.654
0.175	0.426	0.157	0.331	0.213	0.246	0.306	0.185	0.432	0.172	0.555	0.198	0.650
0.180	0.439	0.159	0.343	0.215	0.258	0.307	0.197	0.432	0.185	0.554	0.213	0.646
0.185	0.452	0.161	0.356	0.216	0.271	0.307	0.210	0.432	0.199	0.553	0.228	0.643
0.190	0.464	0.163	0.368	0.218	0.284	0.308	0.222	0.432	0.212	0.552	0.243	0.640
0.195	0.477	0.164	0.380	0.219	0.296	0.309	0.235	0.432	0.225	0.551	0.258	0.636
0.200	0.489	0.166	0.394	0.221	0.309	0.309	0.247	0.432	0.239	0.550	0.273	0.633
0.205	0.502	0.167	0.406	0.222	0.322	0.310	0.260	0.432	0.252	0.549	0.287	0.631
0.210	0.514	0.169	0.419	0.223	0.334	0.311	0.272	0.432	0.265	0.548	0.302	0.628
0.215	0.527	0.170	0.432	0.225	0.347	0.311	0.285	0.432	0.278	0.547	0.316	0.625
0.220	0.539	0.172	0.444	0.226	0.360	0.312	0.297	0.432	0.291	0.547	0.330	0.623
0.225	0.552	0.173	0.457	0.227	0.372	0.313	0.310	0.432	0.304	0.546	0.345	0.621
0.230	0.565	0.175	0.470	0.228	0.385	0.313	0.322	0.432	0.317	0.545	0.359	0.618
0.235	0.577	0.176	0.482	0.230	0.397	0.314	0.335	0.432	0.330	0.545	0.373	0.616
0.240	0.590	0.177	0.495	0.231	0.410	0.314	0.347	0.432	0.343	0.544	0.387	0.614
0.245	0.602	0.179	0.508	0.232	0.423	0.315	0.360	0.432	0.356	0.544	0.401	0.612

TABLE 3_Concrete class C60_Steel class S500_a/h=0.10 (cont.)
Bending and axial force with symmetric reinforcement

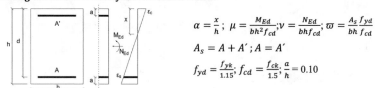

$$\alpha = \frac{x}{h}; \quad \mu = \frac{M_{Ed}}{bh^2 f_{cd}}; \nu = \frac{N_{Ed}}{bh f_{cd}}; \quad \varpi = \frac{A_s f_{yd}}{bh f_{cd}}$$

$$A_s = A + A'; A = A'$$

$$f_{yd} = \frac{f_{yk}}{1.15}; f_{cd} = \frac{f_{ck}}{1.5}; \frac{a}{h} = 0.10$$

μ	ν=0.6 ϖ	α	ν=0.8 ϖ	α	ν=1.0 ϖ	α	ν=1.2 ϖ	α	ν=1.4 ϖ	α	ν=1.6 ϖ	α
0.000	0.000	-	0.000	-	0.000	-	0.201	-	0.400	-	0.600	-
0.005	0.000	-	0.000	-	0.017	5.090	0.216	6.860	0.415	8.180	0.615	8.960
0.010	0.000	-	0.000	-	0.034	3.480	0.231	4.740	0.429	5.810	0.628	6.650
0.015	0.000	-	0.000	-	0.051	2.800	0.246	3.760	0.443	4.630	0.642	5.380
0.020	0.000	-	0.000	-	0.068	2.430	0.261	3.190	0.458	3.920	0.656	4.570
0.025	0.000	-	0.000	-	0.084	2.180	0.276	2.820	0.473	3.430	0.670	4.010
0.030	0.000	-	0.000	-	0.100	2.000	0.291	2.550	0.487	3.090	0.684	3.600
0.035	0.000	-	0.000	-	0.116	1.870	0.306	2.340	0.501	2.810	0.698	3.280
0.040	0.000	-	0.000	-	0.132	1.760	0.322	2.180	0.516	2.610	0.712	3.030
0.045	0.000	-	0.000	-	0.147	1.670	0.336	2.050	0.530	2.440	0.726	2.820
0.050	0.000	-	0.000	-	0.163	1.600	0.351	1.940	0.545	2.290	0.740	2.650
0.055	0.000	-	0.000	-	0.178	1.530	0.366	1.850	0.559	2.180	0.754	2.500
0.060	0.000	-	0.013	1.210	0.193	1.480	0.381	1.770	0.574	2.070	0.768	2.380
0.065	0.000	-	0.028	1.180	0.208	1.430	0.396	1.700	0.588	1.990	0.782	2.270
0.070	0.000	-	0.043	1.160	0.223	1.390	0.410	1.650	0.602	1.910	0.796	2.180
0.075	0.000	-	0.059	1.130	0.238	1.350	0.425	1.590	0.616	1.840	0.811	2.090
0.080	0.000	-	0.074	1.110	0.254	1.320	0.440	1.540	0.631	1.780	0.825	2.020
0.085	0.000	-	0.089	1.090	0.268	1.290	0.454	1.500	0.645	1.730	0.839	1.950
0.090	0.000	-	0.104	1.080	0.283	1.260	0.469	1.460	0.659	1.680	0.853	1.890
0.095	0.000	-	0.119	1.060	0.298	1.240	0.483	1.430	0.673	1.630	0.867	1.830
0.100	0.000	-	0.134	1.040	0.313	1.210	0.498	1.400	0.688	1.590	0.881	1.780
0.105	0.002	0.862	0.149	1.030	0.327	1.190	0.513	1.370	0.702	1.550	0.895	1.740
0.110	0.019	0.851	0.164	1.020	0.342	1.170	0.527	1.340	0.716	1.520	0.909	1.700
0.115	0.037	0.840	0.178	1.010	0.356	1.160	0.541	1.320	0.730	1.490	0.922	1.660
0.120	0.054	0.829	0.193	0.995	0.371	1.140	0.555	1.290	0.744	1.460	0.936	1.620
0.125	0.072	0.818	0.209	0.984	0.385	1.120	0.570	1.270	0.758	1.430	0.950	1.590
0.130	0.089	0.809	0.224	0.974	0.400	1.110	0.584	1.250	0.773	1.400	0.964	1.560
0.135	0.106	0.800	0.239	0.964	0.414	1.090	0.598	1.230	0.787	1.380	0.978	1.530
0.140	0.123	0.791	0.255	0.954	0.429	1.080	0.612	1.220	0.801	1.360	0.992	1.500
0.145	0.140	0.783	0.270	0.944	0.443	1.070	0.627	1.200	0.815	1.340	1.006	1.470
0.150	0.157	0.775	0.286	0.935	0.457	1.060	0.641	1.180	0.829	1.310	1.020	1.450
0.155	0.174	0.767	0.301	0.926	0.472	1.050	0.655	1.170	0.843	1.300	1.034	1.430
0.160	0.190	0.760	0.317	0.917	0.486	1.030	0.669	1.150	0.857	1.280	1.047	1.410
0.165	0.206	0.753	0.333	0.909	0.500	1.020	0.683	1.140	0.871	1.260	1.061	1.390
0.170	0.223	0.747	0.348	0.900	0.514	1.020	0.698	1.130	0.885	1.250	1.075	1.370
0.175	0.239	0.741	0.364	0.892	0.528	1.010	0.711	1.120	0.899	1.230	1.089	1.350
0.180	0.255	0.735	0.380	0.884	0.542	0.998	0.725	1.100	0.912	1.220	1.103	1.330
0.185	0.271	0.729	0.395	0.877	0.557	0.990	0.739	1.090	0.926	1.200	1.116	1.310
0.190	0.286	0.724	0.411	0.869	0.572	0.982	0.753	1.080	0.940	1.190	1.130	1.300
0.195	0.302	0.719	0.426	0.862	0.587	0.974	0.767	1.070	0.954	1.180	1.144	1.280
0.200	0.317	0.714	0.442	0.855	0.601	0.967	0.781	1.060	0.968	1.160	1.158	1.270
0.205	0.333	0.710	0.457	0.849	0.616	0.959	0.795	1.050	0.982	1.150	1.171	1.250
0.210	0.348	0.705	0.473	0.842	0.630	0.952	0.809	1.050	0.995	1.140	1.185	1.240
0.215	0.363	0.701	0.489	0.836	0.645	0.945	0.823	1.040	1.009	1.130	1.199	1.230
0.220	0.379	0.697	0.504	0.830	0.660	0.938	0.837	1.030	1.023	1.120	1.212	1.220
0.225	0.394	0.693	0.519	0.824	0.674	0.932	0.850	1.020	1.037	1.110	1.226	1.200
0.230	0.408	0.689	0.535	0.818	0.690	0.925	0.864	1.010	1.051	1.100	1.240	1.190
0.235	0.424	0.686	0.550	0.812	0.704	0.919	0.878	1.010	1.064	1.090	1.253	1.180
0.240	0.439	0.682	0.565	0.807	0.719	0.912	0.892	1.000	1.078	1.090	1.267	1.170
0.245	0.453	0.679	0.580	0.802	0.734	0.906	0.906	0.993	1.092	1.080	1.281	1.160

TABLE 3_Concrete class C60_Steel class S500_a/h=0.10 (cont.)
Bending and axial force with symmetric reinforcement

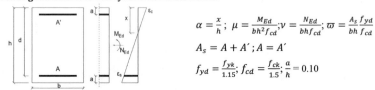

$$\alpha = \frac{x}{h}; \quad \mu = \frac{M_{Ed}}{bh^2 f_{cd}}; \nu = \frac{N_{Ed}}{bh f_{cd}}; \quad \varpi = \frac{A_s}{bh}\frac{f_{yd}}{f_{cd}}$$

$$A_S = A + A'; A = A'$$

$$f_{yd} = \frac{f_{yk}}{1.15}; f_{cd} = \frac{f_{ck}}{1.5}; \frac{a}{h} = 0.10$$

	$\nu=0.0$		$\nu=0.1$		$\nu=0.2$		$\nu=0.3$		$\nu=0.4$		$\nu=0.5$	
μ	ϖ	α	ϖ	α	ϖ	α	ϖ	α	ϖ	α	ϖ	α
0.250	0.615	0.180	0.521	0.233	0.435	0.316	0.372	0.432	0.369	0.543	0.415	0.610
0.255	0.628	0.181	0.533	0.234	0.448	0.316	0.385	0.432	0.382	0.542	0.429	0.608
0.260	0.640	0.183	0.546	0.235	0.461	0.317	0.397	0.432	0.395	0.542	0.443	0.607
0.265	0.653	0.184	0.558	0.236	0.473	0.317	0.410	0.432	0.408	0.541	0.457	0.605
0.270	0.665	0.185	0.571	0.237	0.486	0.318	0.422	0.432	0.421	0.541	0.470	0.603
0.275	0.678	0.186	0.583	0.238	0.499	0.318	0.435	0.432	0.434	0.540	0.484	0.602
0.280	0.690	0.187	0.596	0.239	0.511	0.319	0.447	0.432	0.447	0.540	0.498	0.600
0.285	0.703	0.189	0.608	0.240	0.524	0.319	0.460	0.432	0.460	0.540	0.512	0.599
0.290	0.716	0.190	0.621	0.241	0.536	0.320	0.472	0.432	0.472	0.539	0.525	0.597
0.295	0.728	0.191	0.634	0.242	0.549	0.320	0.485	0.432	0.485	0.539	0.538	0.596
0.300	0.741	0.192	0.647	0.243	0.562	0.321	0.497	0.432	0.498	0.538	0.552	0.595
0.305	0.753	0.193	0.659	0.244	0.574	0.321	0.510	0.432	0.511	0.538	0.565	0.593
0.310	0.766	0.194	0.672	0.245	0.587	0.322	0.522	0.432	0.524	0.538	0.579	0.592
0.315	0.779	0.195	0.685	0.246	0.600	0.322	0.535	0.432	0.537	0.537	0.593	0.591
0.320	0.791	0.196	0.697	0.247	0.612	0.323	0.547	0.432	0.549	0.537	0.606	0.590
0.325	0.804	0.197	0.709	0.248	0.625	0.323	0.560	0.432	0.562	0.537	0.619	0.589
0.330	0.816	0.198	0.722	0.249	0.638	0.324	0.572	0.432	0.575	0.536	0.633	0.587
0.335	0.829	0.200	0.735	0.250	0.650	0.324	0.585	0.432	0.588	0.536	0.646	0.586
0.340	0.842	0.201	0.748	0.251	0.663	0.325	0.597	0.432	0.600	0.536	0.660	0.585
0.345	0.854	0.202	0.760	0.251	0.675	0.325	0.610	0.432	0.613	0.535	0.673	0.584
0.350	0.867	0.202	0.773	0.252	0.688	0.325	0.622	0.432	0.626	0.535	0.686	0.583
0.355	0.879	0.203	0.785	0.253	0.700	0.326	0.635	0.432	0.639	0.535	0.699	0.582
0.360	0.892	0.204	0.799	0.254	0.713	0.326	0.647	0.432	0.651	0.534	0.712	0.581
0.365	0.904	0.205	0.810	0.255	0.726	0.327	0.660	0.432	0.664	0.534	0.726	0.581
0.370	0.917	0.206	0.824	0.255	0.738	0.327	0.672	0.432	0.677	0.534	0.739	0.580
0.375	0.930	0.207	0.836	0.256	0.751	0.327	0.685	0.432	0.690	0.534	0.752	0.579
0.380	0.942	0.208	0.849	0.257	0.763	0.328	0.697	0.432	0.702	0.533	0.765	0.578
0.385	0.955	0.209	0.861	0.258	0.776	0.328	0.710	0.432	0.715	0.533	0.778	0.577
0.390	0.967	0.210	0.874	0.258	0.789	0.329	0.722	0.432	0.728	0.533	0.792	0.576
0.395	0.980	0.211	0.887	0.259	0.801	0.329	0.735	0.432	0.741	0.533	0.805	0.576
0.400	0.993	0.212	0.899	0.260	0.814	0.329	0.747	0.432	0.753	0.533	0.818	0.575
0.410	1.018	0.213	0.925	0.261	0.839	0.330	0.772	0.432	0.779	0.532	0.844	0.573
0.420	1.043	0.215	0.950	0.263	0.864	0.331	0.797	0.432	0.804	0.532	0.870	0.572
0.430	1.068	0.217	0.974	0.264	0.889	0.332	0.822	0.432	0.829	0.531	0.896	0.571
0.440	1.093	0.218	1.000	0.265	0.915	0.332	0.847	0.432	0.855	0.531	0.922	0.570
0.450	1.118	0.220	1.025	0.267	0.940	0.333	0.872	0.432	0.880	0.531	0.948	0.568
0.460	1.144	0.221	1.051	0.268	0.965	0.334	0.897	0.432	0.905	0.530	0.974	0.567
0.470	1.169	0.223	1.076	0.269	0.990	0.334	0.922	0.432	0.930	0.530	1.000	0.566
0.480	1.194	0.224	1.100	0.270	1.015	0.335	0.947	0.432	0.956	0.530	1.025	0.565
0.490	1.219	0.226	1.126	0.271	1.040	0.335	0.972	0.432	0.981	0.529	1.051	0.564
0.500	1.244	0.227	1.152	0.273	1.066	0.336	0.997	0.432	1.006	0.529	1.077	0.563
0.510	1.270	0.228	1.177	0.274	1.091	0.337	1.022	0.432	1.031	0.529	1.103	0.562
0.520	1.295	0.230	1.202	0.275	1.116	0.337	1.047	0.432	1.057	0.528	1.129	0.561
0.530	1.320	0.231	1.227	0.276	1.141	0.338	1.072	0.432	1.082	0.528	1.154	0.560
0.540	1.345	0.232	1.252	0.277	1.166	0.338	1.097	0.432	1.107	0.528	1.180	0.560
0.550			1.277	0.278	1.191	0.339	1.122	0.432	1.132	0.528	1.206	0.559
0.560			1.302	0.279	1.216	0.339	1.147	0.432	1.158	0.527	1.232	0.558
0.570			1.328	0.280	1.242	0.340	1.172	0.432	1.183	0.527	1.257	0.557
0.580			1.353	0.281	1.267	0.340	1.197	0.432	1.208	0.527	1.283	0.557
0.590			1.378	0.282	1.292	0.341	1.222	0.432	1.233	0.527	1.308	0.556

TABLE 3_Concrete class C60_Steel class S500_a/h=0.10 (cont.)
Bending and axial force with symmetric reinforcement

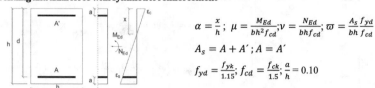

$$\alpha = \frac{x}{h}; \quad \mu = \frac{M_{Ed}}{bh^2 f_{cd}}; \quad \nu = \frac{N_{Ed}}{bh f_{cd}}; \quad \varpi = \frac{A_s f_{yd}}{bh \; f_{cd}}$$

$$A_s = A + A'; A = A'$$

$$f_{yd} = \frac{f_{yk}}{1.15}; \quad f_{cd} = \frac{f_{ck}}{1.5}; \quad \frac{a}{h} = 0.10$$

μ	$\nu=0.6$ ϖ	α	$\nu=0.8$ ϖ	α	$\nu=1.0$ ϖ	α	$\nu=1.2$ ϖ	α	$\nu=1.4$ ϖ	α	$\nu=1.6$ ϖ	α
0.250	0.468	0.676	0.595	0.797	0.748	0.900	0.921	0.987	1.105	1.070	1.294	1.150
0.255	0.482	0.673	0.610	0.792	0.763	0.894	0.935	0.981	1.119	1.060	1.308	1.140
0.260	0.497	0.670	0.625	0.787	0.778	0.889	0.949	0.975	1.133	1.050	1.321	1.130
0.265	0.511	0.667	0.641	0.783	0.793	0.883	0.963	0.969	1.146	1.050	1.335	1.130
0.270	0.526	0.664	0.655	0.778	0.807	0.878	0.977	0.963	1.160	1.040	1.349	1.120
0.275	0.540	0.662	0.670	0.774	0.822	0.872	0.991	0.957	1.173	1.030	1.362	1.110
0.280	0.554	0.659	0.685	0.770	0.837	0.867	1.006	0.951	1.187	1.030	1.376	1.100
0.285	0.569	0.657	0.700	0.766	0.852	0.862	1.020	0.946	1.201	1.020	1.389	1.090
0.290	0.582	0.654	0.715	0.762	0.866	0.857	1.034	0.940	1.214	1.010	1.403	1.090
0.295	0.597	0.652	0.729	0.758	0.881	0.852	1.049	0.935	1.228	1.010	1.416	1.080
0.300	0.611	0.650	0.743	0.754	0.895	0.848	1.063	0.930	1.241	1.000	1.430	1.070
0.305	0.625	0.648	0.758	0.751	0.910	0.843	1.077	0.924	1.255	0.996	1.443	1.070
0.310	0.639	0.646	0.773	0.747	0.925	0.839	1.092	0.919	1.269	0.990	1.457	1.060
0.315	0.653	0.643	0.787	0.744	0.939	0.834	1.106	0.914	1.283	0.985	1.470	1.050
0.320	0.666	0.642	0.802	0.740	0.954	0.830	1.120	0.910	1.297	0.980	1.484	1.050
0.325	0.681	0.640	0.816	0.737	0.968	0.826	1.134	0.905	1.311	0.975	1.497	1.040
0.330	0.694	0.638	0.830	0.734	0.983	0.822	1.149	0.900	1.325	0.970	1.511	1.030
0.335	0.708	0.636	0.845	0.731	0.997	0.818	1.163	0.896	1.339	0.965	1.524	1.030
0.340	0.722	0.634	0.859	0.728	1.011	0.814	1.177	0.891	1.353	0.960	1.538	1.020
0.345	0.736	0.633	0.873	0.725	1.026	0.810	1.191	0.887	1.367	0.955	1.551	1.020
0.350	0.749	0.631	0.888	0.722	1.040	0.806	1.206	0.882	1.381	0.951	1.565	1.010
0.355	0.763	0.629	0.902	0.719	1.054	0.803	1.220	0.878	1.395	0.946	1.578	1.010
0.360	0.777	0.628	0.916	0.717	1.069	0.799	1.234	0.874	1.409	0.941	1.591	1.000
0.365	0.790	0.626	0.930	0.714	1.083	0.796	1.248	0.870	1.423	0.937	1.605	0.997
0.370	0.804	0.625	0.944	0.712	1.098	0.792	1.262	0.866	1.437	0.933	1.619	0.993
0.375	0.817	0.623	0.958	0.709	1.112	0.789	1.277	0.862	1.451	0.928	1.632	0.988
0.380	0.831	0.622	0.973	0.707	1.126	0.786	1.291	0.858	1.465	0.924	1.646	0.984
0.385	0.844	0.621	0.986	0.704	1.140	0.783	1.305	0.855	1.479	0.920	1.660	0.979
0.390	0.858	0.619	1.000	0.702	1.154	0.780	1.319	0.851	1.492	0.916	1.674	0.975
0.395	0.872	0.618	1.014	0.700	1.169	0.777	1.333	0.847	1.506	0.912	1.687	0.971
0.400	0.885	0.617	1.028	0.698	1.183	0.774	1.347	0.844	1.520	0.908	1.701	0.966
0.410	0.912	0.614	1.056	0.693	1.211	0.768	1.375	0.837	1.549	0.900	1.729	0.958
0.420	0.939	0.612	1.083	0.689	1.239	0.762	1.404	0.830	1.577	0.893	1.756	0.950
0.430	0.965	0.610	1.111	0.685	1.267	0.757	1.432	0.824	1.604	0.885	1.784	0.942
0.440	0.992	0.608	1.139	0.682	1.295	0.752	1.460	0.818	1.632	0.879	1.811	0.935
0.450	1.018	0.606	1.166	0.678	1.322	0.747	1.487	0.812	1.660	0.872	1.839	0.927
0.460	1.045	0.604	1.193	0.675	1.351	0.742	1.516	0.806	1.688	0.865	1.866	0.920
0.470	1.071	0.602	1.221	0.672	1.378	0.738	1.543	0.800	1.715	0.859	1.894	0.913
0.480	1.098	0.600	1.248	0.668	1.405	0.734	1.571	0.795	1.743	0.853	1.921	0.907
0.490	1.124	0.598	1.275	0.665	1.433	0.729	1.599	0.790	1.771	0.847	1.949	0.900
0.500	1.150	0.597	1.302	0.663	1.461	0.725	1.626	0.785	1.798	0.841	1.976	0.894
0.510	1.176	0.595	1.329	0.660	1.488	0.722	1.654	0.780	1.826	0.836	2.003	0.888
0.520	1.203	0.594	1.356	0.657	1.515	0.718	1.682	0.776	1.854	0.830	2.031	0.882
0.530	1.229	0.592	1.383	0.654	1.542	0.714	1.709	0.771	1.881	0.825	2.058	0.876
0.540	1.255	0.591	1.409	0.652	1.570	0.711	1.736	0.767	1.908	0.820	2.086	0.870
0.550	1.281	0.589	1.436	0.650	1.597	0.707	1.764	0.763	1.936	0.815	2.113	0.865
0.560	1.307	0.588	1.463	0.647	1.624	0.704	1.791	0.759	1.963	0.810	2.140	0.859
0.570	1.333	0.587	1.489	0.645	1.651	0.701	1.818	0.755	1.991	0.806	2.168	0.854
0.580	1.359	0.586	1.515	0.643	1.678	0.698	1.845	0.751	2.018	0.801	2.195	0.849
0.590	1.385	0.584	1.542	0.641	1.705	0.695	1.872	0.747	2.045	0.797	2.222	0.844

Tab. 5.17

TABLE 3_Concrete class C70_Steel class S500_a/h=0.10

Bending and axial force with symmetric reinforcement

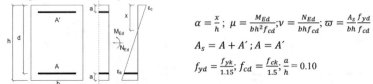

$$\alpha = \frac{x}{h} \; ; \; \mu = \frac{M_{Ed}}{bh^2 f_{cd}} ; v = \frac{N_{Ed}}{bh f_{cd}} ; \varpi = \frac{A_s}{bh}\frac{f_{yd}}{f_{cd}}$$

$$A_s = A + A' \; ; A = A'$$

$$f_{yd} = \frac{f_{yk}}{1.15} ; f_{cd} = \frac{f_{ck}}{1.5} ; \frac{a}{h} = 0.10$$

μ	$v=0.0$ ϖ	α	$v=0.1$ ϖ	α	$v=0.2$ ϖ	α	$v=0.3$ ϖ	α	$v=0.4$ ϖ	α	$v=0.5$ ϖ	α
0.000	0.000	-	0.000	-	0.000	-	0.000	-	0.000	-	0.000	-
0.005	0.010	0.035	0.000	-	0.000	-	0.000	-	0.000	-	0.000	-
0.010	0.021	0.048	0.000	-	0.000	-	0.000	-	0.000	-	0.000	-
0.015	0.032	0.057	0.000	-	0.000	-	0.000	-	0.000	-	0.000	-
0.020	0.044	0.065	0.000	-	0.000	-	0.000	-	0.000	-	0.000	-
0.025	0.055	0.071	0.000	-	0.000	-	0.000	-	0.000	-	0.000	-
0.030	0.067	0.077	0.000	-	0.000	-	0.000	-	0.000	-	0.000	-
0.035	0.079	0.082	0.000	-	0.000	-	0.000	-	0.000	-	0.000	-
0.040	0.091	0.086	0.000	-	0.000	-	0.000	-	0.000	-	0.000	-
0.045	0.103	0.091	0.002	0.158	0.000	-	0.000	-	0.000	-	0.000	-
0.050	0.115	0.096	0.015	0.163	0.000	-	0.000	-	0.000	-	0.000	-
0.055	0.128	0.100	0.027	0.168	0.000	-	0.000	-	0.000	-	0.000	-
0.060	0.140	0.104	0.040	0.172	0.000	-	0.000	-	0.000	-	0.000	-
0.065	0.152	0.108	0.052	0.176	0.000	-	0.000	-	0.000	-	0.000	-
0.070	0.165	0.112	0.064	0.180	0.000	-	0.000	-	0.000	-	0.000	-
0.075	0.177	0.116	0.078	0.183	0.000	-	0.000	-	0.000	-	0.000	-
0.080	0.189	0.119	0.090	0.187	0.007	0.315	0.000	-	0.000	-	0.000	-
0.085	0.202	0.122	0.103	0.190	0.020	0.316	0.000	-	0.000	-	0.000	-
0.090	0.214	0.126	0.115	0.193	0.033	0.318	0.000	-	0.000	-	0.000	-
0.095	0.227	0.129	0.128	0.196	0.046	0.319	0.000	-	0.000	-	0.000	-
0.100	0.239	0.132	0.142	0.199	0.058	0.321	0.003	0.471	0.000	-	0.000	-
0.105	0.252	0.134	0.153	0.202	0.071	0.322	0.016	0.471	0.000	-	0.000	-
0.110	0.264	0.137	0.167	0.205	0.084	0.323	0.028	0.471	0.003	0.627	0.008	0.780
0.115	0.277	0.140	0.179	0.207	0.097	0.325	0.041	0.471	0.020	0.621	0.027	0.768
0.120	0.289	0.143	0.192	0.210	0.109	0.326	0.053	0.471	0.036	0.616	0.045	0.758
0.125	0.302	0.145	0.205	0.212	0.122	0.327	0.066	0.471	0.052	0.611	0.063	0.748
0.130	0.314	0.148	0.217	0.214	0.135	0.328	0.079	0.471	0.068	0.607	0.081	0.739
0.135	0.327	0.150	0.230	0.217	0.148	0.330	0.091	0.471	0.083	0.603	0.099	0.730
0.140	0.339	0.152	0.243	0.219	0.160	0.331	0.104	0.471	0.098	0.599	0.116	0.722
0.145	0.352	0.155	0.255	0.221	0.173	0.332	0.117	0.471	0.113	0.595	0.133	0.714
0.150	0.364	0.157	0.269	0.223	0.186	0.333	0.129	0.471	0.128	0.592	0.150	0.707
0.155	0.377	0.159	0.281	0.225	0.199	0.334	0.142	0.471	0.143	0.589	0.167	0.700
0.160	0.389	0.161	0.294	0.227	0.212	0.335	0.155	0.471	0.158	0.586	0.183	0.694
0.165	0.402	0.163	0.307	0.229	0.224	0.336	0.167	0.471	0.172	0.584	0.199	0.688
0.170	0.414	0.165	0.318	0.231	0.237	0.337	0.180	0.471	0.187	0.581	0.215	0.683
0.175	0.427	0.167	0.331	0.233	0.250	0.338	0.193	0.471	0.201	0.579	0.231	0.677
0.180	0.440	0.169	0.344	0.235	0.263	0.339	0.205	0.471	0.215	0.577	0.247	0.672
0.185	0.452	0.171	0.357	0.236	0.275	0.340	0.218	0.471	0.229	0.574	0.262	0.668
0.190	0.465	0.173	0.370	0.238	0.288	0.341	0.230	0.471	0.243	0.573	0.278	0.663
0.195	0.477	0.175	0.382	0.240	0.301	0.342	0.243	0.471	0.257	0.571	0.293	0.659
0.200	0.490	0.177	0.395	0.241	0.313	0.343	0.256	0.471	0.271	0.569	0.308	0.655
0.205	0.502	0.179	0.407	0.243	0.326	0.344	0.268	0.471	0.285	0.567	0.324	0.651
0.210	0.515	0.180	0.420	0.245	0.339	0.345	0.281	0.471	0.299	0.566	0.338	0.648
0.215	0.528	0.182	0.434	0.246	0.352	0.346	0.294	0.471	0.312	0.564	0.353	0.644
0.220	0.540	0.184	0.446	0.248	0.364	0.347	0.306	0.471	0.326	0.563	0.368	0.641
0.225	0.553	0.185	0.459	0.249	0.377	0.348	0.319	0.471	0.340	0.561	0.383	0.638
0.230	0.565	0.187	0.472	0.251	0.390	0.349	0.332	0.471	0.353	0.560	0.397	0.635
0.235	0.578	0.189	0.484	0.252	0.402	0.350	0.344	0.471	0.367	0.559	0.412	0.632
0.240	0.591	0.190	0.496	0.253	0.415	0.350	0.357	0.471	0.380	0.557	0.426	0.629
0.245	0.603	0.192	0.509	0.255	0.428	0.351	0.369	0.471	0.394	0.556	0.441	0.626

TABLE 3_Concrete class C70_Steel class S500_a/h=0.10 (cont.)
Bending and axial force with symmetric reinforcement

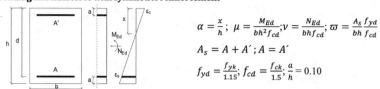

$$\alpha = \frac{x}{h}; \quad \mu = \frac{M_{Ed}}{bh^2 f_{cd}}; \nu = \frac{N_{Ed}}{bhf_{cd}}; \varpi = \frac{A_s}{bh}\frac{f_{yd}}{f_{cd}}$$

$$A_S = A + A'\,; A = A'$$

$$f_{yd} = \frac{f_{yk}}{1.15}; \; f_{cd} = \frac{f_{ck}}{1.5}; \frac{a}{h} = 0.10$$

μ	ν=0.6 ϖ	ν=0.6 α	ν=0.8 ϖ	ν=0.8 α	ν=1.0 ϖ	ν=1.0 α	ν=1.2 ϖ	ν=1.2 α	ν=1.4 ϖ	ν=1.4 α	ν=1.6 ϖ	ν=1.6 α
0.000	0.000	-	0.000	-	0.000	-	0.201	-	0.400	-	0.600	-
0.005	0.000	-	0.000	-	0.020	6.650	0.219	7.320	0.419	7.570	0.619	7.790
0.010	0.000	-	0.000	-	0.040	4.330	0.236	5.250	0.435	5.860	0.634	6.280
0.015	0.000	-	0.000	-	0.059	3.400	0.253	4.220	0.450	4.830	0.649	5.310
0.020	0.000	-	0.000	-	0.077	2.900	0.270	3.590	0.466	4.170	0.664	4.640
0.025	0.000	-	0.000	-	0.096	2.560	0.286	3.170	0.482	3.690	0.679	4.140
0.030	0.000	-	0.000	-	0.113	2.320	0.303	2.860	0.497	3.340	0.694	3.760
0.035	0.000	-	0.000	-	0.130	2.150	0.319	2.620	0.513	3.070	0.709	3.460
0.040	0.000	-	0.000	-	0.148	2.000	0.335	2.440	0.528	2.840	0.723	3.210
0.045	0.000	-	0.000	-	0.165	1.890	0.352	2.280	0.543	2.660	0.738	3.000
0.050	0.000	-	0.005	1.430	0.182	1.790	0.368	2.160	0.559	2.500	0.753	2.830
0.055	0.000	-	0.022	1.380	0.198	1.710	0.384	2.050	0.574	2.370	0.768	2.680
0.060	0.000	-	0.040	1.340	0.215	1.640	0.400	1.950	0.590	2.260	0.783	2.550
0.065	0.000	-	0.057	1.300	0.232	1.580	0.416	1.870	0.605	2.160	0.798	2.440
0.070	0.000	-	0.074	1.260	0.248	1.530	0.431	1.800	0.621	2.070	0.813	2.340
0.075	0.000	-	0.090	1.230	0.264	1.480	0.447	1.740	0.635	2.000	0.828	2.250
0.080	0.000	-	0.106	1.200	0.280	1.440	0.463	1.680	0.651	1.930	0.843	2.170
0.085	0.000	-	0.123	1.180	0.296	1.400	0.478	1.630	0.666	1.870	0.857	2.090
0.090	0.000	-	0.139	1.150	0.312	1.370	0.493	1.590	0.681	1.810	0.872	2.030
0.095	0.000	-	0.154	1.130	0.327	1.340	0.509	1.550	0.696	1.760	0.887	1.970
0.100	0.016	0.928	0.170	1.110	0.343	1.310	0.524	1.510	0.711	1.710	0.901	1.910
0.105	0.034	0.915	0.186	1.090	0.359	1.280	0.539	1.470	0.726	1.670	0.916	1.860
0.110	0.051	0.901	0.202	1.080	0.374	1.260	0.554	1.440	0.741	1.630	0.931	1.820
0.115	0.069	0.888	0.217	1.060	0.389	1.230	0.570	1.410	0.756	1.590	0.945	1.770
0.120	0.087	0.876	0.233	1.050	0.405	1.210	0.585	1.380	0.771	1.560	0.960	1.730
0.125	0.105	0.864	0.248	1.030	0.420	1.190	0.600	1.360	0.785	1.530	0.975	1.690
0.130	0.122	0.853	0.263	1.020	0.436	1.170	0.615	1.340	0.800	1.500	0.989	1.660
0.135	0.140	0.842	0.278	1.010	0.450	1.160	0.630	1.310	0.815	1.470	1.003	1.630
0.140	0.157	0.831	0.294	0.997	0.466	1.140	0.645	1.290	0.830	1.440	1.018	1.600
0.145	0.174	0.821	0.309	0.986	0.481	1.130	0.660	1.270	0.844	1.420	1.032	1.570
0.150	0.191	0.812	0.325	0.976	0.496	1.110	0.675	1.250	0.859	1.400	1.047	1.540
0.155	0.209	0.803	0.341	0.965	0.510	1.100	0.689	1.240	0.873	1.370	1.061	1.510
0.160	0.225	0.794	0.357	0.956	0.526	1.080	0.704	1.220	0.888	1.350	1.076	1.490
0.165	0.242	0.786	0.373	0.946	0.540	1.070	0.719	1.200	0.903	1.330	1.090	1.470
0.170	0.259	0.778	0.388	0.936	0.555	1.060	0.733	1.190	0.917	1.320	1.104	1.450
0.175	0.275	0.771	0.404	0.927	0.569	1.050	0.748	1.170	0.932	1.300	1.119	1.420
0.180	0.292	0.764	0.420	0.918	0.584	1.040	0.763	1.160	0.946	1.280	1.133	1.400
0.185	0.308	0.757	0.436	0.909	0.599	1.030	0.777	1.150	0.960	1.270	1.147	1.390
0.190	0.324	0.751	0.452	0.901	0.614	1.020	0.792	1.130	0.975	1.250	1.162	1.370
0.195	0.340	0.744	0.467	0.893	0.628	1.010	0.806	1.120	0.989	1.240	1.176	1.350
0.200	0.356	0.738	0.483	0.885	0.643	1.000	0.821	1.110	1.003	1.220	1.190	1.330
0.205	0.372	0.733	0.498	0.877	0.657	0.993	0.835	1.100	1.018	1.210	1.204	1.320
0.210	0.387	0.727	0.515	0.869	0.673	0.985	0.850	1.090	1.032	1.200	1.218	1.300
0.215	0.403	0.722	0.530	0.862	0.687	0.977	0.864	1.080	1.046	1.180	1.233	1.290
0.220	0.418	0.717	0.546	0.855	0.702	0.969	0.878	1.070	1.061	1.170	1.247	1.280
0.225	0.434	0.712	0.561	0.848	0.717	0.962	0.893	1.060	1.075	1.160	1.261	1.260
0.230	0.449	0.708	0.577	0.842	0.732	0.954	0.907	1.050	1.089	1.150	1.275	1.250
0.235	0.464	0.704	0.592	0.835	0.747	0.947	0.921	1.040	1.103	1.140	1.289	1.240
0.240	0.480	0.699	0.608	0.829	0.762	0.940	0.936	1.030	1.117	1.130	1.303	1.230
0.245	0.494	0.695	0.623	0.823	0.777	0.933	0.950	1.030	1.132	1.120	1.317	1.210

TABLE 3_Concrete class C70_Steel class S500_a/h=0.10 (cont.)
Bending and axial force with symmetric reinforcement

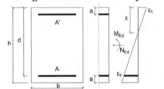

$$\alpha = \frac{x}{h} \; ; \quad \mu = \frac{M_{Ed}}{bh^2 f_{cd}} \; ; \quad \nu = \frac{N_{Ed}}{bh f_{cd}} \; ; \quad \varpi = \frac{A_s\, f_{yd}}{bh\, f_{cd}}$$

$$A_s = A + A' \; ; \quad A = A'$$

$$f_{yd} = \frac{f_{yk}}{1.15} \; ; \quad f_{cd} = \frac{f_{ck}}{1.5} \; ; \quad \frac{a}{h} = 0.10$$

μ	$\nu=0.0$ ϖ	α	$\nu=0.1$ ϖ	α	$\nu=0.2$ ϖ	α	$\nu=0.3$ ϖ	α	$\nu=0.4$ ϖ	α	$\nu=0.5$ ϖ	α
0.250	0.616	0.193	0.521	0.256	0.441	0.352	0.382	0.471	0.407	0.555	0.455	0.624
0.255	0.628	0.195	0.534	0.258	0.453	0.353	0.395	0.471	0.420	0.554	0.469	0.621
0.260	0.641	0.196	0.547	0.259	0.466	0.354	0.407	0.471	0.434	0.553	0.483	0.619
0.265	0.654	0.198	0.560	0.260	0.479	0.355	0.420	0.471	0.447	0.552	0.498	0.617
0.270	0.666	0.199	0.573	0.262	0.491	0.355	0.433	0.471	0.460	0.551	0.512	0.615
0.275	0.679	0.201	0.586	0.263	0.504	0.356	0.445	0.471	0.473	0.550	0.526	0.613
0.280	0.691	0.202	0.598	0.264	0.517	0.357	0.458	0.471	0.487	0.549	0.540	0.611
0.285	0.704	0.203	0.611	0.265	0.529	0.358	0.471	0.471	0.500	0.548	0.554	0.609
0.290	0.717	0.205	0.623	0.266	0.542	0.358	0.483	0.471	0.513	0.547	0.568	0.607
0.295	0.729	0.206	0.636	0.268	0.555	0.359	0.496	0.471	0.526	0.547	0.581	0.605
0.300	0.742	0.208	0.649	0.269	0.567	0.360	0.509	0.471	0.539	0.546	0.595	0.603
0.305	0.755	0.209	0.662	0.270	0.580	0.361	0.521	0.471	0.552	0.545	0.609	0.601
0.310	0.767	0.210	0.675	0.271	0.593	0.361	0.534	0.471	0.566	0.544	0.623	0.600
0.315	0.780	0.211	0.688	0.272	0.605	0.362	0.546	0.471	0.579	0.544	0.636	0.598
0.320	0.792	0.213	0.700	0.273	0.618	0.363	0.559	0.471	0.592	0.543	0.650	0.597
0.325	0.805	0.214	0.712	0.274	0.631	0.363	0.572	0.471	0.605	0.542	0.664	0.595
0.330	0.818	0.215	0.725	0.276	0.644	0.364	0.584	0.471	0.618	0.542	0.677	0.594
0.335	0.830	0.216	0.738	0.277	0.656	0.365	0.597	0.471	0.631	0.541	0.691	0.592
0.340	0.843	0.218	0.750	0.278	0.669	0.365	0.610	0.471	0.644	0.540	0.704	0.591
0.345	0.855	0.219	0.763	0.279	0.681	0.366	0.622	0.471	0.657	0.540	0.718	0.590
0.350	0.868	0.220	0.776	0.280	0.694	0.367	0.635	0.471	0.670	0.539	0.731	0.588
0.355	0.881	0.221	0.788	0.281	0.707	0.367	0.648	0.471	0.683	0.539	0.745	0.587
0.360	0.893	0.222	0.801	0.282	0.719	0.368	0.660	0.471	0.696	0.538	0.758	0.586
0.365	0.906	0.223	0.814	0.283	0.732	0.368	0.673	0.471	0.709	0.537	0.771	0.585
0.370	0.918	0.224	0.826	0.284	0.745	0.369	0.685	0.471	0.721	0.537	0.785	0.584
0.375	0.931	0.225	0.839	0.285	0.758	0.370	0.698	0.471	0.734	0.536	0.798	0.582
0.380	0.943	0.227	0.852	0.286	0.770	0.370	0.711	0.471	0.747	0.536	0.811	0.581
0.385	0.956	0.228	0.864	0.287	0.783	0.371	0.723	0.471	0.760	0.536	0.825	0.580
0.390	0.969	0.229	0.877	0.288	0.795	0.371	0.736	0.471	0.773	0.535	0.838	0.579
0.395	0.981	0.230	0.889	0.288	0.808	0.372	0.749	0.471	0.786	0.535	0.851	0.578
0.400	0.994	0.231	0.902	0.289	0.821	0.373	0.761	0.471	0.799	0.534	0.865	0.577
0.410			0.928	0.291	0.846	0.374	0.787	0.471	0.824	0.533	0.891	0.575
0.420			0.953	0.293	0.871	0.375	0.812	0.471	0.850	0.532	0.918	0.574
0.430			0.978	0.295	0.897	0.376	0.837	0.471	0.876	0.532	0.944	0.572
0.440			1.004	0.296	0.922	0.377	0.862	0.471	0.901	0.531	0.970	0.570
0.450			1.028	0.298	0.947	0.378	0.888	0.471	0.927	0.530	0.996	0.569
0.460			1.054	0.300	0.972	0.379	0.913	0.471	0.953	0.530	1.022	0.567
0.470			1.079	0.301	0.998	0.380	0.938	0.471	0.978	0.529	1.048	0.566
0.480			1.104	0.303	1.023	0.381	0.963	0.471	1.003	0.528	1.075	0.564
0.490			1.129	0.304	1.048	0.382	0.989	0.471	1.029	0.528	1.101	0.563
0.500			1.155	0.306	1.074	0.383	1.014	0.471	1.055	0.527	1.127	0.562
0.510			1.181	0.307	1.099	0.384	1.039	0.471	1.080	0.527	1.153	0.561
0.520			1.206	0.309	1.124	0.385	1.065	0.471	1.105	0.526	1.178	0.559
0.530			1.232	0.310	1.149	0.386	1.090	0.471	1.131	0.526	1.204	0.558
0.540			1.256	0.311	1.175	0.387	1.115	0.471	1.156	0.525	1.230	0.557
0.550			1.281	0.313	1.200	0.388	1.140	0.471	1.181	0.525	1.256	0.556
0.560			1.307	0.314	1.225	0.389	1.166	0.471	1.207	0.524	1.282	0.555
0.570			1.332	0.315	1.250	0.389	1.191	0.471	1.232	0.524	1.308	0.554
0.580			1.357	0.317	1.276	0.390	1.216	0.471	1.258	0.523	1.333	0.553
0.590			1.383	0.318	1.301	0.391	1.241	0.471	1.283	0.523	1.359	0.552

TABLE 3_Concrete class C70_Steel class S500_a/h=0.10 (cont.)
Bending and axial force with symmetric reinforcement

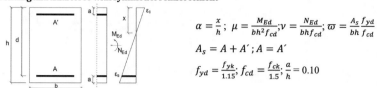

$$\alpha = \frac{x}{h}; \quad \mu = \frac{M_{Ed}}{bh^2 f_{cd}}; \nu = \frac{N_{Ed}}{bh f_{cd}}; \varpi = \frac{A_s}{bh}\frac{f_{yd}}{f_{cd}}$$

$$A_s = A + A'; A = A'$$

$$f_{yd} = \frac{f_{yk}}{1.15}; f_{cd} = \frac{f_{ck}}{1.5}; \frac{a}{h} = 0.10$$

μ	$\nu=0.6$ ϖ	α	$\nu=0.8$ ϖ	α	$\nu=1.0$ ϖ	α	$\nu=1.2$ ϖ	α	$\nu=1.4$ ϖ	α	$\nu=1.6$ ϖ	α
0.250	0.509	0.692	0.638	0.817	0.792	0.926	0.964	1.020	1.146	1.110	1.331	1.200
0.255	0.524	0.688	0.654	0.812	0.807	0.920	0.978	1.010	1.160	1.100	1.345	1.190
0.260	0.539	0.684	0.669	0.806	0.822	0.913	0.992	1.000	1.174	1.090	1.359	1.180
0.265	0.554	0.681	0.684	0.801	0.837	0.907	1.006	0.998	1.188	1.080	1.373	1.170
0.270	0.569	0.678	0.699	0.796	0.852	0.901	1.021	0.991	1.202	1.080	1.387	1.160
0.275	0.583	0.674	0.715	0.791	0.866	0.895	1.035	0.984	1.216	1.070	1.401	1.150
0.280	0.598	0.671	0.729	0.786	0.882	0.889	1.049	0.978	1.230	1.060	1.415	1.140
0.285	0.612	0.668	0.744	0.782	0.896	0.883	1.064	0.972	1.244	1.050	1.429	1.140
0.290	0.626	0.665	0.760	0.777	0.911	0.877	1.078	0.966	1.258	1.050	1.443	1.130
0.295	0.641	0.663	0.774	0.772	0.926	0.872	1.093	0.960	1.272	1.040	1.457	1.120
0.300	0.655	0.660	0.789	0.768	0.941	0.867	1.107	0.954	1.286	1.030	1.471	1.110
0.305	0.670	0.657	0.804	0.764	0.955	0.861	1.121	0.948	1.300	1.030	1.485	1.100
0.310	0.684	0.655	0.818	0.760	0.970	0.856	1.136	0.942	1.313	1.020	1.498	1.100
0.315	0.697	0.652	0.833	0.756	0.985	0.851	1.151	0.937	1.328	1.010	1.512	1.090
0.320	0.712	0.650	0.848	0.752	1.000	0.847	1.165	0.931	1.341	1.010	1.526	1.080
0.325	0.726	0.648	0.862	0.749	1.014	0.842	1.179	0.926	1.355	1.000	1.540	1.070
0.330	0.740	0.645	0.877	0.745	1.029	0.837	1.194	0.920	1.369	0.995	1.554	1.070
0.335	0.754	0.643	0.892	0.742	1.044	0.833	1.208	0.915	1.384	0.989	1.568	1.060
0.340	0.768	0.641	0.906	0.738	1.058	0.828	1.223	0.910	1.397	0.984	1.581	1.050
0.345	0.782	0.639	0.920	0.735	1.073	0.824	1.237	0.905	1.412	0.979	1.595	1.050
0.350	0.796	0.637	0.935	0.732	1.087	0.820	1.251	0.901	1.426	0.974	1.609	1.040
0.355	0.809	0.635	0.949	0.728	1.102	0.816	1.266	0.896	1.440	0.968	1.623	1.040
0.360	0.823	0.633	0.963	0.725	1.116	0.812	1.280	0.891	1.454	0.963	1.636	1.030
0.365	0.837	0.632	0.978	0.722	1.131	0.808	1.295	0.887	1.468	0.958	1.650	1.020
0.370	0.851	0.630	0.992	0.720	1.145	0.804	1.309	0.882	1.482	0.953	1.664	1.020
0.375	0.865	0.628	1.006	0.717	1.159	0.800	1.324	0.878	1.496	0.949	1.677	1.010
0.380	0.878	0.626	1.020	0.714	1.174	0.797	1.338	0.873	1.510	0.944	1.691	1.010
0.385	0.892	0.625	1.035	0.711	1.188	0.793	1.352	0.869	1.525	0.939	1.705	1.000
0.390	0.906	0.623	1.049	0.709	1.203	0.790	1.366	0.865	1.539	0.934	1.719	0.998
0.395	0.919	0.622	1.063	0.706	1.217	0.786	1.380	0.861	1.553	0.930	1.732	0.993
0.400	0.933	0.620	1.077	0.704	1.231	0.783	1.395	0.857	1.567	0.925	1.746	0.988
0.410	0.960	0.617	1.105	0.699	1.259	0.776	1.423	0.849	1.596	0.917	1.774	0.979
0.420	0.987	0.614	1.132	0.694	1.288	0.770	1.452	0.842	1.624	0.908	1.802	0.970
0.430	1.014	0.612	1.161	0.690	1.316	0.764	1.480	0.835	1.652	0.900	1.830	0.961
0.440	1.041	0.609	1.188	0.685	1.344	0.759	1.509	0.828	1.680	0.893	1.857	0.953
0.450	1.067	0.607	1.216	0.681	1.372	0.753	1.537	0.821	1.708	0.885	1.885	0.945
0.460	1.094	0.604	1.244	0.678	1.400	0.748	1.564	0.815	1.736	0.878	1.913	0.937
0.470	1.121	0.602	1.271	0.674	1.428	0.743	1.593	0.809	1.764	0.871	1.941	0.929
0.480	1.148	0.600	1.298	0.670	1.456	0.738	1.621	0.803	1.792	0.864	1.969	0.922
0.490	1.174	0.598	1.325	0.667	1.484	0.733	1.649	0.797	1.819	0.857	1.996	0.914
0.500	1.200	0.596	1.352	0.664	1.511	0.729	1.677	0.792	1.847	0.851	2.024	0.907
0.510	1.227	0.594	1.379	0.661	1.539	0.725	1.705	0.786	1.875	0.845	2.052	0.900
0.520	1.253	0.592	1.407	0.657	1.566	0.721	1.732	0.781	1.903	0.839	2.079	0.894
0.530	1.279	0.591	1.434	0.655	1.594	0.717	1.760	0.776	1.931	0.833	2.107	0.887
0.540	1.306	0.589	1.461	0.652	1.621	0.713	1.787	0.771	1.958	0.828	2.135	0.881
0.550	1.332	0.587	1.487	0.649	1.648	0.709	1.815	0.767	1.986	0.822	2.162	0.875
0.560	1.358	0.586	1.514	0.646	1.676	0.705	1.842	0.762	2.013	0.817	2.190	0.869
0.570	1.384	0.584	1.541	0.644	1.703	0.702	1.870	0.758	2.041	0.812	2.217	0.863
0.580	1.410	0.583	1.568	0.641	1.730	0.698	1.897	0.754	2.069	0.807	2.245	0.858
0.590	1.436	0.581	1.594	0.639	1.757	0.695	1.924	0.750	2.096	0.802	2.272	0.852

Tab. 5.18

TABLE 3_Concrete class C80_Steel class S500_a/h=0.10

Bending and axial force with symmetric reinforcement

$$\alpha = \frac{x}{h}\; ; \quad \mu = \frac{M_{Ed}}{bh^2 f_{cd}}; \nu = \frac{N_{Ed}}{bh f_{cd}}; \quad \varpi = \frac{A_s}{bh}\frac{f_{yd}}{f_{cd}}$$

$$A_s = A + A'\; ; A = A'$$

$$f_{yd} = \frac{f_{yk}}{1.15}; f_{cd} = \frac{f_{ck}}{1.5}; \frac{a}{h} = 0.10$$

	$\nu=0.0$		$\nu=0.1$		$\nu=0.2$		$\nu=0.3$		$\nu=0.4$		$\nu=0.5$	
μ	ϖ	α	ϖ	α	ϖ	α	ϖ	α	ϖ	α	ϖ	α
0.000	0.000	-	0.000	-	0.000	-	0.000	-	0.000	-	0.000	-
0.005	0.010	0.036	0.000	-	0.000	-	0.000	-	0.000	-	0.000	-
0.010	0.021	0.049	0.000	-	0.000	-	0.000	-	0.000	-	0.000	-
0.015	0.032	0.059	0.000	-	0.000	-	0.000	-	0.000	-	0.000	-
0.020	0.044	0.066	0.000	-	0.000	-	0.000	-	0.000	-	0.000	-
0.025	0.055	0.073	0.000	-	0.000	-	0.000	-	0.000	-	0.000	-
0.030	0.067	0.078	0.000	-	0.000	-	0.000	-	0.000	-	0.000	-
0.035	0.079	0.083	0.000	-	0.000	-	0.000	-	0.000	-	0.000	-
0.040	0.091	0.088	0.000	-	0.000	-	0.000	-	0.000	-	0.000	-
0.045	0.103	0.093	0.002	0.168	0.000	-	0.000	-	0.000	-	0.000	-
0.050	0.115	0.098	0.016	0.173	0.000	-	0.000	-	0.000	-	0.000	-
0.055	0.128	0.103	0.028	0.178	0.000	-	0.000	-	0.000	-	0.000	-
0.060	0.140	0.107	0.040	0.182	0.000	-	0.000	-	0.000	-	0.000	-
0.065	0.152	0.111	0.053	0.186	0.000	-	0.000	-	0.000	-	0.000	-
0.070	0.165	0.115	0.066	0.191	0.000	-	0.000	-	0.000	-	0.000	-
0.075	0.177	0.119	0.078	0.194	0.000	-	0.000	-	0.000	-	0.000	-
0.080	0.190	0.123	0.091	0.198	0.010	0.335	0.000	-	0.000	-	0.000	-
0.085	0.202	0.126	0.104	0.201	0.022	0.337	0.000	-	0.000	-	0.000	-
0.090	0.215	0.130	0.117	0.205	0.035	0.338	0.000	-	0.000	-	0.000	-
0.095	0.227	0.133	0.129	0.208	0.048	0.340	0.000	-	0.000	-	0.000	-
0.100	0.239	0.136	0.142	0.211	0.061	0.342	0.009	0.500	0.000	-	0.000	-
0.105	0.252	0.139	0.155	0.214	0.074	0.343	0.022	0.500	0.000	-	0.011	0.826
0.110	0.264	0.142	0.168	0.217	0.087	0.345	0.035	0.500	0.017	0.659	0.030	0.812
0.115	0.277	0.145	0.180	0.219	0.099	0.346	0.048	0.500	0.035	0.652	0.049	0.799
0.120	0.290	0.148	0.193	0.222	0.112	0.348	0.061	0.500	0.052	0.645	0.068	0.787
0.125	0.302	0.151	0.205	0.224	0.125	0.349	0.074	0.500	0.068	0.638	0.086	0.776
0.130	0.315	0.153	0.218	0.227	0.138	0.350	0.087	0.500	0.085	0.632	0.104	0.766
0.135	0.327	0.156	0.231	0.229	0.151	0.352	0.100	0.500	0.101	0.627	0.122	0.756
0.140	0.340	0.158	0.244	0.232	0.163	0.353	0.113	0.500	0.117	0.622	0.140	0.746
0.145	0.352	0.161	0.257	0.234	0.176	0.354	0.126	0.500	0.133	0.617	0.157	0.738
0.150	0.365	0.163	0.269	0.236	0.189	0.356	0.139	0.500	0.148	0.613	0.174	0.730
0.155	0.377	0.166	0.283	0.239	0.202	0.357	0.152	0.500	0.164	0.609	0.191	0.722
0.160	0.390	0.168	0.295	0.241	0.215	0.358	0.166	0.500	0.179	0.605	0.208	0.715
0.165	0.402	0.170	0.307	0.243	0.227	0.360	0.179	0.500	0.194	0.602	0.224	0.708
0.170	0.415	0.172	0.320	0.245	0.240	0.361	0.192	0.500	0.209	0.599	0.241	0.701
0.175	0.428	0.175	0.333	0.247	0.253	0.362	0.205	0.500	0.224	0.596	0.257	0.695
0.180	0.440	0.177	0.346	0.249	0.266	0.363	0.218	0.500	0.239	0.593	0.273	0.690
0.185	0.453	0.179	0.358	0.251	0.279	0.364	0.231	0.500	0.253	0.590	0.289	0.684
0.190	0.465	0.181	0.371	0.253	0.291	0.366	0.244	0.500	0.268	0.588	0.304	0.679
0.195	0.478	0.183	0.383	0.254	0.304	0.367	0.257	0.500	0.283	0.585	0.320	0.674
0.200	0.491	0.185	0.397	0.256	0.317	0.368	0.270	0.500	0.297	0.583	0.336	0.670
0.205	0.503	0.187	0.409	0.258	0.330	0.369	0.283	0.500	0.312	0.581	0.351	0.665
0.210	0.516	0.189	0.422	0.260	0.342	0.370	0.296	0.500	0.326	0.579	0.366	0.661
0.215	0.528	0.190	0.435	0.262	0.355	0.371	0.309	0.500	0.340	0.577	0.381	0.657
0.220	0.541	0.192	0.447	0.263	0.368	0.372	0.322	0.500	0.355	0.575	0.396	0.653
0.225	0.554	0.194	0.460	0.265	0.381	0.373	0.335	0.500	0.369	0.573	0.411	0.650
0.230	0.566	0.196	0.472	0.266	0.393	0.374	0.349	0.500	0.383	0.572	0.426	0.646
0.235	0.579	0.198	0.485	0.268	0.406	0.375	0.362	0.500	0.397	0.570	0.441	0.643
0.240	0.591	0.199	0.499	0.270	0.419	0.376	0.375	0.500	0.411	0.569	0.455	0.640
0.245	0.604	0.201	0.511	0.271	0.432	0.377	0.388	0.500	0.425	0.567	0.470	0.636

TABLE 3_Concrete class C80_Steel class S500_a/h=0.10 (cont.)
Bending and axial force with symmetric reinforcement

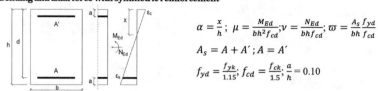

$$\alpha = \frac{x}{h}; \ \mu = \frac{M_{Ed}}{bh^2 f_{cd}}; \nu = \frac{N_{Ed}}{bhf_{cd}}; \varpi = \frac{A_s \, f_{yd}}{bh \, f_{cd}}$$

$$A_s = A + A'; A = A'$$

$$f_{yd} = \frac{f_{yk}}{1.15}; f_{cd} = \frac{f_{ck}}{1.5}, \frac{a}{h} = 0.10$$

	ν=0.6		ν=0.8		ν=1.0		ν=1.2		ν=1.4		ν=1.6	
μ	ϖ	α	ϖ	α	ϖ	α	ϖ	α	ϖ	α	ϖ	α
0.000	0.000	-	0.000	-	0.000	-	0.201	-	0.400	-	0.600	-
0.005	0.000	-	0.000	-	0.023	7.510	0.223	7.430	0.423	7.420	0.623	7.420
0.010	0.000	-	0.000	-	0.045	4.700	0.242	5.270	0.441	5.580	0.640	5.780
0.015	0.000	-	0.000	-	0.066	3.690	0.260	4.310	0.458	4.730	0.656	5.030
0.020	0.000	-	0.000	-	0.086	3.110	0.278	3.720	0.474	4.150	0.672	4.480
0.025	0.000	-	0.000	-	0.106	2.740	0.296	3.290	0.491	3.720	0.687	4.060
0.030	0.000	-	0.000	-	0.125	2.480	0.314	2.970	0.507	3.390	0.703	3.720
0.035	0.000	-	0.000	-	0.144	2.280	0.331	2.730	0.524	3.120	0.719	3.450
0.040	0.000	-	0.000	-	0.162	2.120	0.348	2.540	0.540	2.900	0.734	3.220
0.045	0.000	-	0.006	1.590	0.180	2.000	0.365	2.380	0.556	2.720	0.750	3.030
0.050	0.000	-	0.025	1.520	0.198	1.890	0.382	2.250	0.572	2.570	0.766	2.860
0.055	0.000	-	0.043	1.460	0.216	1.800	0.399	2.130	0.589	2.440	0.781	2.720
0.060	0.000	-	0.062	1.410	0.234	1.720	0.416	2.040	0.604	2.330	0.797	2.590
0.065	0.000	-	0.079	1.360	0.251	1.660	0.432	1.950	0.621	2.230	0.812	2.480
0.070	0.000	-	0.097	1.320	0.268	1.600	0.449	1.880	0.636	2.140	0.828	2.380
0.075	0.000	-	0.114	1.290	0.285	1.550	0.465	1.810	0.652	2.060	0.843	2.300
0.080	0.000	-	0.131	1.260	0.301	1.500	0.482	1.750	0.668	1.990	0.858	2.220
0.085	0.000	-	0.149	1.230	0.318	1.460	0.498	1.700	0.684	1.930	0.874	2.140
0.090	0.011	0.991	0.166	1.200	0.335	1.420	0.514	1.650	0.699	1.870	0.889	2.080
0.095	0.028	0.976	0.182	1.180	0.351	1.390	0.530	1.610	0.715	1.820	0.904	2.020
0.100	0.045	0.961	0.199	1.150	0.368	1.360	0.546	1.560	0.730	1.770	0.919	1.960
0.105	0.063	0.945	0.215	1.130	0.384	1.330	0.562	1.530	0.746	1.720	0.935	1.910
0.110	0.080	0.932	0.232	1.110	0.400	1.300	0.578	1.490	0.762	1.680	0.950	1.860
0.115	0.098	0.917	0.248	1.100	0.416	1.280	0.593	1.460	0.777	1.640	0.965	1.820
0.120	0.115	0.905	0.263	1.080	0.432	1.260	0.609	1.430	0.792	1.610	0.980	1.780
0.125	0.133	0.891	0.280	1.060	0.448	1.230	0.624	1.410	0.808	1.580	0.995	1.740
0.130	0.150	0.879	0.296	1.050	0.463	1.210	0.640	1.380	0.823	1.540	1.010	1.710
0.135	0.168	0.867	0.311	1.040	0.479	1.190	0.655	1.360	0.838	1.520	1.025	1.670
0.140	0.185	0.856	0.327	1.020	0.495	1.180	0.671	1.330	0.853	1.490	1.040	1.640
0.145	0.203	0.845	0.342	1.010	0.510	1.160	0.686	1.310	0.868	1.460	1.055	1.610
0.150	0.220	0.835	0.358	1.000	0.526	1.140	0.701	1.290	0.883	1.440	1.069	1.580
0.155	0.237	0.825	0.374	0.989	0.541	1.130	0.717	1.270	0.898	1.410	1.084	1.560
0.160	0.254	0.816	0.390	0.979	0.557	1.110	0.732	1.250	0.913	1.390	1.099	1.530
0.165	0.271	0.807	0.405	0.968	0.571	1.100	0.747	1.240	0.929	1.370	1.114	1.510
0.170	0.287	0.798	0.421	0.958	0.587	1.090	0.762	1.220	0.943	1.350	1.128	1.480
0.175	0.304	0.790	0.437	0.948	0.602	1.080	0.777	1.210	0.958	1.330	1.143	1.460
0.180	0.321	0.782	0.453	0.939	0.617	1.060	0.792	1.190	0.973	1.320	1.158	1.440
0.185	0.337	0.774	0.468	0.929	0.632	1.050	0.807	1.180	0.988	1.300	1.173	1.420
0.190	0.353	0.767	0.484	0.920	0.647	1.040	0.822	1.160	1.002	1.280	1.187	1.400
0.195	0.369	0.761	0.500	0.911	0.662	1.030	0.837	1.150	1.017	1.270	1.202	1.390
0.200	0.385	0.754	0.516	0.903	0.677	1.020	0.852	1.140	1.032	1.250	1.216	1.370
0.205	0.402	0.748	0.531	0.895	0.692	1.010	0.866	1.130	1.047	1.240	1.231	1.350
0.210	0.417	0.742	0.547	0.886	0.707	1.010	0.881	1.120	1.061	1.230	1.245	1.340
0.215	0.433	0.736	0.563	0.879	0.722	0.997	0.896	1.100	1.076	1.210	1.260	1.320
0.220	0.449	0.731	0.579	0.871	0.736	0.988	0.911	1.090	1.090	1.200	1.274	1.310
0.225	0.464	0.725	0.594	0.864	0.752	0.980	0.925	1.080	1.105	1.190	1.289	1.290
0.230	0.480	0.720	0.610	0.857	0.766	0.972	0.940	1.070	1.120	1.180	1.303	1.280
0.235	0.495	0.716	0.625	0.850	0.781	0.964	0.955	1.060	1.134	1.170	1.318	1.270
0.240	0.510	0.711	0.641	0.843	0.797	0.957	0.969	1.060	1.148	1.150	1.332	1.250
0.245	0.525	0.706	0.657	0.837	0.811	0.950	0.984	1.050	1.163	1.140	1.346	1.240

TABLE 3_Concrete class C80_Steel class S500_a/h=0.10 (cont.)
Bending and axial force with symmetric reinforcement

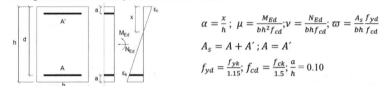

$$\alpha = \frac{x}{h}; \quad \mu = \frac{M_{Ed}}{bh^2 f_{cd}}; \nu = \frac{N_{Ed}}{bhf_{cd}}; \varpi = \frac{A_s\, f_{yd}}{bh\, f_{cd}}$$

$$A_S = A + A'; A = A'$$

$$f_{yd} = \frac{f_{yk}}{1.15}; f_{cd} = \frac{f_{ck}}{1.5}; \frac{a}{h} = 0.10$$

	$\nu=0.0$		$\nu=0.1$		$\nu=0.2$		$\nu=0.3$		$\nu=0.4$		$\nu=0.5$	
μ	ϖ	α	ϖ	α	ϖ	α	ϖ	α	ϖ	α	ϖ	α
0.250	0.616	0.203	0.524	0.273	0.444	0.378	0.401	0.500	0.439	0.566	0.484	0.634
0.255	0.629	0.204	0.536	0.274	0.457	0.379	0.414	0.500	0.453	0.564	0.498	0.631
0.260	0.642	0.206	0.549	0.276	0.470	0.380	0.427	0.500	0.467	0.563	0.513	0.628
0.265	0.654	0.208	0.562	0.277	0.483	0.381	0.440	0.500	0.481	0.562	0.527	0.625
0.270	0.667	0.209	0.574	0.279	0.495	0.382	0.453	0.500	0.495	0.561	0.541	0.623
0.275	0.680	0.211	0.587	0.280	0.508	0.383	0.466	0.500	0.508	0.560	0.555	0.621
0.280	0.692	0.212	0.599	0.281	0.521	0.384	0.479	0.500	0.522	0.558	0.570	0.618
0.285	0.705	0.214	0.612	0.283	0.534	0.385	0.492	0.500	0.536	0.557	0.584	0.616
0.290	0.717	0.215	0.625	0.284	0.546	0.386	0.505	0.500	0.550	0.556	0.598	0.614
0.295	0.730	0.217	0.638	0.285	0.559	0.387	0.518	0.500	0.563	0.555	0.612	0.612
0.300	0.743	0.218	0.651	0.287	0.572	0.387	0.532	0.500	0.577	0.554	0.625	0.610
0.305	0.755	0.220	0.663	0.288	0.585	0.388	0.545	0.500	0.591	0.553	0.640	0.608
0.310	0.768	0.221	0.676	0.289	0.597	0.389	0.558	0.500	0.605	0.553	0.654	0.606
0.315	0.780	0.222	0.688	0.291	0.610	0.390	0.571	0.500	0.618	0.552	0.668	0.604
0.320	0.793	0.224	0.701	0.292	0.623	0.391	0.584	0.500	0.632	0.551	0.682	0.603
0.325	0.806	0.225	0.714	0.293	0.635	0.392	0.597	0.500	0.645	0.550	0.696	0.601
0.330	0.818	0.227	0.727	0.294	0.648	0.392	0.610	0.500	0.659	0.549	0.710	0.599
0.335	0.831	0.228	0.740	0.296	0.661	0.393	0.623	0.500	0.672	0.549	0.724	0.598
0.340	0.844	0.229	0.752	0.297	0.673	0.394	0.636	0.500	0.686	0.548	0.738	0.596
0.345	0.856	0.231	0.765	0.298	0.686	0.395	0.649	0.500	0.700	0.547	0.752	0.595
0.350	0.869	0.232	0.778	0.299	0.699	0.396	0.662	0.500	0.713	0.546	0.766	0.593
0.355			0.790	0.300	0.712	0.396	0.675	0.500	0.727	0.546	0.780	0.592
0.360			0.803	0.301	0.724	0.397	0.688	0.500	0.740	0.545	0.794	0.591
0.365			0.815	0.302	0.737	0.398	0.701	0.500	0.754	0.544	0.808	0.590
0.370			0.828	0.304	0.750	0.399	0.715	0.500	0.767	0.544	0.822	0.588
0.375			0.841	0.305	0.762	0.399	0.728	0.500	0.781	0.543	0.836	0.587
0.380			0.854	0.306	0.775	0.400	0.741	0.500	0.794	0.543	0.849	0.586
0.385			0.867	0.307	0.788	0.401	0.754	0.500	0.807	0.542	0.863	0.585
0.390			0.879	0.308	0.801	0.402	0.767	0.500	0.821	0.542	0.877	0.584
0.395			0.892	0.309	0.813	0.402	0.780	0.500	0.834	0.541	0.890	0.583
0.400			0.905	0.310	0.826	0.403	0.793	0.500	0.848	0.541	0.904	0.581
0.410			0.930	0.312	0.851	0.404	0.819	0.500	0.875	0.540	0.932	0.579
0.420			0.956	0.314	0.877	0.406	0.845	0.500	0.902	0.539	0.959	0.577
0.430			0.981	0.316	0.902	0.407	0.871	0.500	0.928	0.538	0.987	0.576
0.440			1.006	0.318	0.928	0.409	0.898	0.500	0.955	0.537	1.014	0.574
0.450			1.031	0.320	0.953	0.410	0.924	0.500	0.982	0.536	1.041	0.572
0.460			1.057	0.322	0.978	0.411	0.950	0.500	1.008	0.535	1.068	0.571
0.470			1.083	0.324	1.004	0.412	0.976	0.500	1.035	0.534	1.095	0.569
0.480			1.108	0.325	1.029	0.414	1.002	0.500	1.061	0.534	1.123	0.568
0.490			1.133	0.327	1.054	0.415	1.028	0.500	1.088	0.533	1.150	0.566
0.500			1.159	0.329	1.080	0.416	1.054	0.500	1.115	0.532	1.176	0.565
0.510			1.183	0.331	1.105	0.417	1.081	0.500	1.141	0.532	1.203	0.564
0.520			1.209	0.332	1.130	0.418	1.107	0.500	1.168	0.531	1.230	0.562
0.530			1.234	0.334	1.156	0.420	1.133	0.500	1.194	0.530	1.257	0.561
0.540			1.259	0.335	1.181	0.421	1.159	0.500	1.221	0.530	1.284	0.560
0.550			1.284	0.337	1.206	0.422	1.185	0.500	1.247	0.529	1.311	0.559
0.560			1.310	0.339	1.232	0.423	1.211	0.500	1.274	0.529	1.338	0.558
0.570			1.336	0.340	1.257	0.424	1.237	0.500	1.300	0.528	1.365	0.557
0.580			1.361	0.342	1.282	0.425	1.264	0.500	1.327	0.528	1.392	0.556
0.590			1.386	0.343	1.308	0.426	1.290	0.500	1.353	0.527	1.418	0.555

TABLE 3_Concrete class C80_Steel class S500_a/h=0.10 (cont.)
Bending and Axial Force with Symmetric Reinforcement

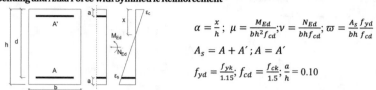

$$\alpha = \frac{x}{h}; \quad \mu = \frac{M_{Ed}}{bh^2 f_{cd}}; \nu = \frac{N_{Ed}}{bh f_{cd}}; \varpi = \frac{A_s\, f_{yd}}{bh\, f_{cd}}$$

$$A_s = A + A'\, ; A = A'$$

$$f_{yd} = \frac{f_{yk}}{1.15}; \; f_{cd} = \frac{f_{ck}}{1.5}; \frac{a}{h} = 0.10$$

	ν=0.6		ν=0.8		ν=1.0		ν=1.2		ν=1.4		ν=1.6	
μ	ϖ	α	ϖ	α	ϖ	α	ϖ	α	ϖ	α	ϖ	α
0.250	0.540	0.702	0.672	0.830	0.826	0.942	0.998	1.040	1.177	1.130	1.361	1.230
0.255	0.555	0.698	0.687	0.825	0.841	0.935	1.013	1.030	1.192	1.120	1.375	1.220
0.260	0.570	0.694	0.703	0.818	0.856	0.928	1.027	1.020	1.206	1.120	1.389	1.210
0.265	0.585	0.690	0.718	0.813	0.871	0.921	1.041	1.020	1.221	1.110	1.404	1.200
0.270	0.599	0.687	0.733	0.807	0.886	0.915	1.056	1.010	1.235	1.100	1.418	1.190
0.275	0.614	0.683	0.748	0.802	0.901	0.908	1.070	1.000	1.249	1.090	1.432	1.180
0.280	0.629	0.680	0.763	0.797	0.916	0.902	1.085	0.994	1.264	1.080	1.446	1.170
0.285	0.643	0.676	0.778	0.792	0.931	0.896	1.099	0.988	1.278	1.070	1.461	1.160
0.290	0.658	0.673	0.793	0.787	0.946	0.890	1.113	0.981	1.292	1.060	1.475	1.150
0.295	0.673	0.670	0.808	0.782	0.961	0.884	1.128	0.975	1.306	1.060	1.489	1.140
0.300	0.687	0.667	0.823	0.778	0.976	0.878	1.142	0.968	1.320	1.050	1.503	1.130
0.305	0.701	0.664	0.838	0.773	0.990	0.873	1.157	0.962	1.334	1.040	1.517	1.120
0.310	0.716	0.662	0.852	0.769	1.005	0.867	1.172	0.956	1.349	1.040	1.531	1.120
0.315	0.730	0.659	0.868	0.765	1.020	0.862	1.186	0.950	1.363	1.030	1.545	1.110
0.320	0.744	0.656	0.882	0.760	1.035	0.857	1.200	0.944	1.377	1.020	1.559	1.100
0.325	0.758	0.654	0.897	0.756	1.049	0.852	1.215	0.939	1.391	1.020	1.573	1.090
0.330	0.772	0.651	0.911	0.753	1.064	0.847	1.230	0.933	1.405	1.010	1.588	1.090
0.335	0.786	0.649	0.926	0.749	1.079	0.842	1.244	0.928	1.419	1.000	1.602	1.080
0.340	0.800	0.647	0.940	0.745	1.094	0.838	1.258	0.922	1.434	0.999	1.616	1.070
0.345	0.814	0.644	0.955	0.742	1.108	0.833	1.273	0.917	1.448	0.993	1.630	1.070
0.350	0.828	0.642	0.969	0.738	1.123	0.828	1.287	0.912	1.462	0.987	1.644	1.060
0.355	0.842	0.640	0.984	0.735	1.137	0.824	1.302	0.907	1.476	0.982	1.657	1.050
0.360	0.856	0.638	0.998	0.731	1.152	0.820	1.316	0.902	1.490	0.976	1.671	1.050
0.365	0.870	0.636	1.012	0.728	1.166	0.816	1.331	0.897	1.504	0.971	1.685	1.040
0.370	0.884	0.634	1.027	0.725	1.180	0.812	1.345	0.892	1.518	0.966	1.699	1.030
0.375	0.898	0.632	1.041	0.722	1.195	0.808	1.360	0.887	1.532	0.961	1.713	1.030
0.380	0.911	0.630	1.055	0.719	1.209	0.804	1.374	0.883	1.547	0.956	1.727	1.020
0.385	0.925	0.628	1.069	0.716	1.224	0.800	1.388	0.878	1.561	0.951	1.741	1.020
0.390	0.939	0.627	1.084	0.713	1.238	0.796	1.403	0.874	1.575	0.946	1.755	1.010
0.395	0.952	0.625	1.097	0.711	1.252	0.793	1.417	0.870	1.589	0.941	1.769	1.010
0.400	0.966	0.623	1.112	0.708	1.267	0.789	1.431	0.865	1.603	0.936	1.783	1.000
0.410	0.993	0.620	1.140	0.703	1.295	0.782	1.459	0.857	1.632	0.927	1.810	0.992
0.420	1.021	0.617	1.168	0.698	1.324	0.776	1.488	0.849	1.660	0.918	1.838	0.982
0.430	1.047	0.614	1.196	0.693	1.352	0.769	1.517	0.842	1.688	0.910	1.866	0.973
0.440	1.074	0.611	1.224	0.689	1.380	0.763	1.545	0.835	1.716	0.901	1.894	0.964
0.450	1.102	0.609	1.251	0.684	1.408	0.758	1.573	0.827	1.744	0.893	1.922	0.955
0.460	1.130	0.606	1.279	0.680	1.436	0.752	1.601	0.821	1.772	0.886	1.950	0.947
0.470	1.157	0.604	1.306	0.676	1.464	0.747	1.629	0.814	1.801	0.878	1.978	0.939
0.480	1.184	0.602	1.334	0.672	1.492	0.742	1.657	0.808	1.829	0.871	2.005	0.931
0.490	1.212	0.600	1.361	0.669	1.520	0.737	1.686	0.802	1.857	0.864	2.033	0.923
0.500	1.239	0.598	1.388	0.665	1.548	0.732	1.713	0.796	1.885	0.857	2.061	0.915
0.510	1.267	0.596	1.415	0.662	1.575	0.727	1.741	0.790	1.913	0.851	2.089	0.908
0.520	1.294	0.594	1.442	0.659	1.603	0.723	1.769	0.785	1.940	0.844	2.116	0.901
0.530	1.321	0.592	1.469	0.656	1.630	0.719	1.797	0.780	1.968	0.838	2.144	0.894
0.540	1.348	0.590	1.497	0.653	1.658	0.715	1.824	0.775	1.996	0.832	2.172	0.888
0.550	1.376	0.588	1.523	0.650	1.685	0.711	1.852	0.770	2.024	0.827	2.199	0.881
0.560	1.403	0.587	1.550	0.647	1.712	0.707	1.880	0.765	2.051	0.821	2.227	0.875
0.570	1.430	0.585	1.577	0.644	1.740	0.703	1.907	0.760	2.078	0.816	2.255	0.869
0.580	1.457	0.584	1.604	0.642	1.767	0.700	1.934	0.756	2.106	0.811	2.282	0.863
0.590	1.484	0.582	1.630	0.639	1.794	0.696	1.961	0.752	2.134	0.806	2.310	0.857

Tab. 5.19

TABLE 3_Concrete class C90_Steel class S500_a/h=0.10

Bending and axial force with symmetric reinforcement

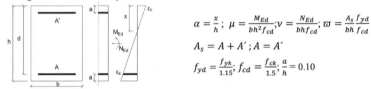

$$\alpha = \frac{x}{h}\,; \quad \mu = \frac{M_{Ed}}{bh^2 f_{cd}}; \nu = \frac{N_{Ed}}{bhf_{cd}}; \quad \varpi = \frac{A_s}{bh}\frac{f_{yd}}{f_{cd}}$$

$$A_s = A + A'\,; A = A'$$

$$f_{yd} = \frac{f_{yk}}{1.15}; f_{cd} = \frac{f_{ck}}{1.5}, \frac{a}{h} = 0.10$$

	$\nu=0.0$		$\nu=0.1$		$\nu=0.2$		$\nu=0.3$		$\nu=0.4$		$\nu=0.5$	
μ	ϖ	α	ϖ	α	ϖ	α	ϖ	α	ϖ	α	ϖ	α
0.000	0.000	-	0.000	-	0.000	-	0.000	-	0.000	-	0.000	-
0.005	0.010	0.036	0.000	-	0.000	-	0.000	-	0.000	-	0.000	-
0.010	0.021	0.050	0.000	-	0.000	-	0.000	-	0.000	-	0.000	-
0.015	0.033	0.059	0.000	-	0.000	-	0.000	-	0.000	-	0.000	-
0.020	0.044	0.067	0.000	-	0.000	-	0.000	-	0.000	-	0.000	-
0.025	0.055	0.073	0.000	-	0.000	-	0.000	-	0.000	-	0.000	-
0.030	0.067	0.079	0.000	-	0.000	-	0.000	-	0.000	-	0.000	-
0.035	0.079	0.084	0.000	-	0.000	-	0.000	-	0.000	-	0.000	-
0.040	0.091	0.089	0.000	-	0.000	-	0.000	-	0.000	-	0.000	-
0.045	0.104	0.095	0.003	0.173	0.000	-	0.000	-	0.000	-	0.000	-
0.050	0.116	0.100	0.015	0.178	0.000	-	0.000	-	0.000	-	0.000	-
0.055	0.128	0.104	0.028	0.183	0.000	-	0.000	-	0.000	-	0.000	-
0.060	0.140	0.109	0.041	0.187	0.000	-	0.000	-	0.000	-	0.000	-
0.065	0.153	0.113	0.053	0.191	0.000	-	0.000	-	0.000	-	0.000	-
0.070	0.165	0.117	0.066	0.195	0.000	-	0.000	-	0.000	-	0.000	-
0.075	0.177	0.121	0.079	0.199	0.000	-	0.000	-	0.000	-	0.000	-
0.080	0.190	0.124	0.091	0.202	0.011	0.344	0.000	-	0.000	-	0.000	-
0.085	0.202	0.128	0.104	0.206	0.024	0.346	0.000	-	0.000	-	0.000	-
0.090	0.215	0.131	0.117	0.209	0.037	0.348	0.000	-	0.000	-	0.000	-
0.095	0.227	0.135	0.130	0.212	0.049	0.349	0.000	-	0.000	-	0.000	-
0.100	0.240	0.138	0.142	0.215	0.062	0.351	0.012	0.514	0.000	-	0.005	0.853
0.105	0.252	0.141	0.155	0.218	0.075	0.352	0.025	0.514	0.007	0.682	0.024	0.839
0.110	0.265	0.144	0.167	0.221	0.088	0.354	0.039	0.514	0.025	0.673	0.043	0.824
0.115	0.277	0.147	0.180	0.224	0.101	0.355	0.052	0.514	0.043	0.665	0.061	0.812
0.120	0.290	0.150	0.193	0.226	0.114	0.356	0.066	0.514	0.060	0.657	0.080	0.799
0.125	0.302	0.152	0.206	0.229	0.126	0.358	0.079	0.514	0.077	0.650	0.098	0.787
0.130	0.315	0.155	0.218	0.231	0.139	0.359	0.093	0.514	0.093	0.644	0.116	0.776
0.135	0.327	0.158	0.231	0.234	0.152	0.361	0.106	0.514	0.110	0.638	0.134	0.766
0.140	0.340	0.160	0.244	0.236	0.165	0.362	0.120	0.514	0.126	0.632	0.152	0.757
0.145	0.352	0.163	0.258	0.239	0.177	0.363	0.133	0.514	0.142	0.627	0.169	0.748
0.150	0.365	0.165	0.270	0.241	0.190	0.364	0.146	0.514	0.157	0.623	0.186	0.739
0.155	0.378	0.168	0.283	0.243	0.203	0.366	0.160	0.514	0.173	0.618	0.203	0.731
0.160	0.390	0.170	0.295	0.245	0.216	0.367	0.173	0.514	0.188	0.614	0.220	0.724
0.165	0.403	0.172	0.307	0.247	0.229	0.368	0.187	0.514	0.203	0.610	0.236	0.717
0.170	0.415	0.174	0.321	0.249	0.241	0.369	0.200	0.514	0.218	0.607	0.252	0.710
0.175	0.428	0.177	0.333	0.251	0.254	0.370	0.214	0.514	0.233	0.604	0.269	0.704
0.180	0.440	0.179	0.346	0.253	0.267	0.372	0.227	0.514	0.248	0.601	0.285	0.698
0.185	0.453	0.181	0.359	0.255	0.280	0.373	0.240	0.514	0.263	0.598	0.301	0.692
0.190	0.466	0.183	0.371	0.257	0.293	0.374	0.254	0.514	0.278	0.595	0.316	0.687
0.195	0.478	0.185	0.384	0.259	0.305	0.375	0.267	0.514	0.292	0.592	0.332	0.682
0.200	0.491	0.187	0.397	0.261	0.318	0.376	0.281	0.514	0.307	0.590	0.348	0.677
0.205	0.503	0.189	0.409	0.262	0.331	0.377	0.294	0.514	0.322	0.588	0.363	0.672
0.210	0.516	0.191	0.422	0.264	0.344	0.378	0.308	0.514	0.336	0.585	0.378	0.668
0.215	0.529	0.193	0.435	0.266	0.356	0.379	0.321	0.514	0.350	0.583	0.393	0.664
0.220	0.541	0.194	0.448	0.268	0.369	0.380	0.334	0.514	0.364	0.581	0.408	0.660
0.225	0.554	0.196	0.461	0.269	0.382	0.381	0.348	0.514	0.379	0.579	0.423	0.656
0.230	0.566	0.198	0.473	0.271	0.394	0.382	0.361	0.514	0.393	0.578	0.438	0.652
0.235	0.579	0.200	0.486	0.273	0.407	0.383	0.375	0.514	0.407	0.576	0.452	0.649
0.240	0.592	0.202	0.499	0.274	0.420	0.384	0.388	0.514	0.421	0.574	0.467	0.646
0.245	0.604	0.203	0.512	0.276	0.433	0.385	0.402	0.514	0.435	0.573	0.482	0.642

TABLE 3_Concrete class C90_Steel class S500_a/h=0.10 (cont.)
Bending and axial force with symmetric reinforcement

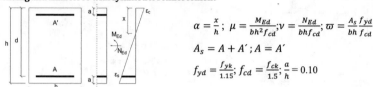

$$\alpha = \frac{x}{h}; \quad \mu = \frac{M_{Ed}}{bh^2 f_{cd}}; \nu = \frac{N_{Ed}}{bh f_{cd}}; \varpi = \frac{A_s}{bh} \frac{f_{yd}}{f_{cd}}$$

$$A_s = A + A'; A = A'$$

$$f_{yd} = \frac{f_{yk}}{1.15}; f_{cd} = \frac{f_{ck}}{1.5}; \frac{a}{h} = 0.10$$

	ν=0.6		ν=0.8		ν=1.0		ν=1.2		ν=1.4		ν=1.6	
μ	ϖ	α	ϖ	α	ϖ	α	ϖ	α	ϖ	α	ϖ	α
0.000	0.000	-	0.000	-	0.000	-	0.201	-	0.400	-	0.600	-
0.005	0.000	-	0.000	-	0.024	7.620	0.225	7.540	0.425	7.570	0.624	7.600
0.010	0.000	-	0.000	-	0.047	4.770	0.247	4.980	0.446	5.120	0.645	5.190
0.015	0.000	-	0.000	-	0.070	3.710	0.265	4.160	0.463	4.430	0.661	4.610
0.020	0.000	-	0.000	-	0.091	3.130	0.283	3.620	0.480	3.940	0.677	4.170
0.025	0.000	-	0.000	-	0.111	2.760	0.302	3.220	0.497	3.560	0.694	3.820
0.030	0.000	-	0.000	-	0.131	2.500	0.320	2.930	0.514	3.270	0.710	3.540
0.035	0.000	-	0.000	-	0.150	2.300	0.338	2.700	0.530	3.030	0.726	3.300
0.040	0.000	-	0.000	-	0.170	2.140	0.355	2.520	0.547	2.840	0.742	3.100
0.045	0.000	-	0.015	1.620	0.188	2.010	0.373	2.370	0.564	2.670	0.757	2.930
0.050	0.000	-	0.035	1.550	0.207	1.910	0.390	2.240	0.580	2.530	0.773	2.780
0.055	0.000	-	0.054	1.490	0.225	1.820	0.407	2.130	0.597	2.410	0.789	2.650
0.060	0.000	-	0.073	1.430	0.243	1.740	0.425	2.040	0.613	2.300	0.805	2.540
0.065	0.000	-	0.091	1.380	0.261	1.670	0.442	1.950	0.629	2.210	0.821	2.440
0.070	0.000	-	0.109	1.340	0.278	1.620	0.459	1.880	0.645	2.120	0.836	2.350
0.075	0.000	-	0.127	1.300	0.296	1.560	0.475	1.810	0.661	2.050	0.852	2.260
0.080	0.000	-	0.145	1.270	0.313	1.520	0.492	1.760	0.678	1.980	0.868	2.190
0.085	0.009	1.020	0.162	1.240	0.330	1.480	0.508	1.700	0.694	1.920	0.883	2.120
0.090	0.026	1.000	0.179	1.220	0.347	1.440	0.525	1.660	0.710	1.860	0.899	2.060
0.095	0.043	0.988	0.196	1.190	0.364	1.400	0.541	1.610	0.725	1.810	0.914	2.000
0.100	0.060	0.972	0.213	1.170	0.380	1.370	0.557	1.570	0.741	1.770	0.930	1.950
0.105	0.078	0.957	0.229	1.150	0.397	1.340	0.574	1.530	0.757	1.720	0.945	1.900
0.110	0.095	0.943	0.246	1.130	0.413	1.310	0.589	1.500	0.773	1.680	0.960	1.850
0.115	0.113	0.928	0.263	1.110	0.429	1.290	0.605	1.470	0.788	1.640	0.976	1.810
0.120	0.130	0.915	0.279	1.090	0.445	1.270	0.621	1.440	0.804	1.610	0.991	1.770
0.125	0.147	0.902	0.295	1.080	0.462	1.240	0.637	1.410	0.820	1.580	1.006	1.740
0.130	0.165	0.889	0.311	1.060	0.477	1.220	0.653	1.390	0.835	1.550	1.021	1.700
0.135	0.182	0.877	0.327	1.050	0.494	1.200	0.668	1.360	0.850	1.520	1.036	1.670
0.140	0.199	0.866	0.343	1.030	0.509	1.190	0.684	1.340	0.865	1.490	1.051	1.640
0.145	0.217	0.855	0.359	1.020	0.525	1.170	0.700	1.320	0.881	1.470	1.066	1.610
0.150	0.234	0.844	0.375	1.010	0.541	1.150	0.715	1.300	0.896	1.440	1.081	1.580
0.155	0.251	0.834	0.391	0.997	0.556	1.140	0.730	1.280	0.912	1.420	1.096	1.560
0.160	0.268	0.824	0.406	0.987	0.572	1.120	0.746	1.260	0.926	1.400	1.111	1.530
0.165	0.285	0.815	0.422	0.976	0.587	1.110	0.762	1.240	0.942	1.380	1.126	1.510
0.170	0.301	0.806	0.438	0.966	0.603	1.100	0.777	1.230	0.957	1.360	1.141	1.490
0.175	0.318	0.798	0.453	0.956	0.618	1.080	0.792	1.210	0.972	1.340	1.156	1.460
0.180	0.334	0.790	0.469	0.946	0.633	1.070	0.807	1.200	0.987	1.320	1.171	1.440
0.185	0.351	0.782	0.485	0.937	0.649	1.060	0.822	1.180	1.002	1.310	1.186	1.420
0.190	0.367	0.775	0.500	0.927	0.664	1.050	0.837	1.170	1.017	1.290	1.201	1.410
0.195	0.383	0.768	0.516	0.918	0.679	1.040	0.852	1.160	1.031	1.270	1.215	1.390
0.200	0.399	0.761	0.532	0.910	0.694	1.030	0.867	1.150	1.047	1.260	1.230	1.370
0.205	0.415	0.755	0.547	0.901	0.709	1.020	0.882	1.130	1.061	1.240	1.245	1.360
0.210	0.431	0.749	0.564	0.893	0.724	1.010	0.897	1.120	1.076	1.230	1.259	1.340
0.215	0.446	0.743	0.579	0.885	0.739	1.000	0.912	1.110	1.091	1.220	1.274	1.320
0.220	0.462	0.737	0.595	0.877	0.754	0.995	0.927	1.100	1.106	1.210	1.289	1.310
0.225	0.478	0.732	0.610	0.870	0.769	0.986	0.942	1.090	1.120	1.190	1.303	1.300
0.230	0.493	0.727	0.626	0.863	0.784	0.978	0.956	1.080	1.135	1.180	1.318	1.280
0.235	0.508	0.722	0.641	0.856	0.799	0.970	0.971	1.070	1.150	1.170	1.332	1.270
0.240	0.524	0.717	0.657	0.849	0.813	0.963	0.986	1.060	1.164	1.160	1.347	1.260
0.245	0.539	0.712	0.672	0.842	0.829	0.955	1.000	1.050	1.179	1.150	1.361	1.250

TABLE 3_Concrete class C90_Steel class S500_a/h=0.10 (cont.)
Bending and Axial Force with Symmetric Reinforcement

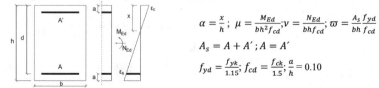

$$\alpha = \frac{x}{h}; \quad \mu = \frac{M_{Ed}}{bh^2 f_{cd}}; \nu = \frac{N_{Ed}}{bh f_{cd}}; \quad \varpi = \frac{A_s}{bh}\frac{f_{yd}}{f_{cd}}$$

$$A_S = A + A'; A = A'$$

$$f_{yd} = \frac{f_{yk}}{1.15}; f_{cd} = \frac{f_{ck}}{1.5}; \frac{a}{h} = 0.10$$

	$\nu=0.0$		$\nu=0.1$		$\nu=0.2$		$\nu=0.3$		$\nu=0.4$		$\nu=0.5$	
μ	ϖ	α	ϖ	α	ϖ	α	ϖ	α	ϖ	α	ϖ	α
0.250	0.617	0.205	0.525	0.277	0.445	0.386	0.415	0.514	0.449	0.571	0.496	0.639
0.255	0.629	0.207	0.537	0.279	0.458	0.387	0.428	0.514	0.463	0.570	0.511	0.636
0.260	0.642	0.208	0.550	0.280	0.471	0.388	0.442	0.514	0.477	0.568	0.525	0.634
0.265	0.655	0.210	0.562	0.282	0.484	0.389	0.455	0.514	0.491	0.567	0.540	0.631
0.270	0.667	0.211	0.575	0.283	0.496	0.390	0.469	0.514	0.505	0.566	0.553	0.628
0.275	0.680	0.213	0.587	0.284	0.509	0.391	0.482	0.514	0.519	0.564	0.568	0.626
0.280	0.692	0.215	0.601	0.286	0.522	0.392	0.496	0.514	0.533	0.563	0.582	0.623
0.285	0.705	0.216	0.613	0.287	0.535	0.393	0.509	0.514	0.547	0.562	0.596	0.621
0.290	0.718	0.218	0.626	0.289	0.547	0.393	0.522	0.514	0.560	0.561	0.610	0.619
0.295	0.730	0.219	0.638	0.290	0.560	0.394	0.536	0.514	0.574	0.560	0.624	0.616
0.300	0.743	0.221	0.651	0.291	0.573	0.395	0.549	0.514	0.588	0.559	0.638	0.614
0.305	0.756	0.222	0.663	0.292	0.586	0.396	0.563	0.514	0.602	0.558	0.652	0.612
0.310	0.768	0.223	0.677	0.294	0.598	0.397	0.576	0.514	0.615	0.557	0.666	0.610
0.315	0.781	0.225	0.690	0.295	0.611	0.398	0.590	0.514	0.629	0.556	0.680	0.609
0.320	0.794	0.226	0.702	0.296	0.624	0.399	0.603	0.514	0.643	0.555	0.694	0.607
0.325	0.806	0.228	0.715	0.298	0.636	0.399	0.617	0.514	0.656	0.554	0.708	0.605
0.330	0.819	0.229	0.727	0.299	0.649	0.400	0.630	0.514	0.670	0.553	0.722	0.603
0.335	0.831	0.230	0.740	0.300	0.662	0.401	0.643	0.514	0.683	0.553	0.736	0.602
0.340	0.844	0.232	0.753	0.301	0.675	0.402	0.657	0.514	0.697	0.552	0.750	0.600
0.345			0.766	0.302	0.687	0.402	0.670	0.514	0.710	0.551	0.764	0.599
0.350			0.778	0.303	0.700	0.403	0.684	0.514	0.724	0.550	0.778	0.597
0.355			0.790	0.305	0.713	0.404	0.697	0.514	0.738	0.549	0.792	0.596
0.360			0.803	0.306	0.725	0.405	0.711	0.514	0.751	0.549	0.806	0.595
0.365			0.816	0.307	0.738	0.405	0.724	0.514	0.765	0.548	0.820	0.593
0.370			0.828	0.308	0.751	0.406	0.737	0.514	0.778	0.547	0.834	0.592
0.375			0.841	0.309	0.763	0.407	0.751	0.514	0.792	0.547	0.847	0.591
0.380			0.855	0.310	0.776	0.408	0.764	0.514	0.805	0.546	0.861	0.589
0.385			0.867	0.311	0.789	0.408	0.778	0.514	0.819	0.546	0.875	0.588
0.390			0.880	0.312	0.801	0.409	0.791	0.514	0.832	0.545	0.889	0.587
0.395			0.892	0.313	0.814	0.410	0.805	0.514	0.846	0.544	0.903	0.586
0.400			0.905	0.314	0.827	0.410	0.818	0.514	0.859	0.544	0.916	0.585
0.410			0.930	0.316	0.852	0.412	0.845	0.514	0.886	0.543	0.944	0.583
0.420			0.955	0.318	0.878	0.413	0.872	0.514	0.913	0.542	0.971	0.581
0.430			0.981	0.320	0.903	0.414	0.899	0.514	0.939	0.541	0.998	0.579
0.440			1.006	0.322	0.928	0.416	0.925	0.514	0.966	0.540	1.026	0.577
0.450			1.031	0.324	0.954	0.417	0.952	0.514	0.993	0.539	1.053	0.575
0.460			1.057	0.326	0.979	0.418	0.979	0.514	1.020	0.538	1.080	0.573
0.470			1.083	0.328	1.004	0.420	1.006	0.514	1.046	0.537	1.107	0.572
0.480			1.108	0.330	1.030	0.421	1.033	0.514	1.073	0.536	1.134	0.570
0.490			1.133	0.331	1.055	0.422	1.060	0.514	1.099	0.536	1.162	0.569
0.500			1.159	0.333	1.081	0.423	1.087	0.514	1.126	0.535	1.188	0.567
0.510			1.183	0.335	1.106	0.424	1.113	0.514	1.153	0.534	1.215	0.566
0.520			1.210	0.337	1.131	0.425	1.140	0.514	1.179	0.534	1.242	0.565
0.530			1.235	0.338	1.156	0.427	1.167	0.514	1.206	0.533	1.269	0.564
0.540			1.260	0.340	1.182	0.428	1.194	0.514	1.232	0.532	1.296	0.562
0.550			1.285	0.341	1.207	0.429	1.221	0.514	1.259	0.532	1.323	0.561
0.560			1.310	0.343	1.232	0.430	1.248	0.514	1.285	0.531	1.350	0.560
0.570			1.335	0.344	1.258	0.431	1.275	0.514	1.312	0.531	1.377	0.559
0.580			1.361	0.346	1.283	0.432	1.302	0.514	1.339	0.530	1.404	0.558
0.590			1.386	0.347	1.308	0.433	1.328	0.514	1.365	0.530	1.431	0.557

TABLE 3_Concrete class C90_Steel class S500_a/h=0.10 (cont.)
Bending and axial force with symmetric reinforcement

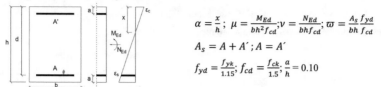

$$\alpha = \frac{x}{h}; \quad \mu = \frac{M_{Ed}}{bh^2 f_{cd}}; \nu = \frac{N_{Ed}}{bh f_{cd}}; \varpi = \frac{A_s}{bh} \frac{f_{yd}}{f_{cd}}$$

$$A_s = A + A'; A = A'$$

$$f_{yd} = \frac{f_{yk}}{1.15}; f_{cd} = \frac{f_{ck}}{1.5}; \frac{a}{h} = 0.10$$

μ	$\nu=0.6$ ϖ	α	$\nu=0.8$ ϖ	α	$\nu=1.0$ ϖ	α	$\nu=1.2$ ϖ	α	$\nu=1.4$ ϖ	α	$\nu=1.6$ ϖ	α
0.250	0.554	0.708	0.687	0.836	0.843	0.948	1.015	1.040	1.193	1.140	1.376	1.230
0.255	0.569	0.704	0.703	0.830	0.858	0.941	1.030	1.040	1.208	1.130	1.390	1.220
0.260	0.584	0.700	0.718	0.824	0.873	0.934	1.044	1.030	1.222	1.120	1.404	1.210
0.265	0.598	0.696	0.733	0.818	0.888	0.927	1.059	1.020	1.237	1.110	1.419	1.200
0.270	0.613	0.692	0.748	0.813	0.903	0.920	1.074	1.010	1.251	1.100	1.433	1.190
0.275	0.628	0.688	0.763	0.807	0.918	0.913	1.088	1.010	1.265	1.090	1.448	1.180
0.280	0.643	0.685	0.779	0.802	0.933	0.907	1.102	1.000	1.280	1.090	1.462	1.170
0.285	0.657	0.681	0.793	0.797	0.948	0.901	1.117	0.993	1.295	1.080	1.476	1.160
0.290	0.671	0.678	0.808	0.792	0.963	0.895	1.131	0.986	1.309	1.070	1.490	1.150
0.295	0.686	0.675	0.824	0.787	0.977	0.889	1.145	0.980	1.323	1.060	1.505	1.140
0.300	0.701	0.672	0.838	0.782	0.993	0.883	1.160	0.973	1.338	1.060	1.519	1.140
0.305	0.715	0.669	0.853	0.778	1.007	0.878	1.174	0.967	1.352	1.050	1.533	1.130
0.310	0.729	0.666	0.868	0.773	1.022	0.872	1.189	0.961	1.366	1.040	1.547	1.120
0.315	0.744	0.663	0.883	0.769	1.037	0.867	1.204	0.955	1.380	1.030	1.562	1.110
0.320	0.758	0.661	0.897	0.765	1.051	0.861	1.218	0.949	1.394	1.030	1.576	1.100
0.325	0.772	0.658	0.912	0.761	1.066	0.856	1.232	0.943	1.409	1.020	1.590	1.100
0.330	0.786	0.655	0.926	0.757	1.081	0.851	1.247	0.937	1.423	1.020	1.604	1.090
0.335	0.800	0.653	0.941	0.753	1.095	0.846	1.261	0.932	1.437	1.010	1.618	1.080
0.340	0.814	0.651	0.955	0.749	1.110	0.842	1.276	0.926	1.451	1.000	1.632	1.080
0.345	0.828	0.648	0.970	0.746	1.125	0.837	1.290	0.921	1.466	0.997	1.647	1.070
0.350	0.842	0.646	0.984	0.742	1.139	0.833	1.305	0.916	1.480	0.992	1.661	1.060
0.355	0.856	0.644	0.999	0.739	1.154	0.828	1.319	0.911	1.494	0.986	1.675	1.060
0.360	0.870	0.642	1.013	0.735	1.168	0.824	1.333	0.906	1.508	0.981	1.689	1.050
0.365	0.884	0.640	1.027	0.732	1.182	0.820	1.348	0.901	1.522	0.975	1.703	1.040
0.370	0.898	0.638	1.042	0.729	1.197	0.815	1.362	0.896	1.536	0.970	1.717	1.040
0.375	0.911	0.636	1.056	0.726	1.211	0.811	1.376	0.891	1.550	0.965	1.731	1.030
0.380	0.925	0.634	1.070	0.723	1.226	0.808	1.391	0.887	1.565	0.959	1.745	1.030
0.385	0.939	0.632	1.084	0.720	1.240	0.804	1.405	0.882	1.578	0.955	1.759	1.020
0.390	0.952	0.630	1.098	0.717	1.254	0.800	1.420	0.878	1.593	0.950	1.773	1.020
0.395	0.966	0.628	1.112	0.714	1.269	0.796	1.434	0.873	1.607	0.945	1.787	1.010
0.400	0.980	0.627	1.127	0.712	1.283	0.793	1.448	0.869	1.621	0.940	1.801	1.010
0.410	1.007	0.623	1.155	0.706	1.311	0.786	1.476	0.861	1.649	0.931	1.828	0.996
0.420	1.034	0.620	1.183	0.701	1.340	0.779	1.505	0.853	1.677	0.922	1.856	0.986
0.430	1.061	0.617	1.211	0.696	1.368	0.773	1.534	0.845	1.705	0.913	1.884	0.977
0.440	1.088	0.614	1.238	0.692	1.396	0.767	1.562	0.838	1.734	0.905	1.912	0.967
0.450	1.115	0.612	1.266	0.687	1.424	0.761	1.590	0.831	1.762	0.897	1.940	0.959
0.460	1.142	0.609	1.293	0.683	1.453	0.755	1.618	0.824	1.790	0.889	1.968	0.950
0.470	1.169	0.607	1.321	0.679	1.480	0.750	1.646	0.817	1.818	0.881	1.995	0.942
0.480	1.197	0.604	1.348	0.675	1.508	0.744	1.674	0.811	1.846	0.874	2.023	0.934
0.490	1.224	0.602	1.376	0.672	1.536	0.740	1.702	0.805	1.874	0.867	2.051	0.926
0.500	1.251	0.600	1.403	0.668	1.563	0.735	1.730	0.799	1.902	0.860	2.079	0.918
0.510	1.279	0.598	1.430	0.665	1.591	0.730	1.758	0.793	1.930	0.854	2.106	0.911
0.520	1.306	0.596	1.457	0.661	1.618	0.726	1.785	0.788	1.957	0.847	2.134	0.904
0.530	1.334	0.594	1.484	0.658	1.646	0.721	1.813	0.782	1.985	0.841	2.162	0.897
0.540	1.361	0.593	1.511	0.655	1.673	0.717	1.841	0.777	2.013	0.835	2.189	0.890
0.550	1.388	0.591	1.538	0.652	1.701	0.713	1.868	0.772	2.040	0.829	2.217	0.884
0.560	1.415	0.589	1.565	0.649	1.728	0.709	1.896	0.768	2.068	0.824	2.244	0.878
0.570	1.442	0.588	1.591	0.647	1.755	0.706	1.923	0.763	2.096	0.818	2.272	0.872
0.580	1.469	0.586	1.618	0.644	1.782	0.702	1.951	0.758	2.123	0.813	2.299	0.866
0.590	1.496	0.585	1.645	0.641	1.809	0.699	1.978	0.754	2.150	0.808	2.327	0.860

5.3.2 Simply Reinforced Sections—TABLE 4

Tab 5.20

TABLE 4_Concrete classes C12-C50_Steel class S500

Bending and axial force of simply reinforced rectangular sections

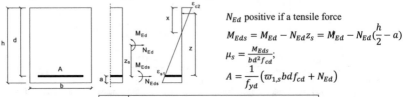

N_{Ed} positive if a tensile force

$$M_{Eds} = M_{Ed} - N_{Ed}z_s = M_{Ed} - N_{Ed}\left(\frac{h}{2} - a\right)$$

$$\mu_s = \frac{M_{Eds}}{bd^2 f_{cd}};$$

$$A = \frac{1}{f_{yd}}\left(\varpi_{1,s}bdf_{cd} + N_{Ed}\right)$$

μ_s	$\varpi_{1,s}$	$\alpha = \dfrac{x}{d}$	$\varsigma = \dfrac{z}{d}$	ε_{c2} ‰	ε_{s1} ‰	σ_{sd} MPa
0.01	0.0101	0.030	0.993	-0,77	25.00	435
0.02	0.0203	0.044	0.984	-1,15	25.00	435
0.03	0.0305	0.055	0.982	-1,46	25.00	435
0.04	0.0410	0.066	0.976	-1,76	25.00	435
0.05	0.0515	0.076	0.971	-2.06	25.00	435
0.06	0.0621	0.087	0.966	-2.37	25.00	435
0.07	0.0728	0.097	0.961	-2.68	25.00	435
0.08	0.0836	0.107	0.957	-3.01	25.00	435
0.09	0.0946	0.118	0.952	-3.35	25.00	435
0.10	0.1057	0.131	0.946	-3.50	23.30	435
0.11	0.1170	0.144	0.940	-3.50	20.70	435
0.12	0.1284	0.159	0.934	-3.50	18.60	435
0.13	0.1400	0.173	0.929	-3.50	16.70	435
0.14	0.1518	0.188	0.922	-3.50	15,20	435
0.15	0.1637	0.202	0.916	-3.50	13.80	435
0.16	0.1759	0.217	0.910	-3.50	12.60	435
0.17	0.1881	0.232	0.904	-3.50	11,60	435
0.18	0.2007	0.248	0.897	-3.50	10.60	435
0.19	0.2134	0.264	0.891	-3.50	9.78	435
0.20	0.2263	0.280	0.884	-3.50	9.02	435
0.21	0.2394	0.296	0.877	-3.50	8.33	435
0.22	0.2528	0.312	0.870	-3.50	7.71	435
0.23	0.2665	0.329	0.863	-3.50	7.13	435
0.24	0.2804	0.346	0.856	-3.50	6.61	435
0.25	0.2945	0.364	0.849	-3.50	6.12	435
0.26	0.3090	0.382	0.841	-3.50	5,67	435
0.27	0.3238	0.400	0.834	-3.50	5,25	435
0.28	0.3390	0.419	0.826	-3.50	4,86	435
0.29	0.3545	0.438	0.818	-3.50	4,49	435
0.30	0.3704	0.458	0.810	-3.50	4,15	435
0.31	0.3868	0.478	0.801	-3.50	3.83	435
0.32	0.4037	0.499	0.793	-3.50	3.52	435
0.33	0.4211	0.520	0.784	-3.50	3.23	435
0.34	0.4389	0.542	0.775	-3.50	2.96	435
0.35	0.4575	0.565	0.765	-3.50	2.69	435
0.36	0.4767	0.589	0.755	-3.50	2.44	435
0.37	0.4966	0.613	0.745	-3.50	2.20	435
0.38	0.5176	0.639	0.734	-3.50	1.97	395
0.39	0.5395	0.666	0.723	-3.50	1.75	350
0.40	0.5625	0.695	0.711	-3.50	1.54	307
0.41	0.5869	0.725	0.699	-3.50	1.33	266
0.42	0.6131	0.757	0.685	-3.50	1.12	224
0.43	0.6411	0.792	0.671	-3.50	0.92	184
0.44	0.6718	0.830	0.655	-3.50	0.72	143
0.45	0.7061	0.872	0.637	-3.50	0.51	103

Tab 5.21

TABLE 4_Concrete classes C12-C50_Steel class S400

Bending and axial force of simply reinforced rectangular sections

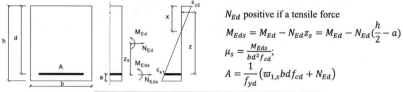

N_{Ed} positive if a tensile force

$$M_{Eds} = M_{Ed} - N_{Ed}z_s = M_{Ed} - N_{Ed}\left(\frac{h}{2} - a\right)$$

$$\mu_s = \frac{M_{Eds}}{bd^2 f_{cd}};$$

$$A = \frac{1}{f_{yd}}\left(\varpi_{1,s}bd f_{cd} + N_{Ed}\right)$$

μ_s	$\varpi_{1,s}$	$\alpha = \dfrac{x}{d}$	$\varsigma = \dfrac{z}{d}$	ε_{c2} ‰	ε_{s1} ‰	σ_{sd} MPa
0.01	0.0101	0.030	0.993	-0.77	25.00	348
0.02	0.0203	0.044	0.984	-1,15	25.00	348
0.03	0.0305	0.055	0.982	-1,46	25.00	348
0.04	0.0410	0.066	0.976	-1,76	25.00	348
0.05	0.0515	0.076	0.971	-2.06	25.00	348
0.06	0.0621	0.087	0.966	-2.37	25.00	348
0.07	0.0728	0.097	0.961	-2.68	25.00	348
0.08	0.0836	0.107	0.957	-3.01	25.00	348
0.09	0.0946	0.118	0.952	-3.35	25.00	348
0.10	0.1057	0.131	0.946	-3.50	23.30	348
0.11	0.1170	0.144	0.940	-3.50	20.70	348
0.12	0.1284	0.159	0.934	-3.50	18.60	348
0.13	0.1400	0.173	0.929	-3.50	16.70	348
0.14	0.1518	0.188	0.922	-3.50	15,20	348
0.15	0.1637	0.202	0.916	-3.50	13.80	348
0.16	0.1759	0.217	0.910	-3.50	12.60	348
0.17	0.1881	0.232	0.904	-3.50	11,60	348
0.18	0.2007	0.248	0.897	-3.50	10.60	348
0.19	0.2134	0.264	0.891	-3.50	9.78	348
0.20	0.2263	0.280	0.884	-3.50	9.02	348
0.21	0.2394	0.296	0.877	-3.50	8.33	348
0.22	0.2528	0.312	0.870	-3.50	7.71	348
0.23	0.2665	0.329	0.863	-3.50	7.13	348
0.24	0.2804	0.346	0.856	-3.50	6.61	348
0.25	0.2945	0.364	0.849	-3.50	6.12	348
0.26	0.3090	0.382	0.841	-3.50	5,67	348
0.27	0.3238	0.400	0.834	-3.50	5,25	348
0.28	0.3390	0.419	0.826	-3.50	4,86	348
0.29	0.3545	0.438	0.818	-3.50	4,49	348
0.30	0.3704	0.458	0.810	-3.50	4,15	348
0.31	0.3868	0.478	0.801	-3.50	3.83	348
0.32	0.4037	0.499	0.793	-3.50	3.52	348
0.33	0.4211	0.520	0.784	-3.50	3.23	348
0.34	0.4389	0.542	0.775	-3.50	2.96	348
0.35	0.4575	0.565	0.765	-3.50	2.69	348
0.36	0.4767	0.589	0.755	-3.50	2.44	348
0.37	0.4966	0.613	0.745	-3.50	2.20	348
0.38	0.5176	0.639	0.734	-3.50	1,97	348
0.39	0.5395	0.666	0.723	-3.50	1,75	348
0.40	0.5625	0.695	0.711	-3.50	1,54	307
0.41	0.5869	0.725	0.699	-3.50	1,33	266
0.42	0.6131	0.757	0.685	-3.50	1,12	224
0.43	0.6411	0.792	0.671	-3.50	0.92	184
0.44	0.6718	0.830	0.655	-3.50	0.72	143
0.45	0.7061	0.872	0.637	-3.50	0.51	103

Tab 5.22

TABLE 4_Concrete classes C55_Steel class S500

Bending and axial force of simply reinforced rectangular sections

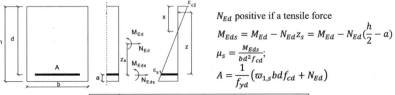

N_{Ed} positive if a tensile force

$$M_{Eds} = M_{Ed} - N_{Ed}z_s = M_{Ed} - N_{Ed}\left(\frac{h}{2} - a\right)$$

$$\mu_s = \frac{M_{Eds}}{bd^2 f_{cd}};$$

$$A = \frac{1}{f_{yd}}\left(\varpi_{1,s}bd f_{cd} + N_{Ed}\right)$$

μ_s	$\varpi_{1,s}$	$\alpha = \dfrac{x}{d}$	$\varsigma = \dfrac{z}{d}$	ε_{c2} ‰	ε_{s1} ‰	σ_{sd} MPa
0.01	0.0101	0.033	0.993	-0.85	25.00	435
0.02	0.0203	0.048	0.984	-1,26	25.00	435
0.03	0.0306	0.060	0.981	-1,59	25.00	435
0.04	0.0410	0.071	0.975	-1,90	25.00	435
0.05	0.0516	0.081	0.970	-2.20	25.00	435
0.06	0.0621	0.091	0.967	-2.51	25.00	435
0.07	0.0728	0.102	0.961	-2.83	25.00	435
0.08	0.0836	0.113	0.957	-3.10	24.40	435
0.09	0.0947	0.128	0.951	-3.10	21.20	435
0.10	0.1059	0.143	0.945	-3.10	18.60	435
0.11	0.1172	0.158	0.939	-3.10	16.50	435
0.12	0.1287	0.173	0.932	-3.10	14.80	435
0.13	0.1404	0.189	0.926	-3.10	13.30	435
0.14	0.1522	0.205	0.920	-3.10	12.00	435
0.15	0.1642	0.221	0.913	-3.10	10.90	435
0.16	0.1764	0.238	0.907	-3.10	9.94	435
0.17	0.1888	0.254	0.900	-3.10	9.08	435
0.18	0.2014	0.271	0.894	-3.10	8.32	435
0.19	0.2142	0.289	0.887	-3.10	7.64	435
0.20	0.2272	0.306	0.880	-3.10	7.02	435
0.21	0.2405	0.324	0.873	-3.10	6.46	435
0.22	0.2540	0.342	0.866	-3.10	5.96	435
0.23	0.2679	0.361	0.859	-3.10	5.49	435
0.24	0.2820	0.380	0.851	-3.10	5.06	435
0.25	0.2963	0.399	0.844	-3.10	4.66	435
0.26	0.3110	0.419	0.836	-3.10	4.29	435
0.27	0.3261	0.440	0.828	-3.10	3.95	435
0.28	0.3415	0.460	0.820	-3.10	3.63	435
0.29	0.3574	0.482	0.811	-3.10	3.33	435
0.30	0.3737	0.504	0.803	-3.10	3.06	435
0.31	0.3905	0.526	0.794	-3.10	2.79	435
0.32	0.4077	0.550	0.785	-3.10	2.54	435
0.33	0.4256	0.574	0.775	-3.10	2.30	435
0.34	0.4442	0.599	0.765	-3.10	2.08	416
0.35	0.4633	0.624	0.755	-3.10	1.86	373
0.36	0.4833	0.651	0.745	-3.10	1.66	332
0.37	0.5043	0.680	0.734	-3.10	1.46	292
0.38	0.5262	0.709	0.722	-3.10	1.27	254
0.39	0.5493	0.740	0.710	-3.10	1.09	217
0.40	0.5739	0.774	0.697	-3.10	0.91	182
0.41	0.6002	0.809	0.683	-3.10	0.73	146
0.42	0.6287	0.847	0.668	-3.10	0.56	112
0.43	0.6600	0.890	0.651	-3.10	0.39	77
0.44	0.6952	0.937	0.633	-3.10	0.21	42

5.4 Bending and Axial Force of Rectangular Sections with Prescribed Neutral Axis Depth

Purpose of the Tables in This Section

This section contains tables to be used in the design of reinforced rectangular sections under bending moment and axial load with prescribed neutral axis position. Enforcing the neutral axis position is equivalent to prescribe the ductility. The different tables give different steel areas for the corresponding ductility. By this procedure it is possible to choose the apropriate solutions with higher ductility or with less amount of reinforcement.

How to Use the Tables in This Section

Example 3 of Chapter 4 shows how these tables are applied and shows a comparison of different solutions.

Cautions When Using the Tables in This Section

To use these tables note that both bending moment and axial load applied at the centroid of the section must be relocated to the tensile steel position. Each table is valid for a given range of the reduced bending moment μ_s. When a value is not available in the table range the previous table has to be chosen, as indicated at the bottom of each Table

5.4.1 Prescribed Neutral Axis Depth x/d– TABLES 5 to 8

Tab 5.23

TABLE 5_Concrete classes C12-C50_Steel class S500_x/d= 0.250

Bending and axial force of rectangular sections doubly reinforced with prescribed neutral axis depth

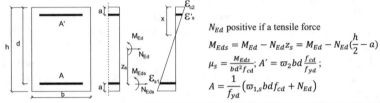

N_{Ed} positive if a tensile force

$$M_{Eds} = M_{Ed} - N_{Ed}z_s = M_{Ed} - N_{Ed}(\frac{h}{2} - a)$$

$$\mu_s = \frac{M_{Eds}}{bd^2 f_{cd}}; \quad A' = \varpi_2 bd \frac{f_{cd}}{f_{yd}};$$

$$A = \frac{1}{f_{yd}}(\varpi_{1,s} bd f_{cd} + N_{Ed})$$

	Prescribed neutral axis position, x/d= 0.350							
	$a/d = 0.05$		$a/d = 0.10$		$a/d = 0.15$		$a/d = 0.20$	
μ_s	$\varpi_{1,s}$	ϖ_2	$\varpi_{1,s}$	ϖ_2	$\varpi_{1,s}$	ϖ_2	$\varpi_{1,s}$	ϖ_2
0.19	0.212	0.009	0.212	0.010	0.213	0.016	0.213	0.034
0.20	0.222	0.020	0.223	0.022	0.224	0.034	0.226	0.072
0.21	0.233	0.030	0.234	0.033	0.236	0.052	0.238	0.111
0.22	0.243	0.041	0.245	0.045	0.248	0.071	0.251	0.150
0.23	0.254	0.051	0.256	0.056	0.260	0.089	0.263	0.189
0.24	0.264	0.062	0.268	0.068	0.271	0.107	0.276	0.228
0.25	0.275	0.072	0.279	0.079	0.283	0.125	0.288	0.267
0.26	0.285	0.083	0.290	0.091	0.295	0.144	0.301	0.305
0.27	0.296	0.093	0.301	0.102	0.307	0.162	0.313	0.344
0.28	0.306	0.104	0.312	0.113	0.318	0.180	0.326	0.383
0.29	0.317	0.114	0.323	0.125	0.330	0.199	0.338	0.422
0.30	0.327	0.125	0.334	0.136	0.342	0.217	0.351	0.461
0.31	0.338	0.135	0.345	0.148	0.354	0.235	0.363	0.500
0.32	0.348	0.146	0.356	0.159	0.366	0.253	0.376	0.538
0.33	0.359	0.156	0.368	0.171	0.377	0.272	0.388	0.577
0.34	0.369	0.167	0.379	0.182	0.389	0.290	0.401	0.616
0.35	0.380	0.178	0.390	0.194	0.401	0.308	0.413	0.655
0.36	0.390	0.188	0.401	0.206	0.413	0.326	0.426	0.694
0.37	0.401	0.199	0.412	0.217	0.424	0.345	0.438	0.732
0.38	0.412	0.209	0.423	0.229	0.436	0.363	0.451	0.771
0.39	0.422	0.220	0.434	0.240	0.448	0.381	0.463	0.810
0.40	0.433	0.230	0.445	0.252	0.460	0.399	0.476	0.849
0.41	0.443	0.241	0.456	0.263	0.471	0.418	0.488	0.888
0.42	0.454	0.251	0.468	0.275	0.483	0.436	0.501	0.927
0.43	0.464	0.262	0.479	0.286	0.495	0.454	0.513	0.965
0.44	0.475	0.272	0.490	0.298	0.507	0.473	0.526	1.004
0.45	0.485	0.283	0.501	0.309	0.518	0.491	0.538	1.043
0.46	0.496	0.293	0.512	0.321	0.530	0.509	0.551	1.082
0.47	0.506	0.304	0.523	0.332	0.542	0.527	0.563	1.121
0.48	0.517	0.314	0.534	0.344	0.554	0.546	0.576	1.159
0.49	0.527	0.325	0.545	0.355	0.566	0.564	0.588	1.198
0.50	0.538	0.335	0.556	0.367	0.577	0.582	0.601	1.237

Note: With values of μ_s less than the smaller value in this table see Tab.5.20 Table 4_Concrete classes C12-C50_Steel class S500.

Tab 5.24

TABLE 5_Concrete classes C12-C50_Steel class S400_x/d = 0.250

Bending and axial force of rectangular sections doubly reinforced with prescribed neutral axis depth

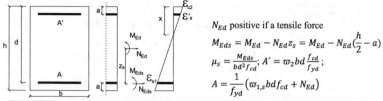

N_{Ed} positive if a tensile force

$$M_{Eds} = M_{Ed} - N_{Ed}z_s = M_{Ed} - N_{Ed}(\frac{h}{2} - a)$$

$$\mu_s = \frac{M_{Eds}}{bd^2 f_{cd}}; \; A' = \varpi_2 bd \frac{f_{cd}}{f_{yd}};$$

$$A = \frac{1}{f_{yd}}(\varpi_{1,s}bdf_{cd} + N_{Ed})$$

	Prescribed neutral axis position, x/d = 0.350							
	a/d = 0.05		a/d = 0.10		a/d = 0.15		a/d = 0.20	
μ_s	$\varpi_{1,s}$	ϖ_2	$\varpi_{1,s}$	ϖ_2	μ_{Sds}	$\varpi_{1,s}$	ϖ_2	$\varpi_{1,s}$
0.190	0.212	0.009	0.212	0.010	0.213	0.013	0.213	0.027
0.200	0.222	0.020	0.223	0.021	0.224	0.027	0.226	0.058
0.210	0.233	0.030	0.234	0.032	0.236	0.042	0.238	0.089
0.220	0.243	0.041	0.245	0.043	0.248	0.057	0.251	0.120
0.230	0.254	0.051	0.256	0.054	0.260	0.071	0.263	0.151
0.240	0.264	0.062	0.268	0.065	0.271	0.086	0.276	0.182
0.250	0.275	0.072	0.279	0.076	0.283	0.100	0.288	0.213
0.260	0.285	0.083	0.290	0.087	0.295	0.115	0.301	0.244
0.270	0.296	0.093	0.301	0.099	0.307	0.130	0.313	0.275
0.280	0.306	0.104	0.312	0.110	0.318	0.144	0.326	0.306
0.290	0.317	0.114	0.323	0.121	0.330	0.159	0.338	0.338
0.300	0.327	0.125	0.334	0.132	0.342	0.173	0.351	0.369
0.310	0.338	0.135	0.345	0.143	0.354	0.188	0.363	0.400
0.320	0.348	0.146	0.356	0.154	0.366	0.203	0.376	0.431
0.330	0.359	0.156	0.368	0.165	0.377	0.217	0.388	0.462
0.340	0.369	0.167	0.379	0.176	0.389	0.232	0.401	0.493
0.350	0.380	0.178	0.390	0.187	0.401	0.246	0.413	0.524
0.360	0.390	0.188	0.401	0.199	0.413	0.261	0.426	0.555
0.370	0.401	0.199	0.412	0.210	0.424	0.276	0.438	0.586
0.380	0.412	0.209	0.423	0.221	0.436	0.290	0.451	0.617
0.390	0.422	0.220	0.434	0.232	0.448	0.305	0.463	0.648
0.400	0.433	0.230	0.445	0.243	0.460	0.320	0.476	0.679
0.410	0.443	0.241	0.456	0.254	0.471	0.334	0.488	0.710
0.420	0.454	0.251	0.468	0.265	0.483	0.349	0.501	0.741
0.430	0.464	0.262	0.479	0.276	0.495	0.363	0.513	0.772
0.440	0.475	0.272	0.490	0.287	0.507	0.378	0.526	0.803
0.450	0.485	0.283	0.501	0.299	0.518	0.393	0.538	0.834
0.460	0.496	0.293	0.512	0.310	0.530	0.407	0.551	0.865
0.470	0.506	0.304	0.523	0.321	0.542	0.422	0.563	0.897
0.480	0.517	0.314	0.534	0.332	0.554	0.436	0.576	0.928
0.490	0.527	0.325	0.545	0.343	0.566	0.451	0.588	0.959
0.500	0.538	0.335	0.556	0.354	0.577	0.466	0.601	0.990

Note: With values of μ_s less than the smaller value in this table see Tab. 5.21 Table 4_Concrete classes C12-C50_Steel class S400.

Tab 5.25

TABLE 5_Concrete classes C55_Steel class S500_x/d = 0.250

Bending and axial force of rectangular sections doubly reinforced with prescribed neutral axis depth

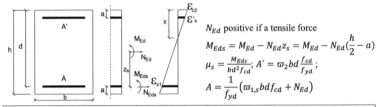

N_{Ed} positive if a tensile force

$$M_{Eds} = M_{Ed} - N_{Ed}z_s = M_{Ed} - N_{Ed}\left(\frac{h}{2} - a\right)$$

$$\mu_s = \frac{M_{Eds}}{bd^2 f_{cd}}; A' = \varpi_2 bd \frac{f_{cd}}{f_{yd}};$$

$$A = \frac{1}{f_{yd}}\left(\varpi_{1,s}bd f_{cd} + N_{Ed}\right)$$

μ_s	\multicolumn{2}{c}{$a/d = 0.05$}		\multicolumn{2}{c}{$a/d = 0.10$}		\multicolumn{2}{c}{$a/d = 0.15$}		$a/d = 0.20$
	$\varpi_{1,s}$	ϖ_2	$\varpi_{1,s}$	ϖ_2	μ_{Sds}	$\varpi_{1,s}$	ϖ_2	$\varpi_{1,s}$
0.17	0.188	0.003	0.188	0.003	0.189	0.006	0.189	0.012
0.18	0.199	0.013	0.200	0.017	0.200	0.026	0.201	0.056
0.19	0.209	0.024	0.211	0.030	0.212	0.047	0.214	0.099
0.20	0.220	0.034	0.222	0.043	0.224	0.067	0.226	0.143
0.21	0.230	0.045	0.233	0.055	0.236	0.088	0.239	0.187
0.22	0.241	0.056	0.244	0.068	0.247	0.109	0.251	0.231
0.23	0.251	0.066	0.255	0.081	0.259	0.129	0.264	0.275
0.24	0.262	0.077	0.266	0.094	0.271	0.150	0.276	0.319
0.25	0.273	0.087	0.277	0.107	0.283	0.171	0.289	0.362
0.26	0.283	0.098	0.288	0.120	0.295	0.191	0.301	0.406
0.27	0.294	0.108	0.300	0.133	0.306	0.212	0.314	0.450
0.28	0.304	0.119	0.311	0.146	0.318	0.232	0.326	0.494
0.29	0.315	0.129	0.322	0.159	0.330	0.253	0.339	0.538
0.30	0.325	0.140	0.333	0.172	0.342	0.274	0.351	0.582
0.31	0.336	0.150	0.344	0.185	0.353	0.294	0.364	0.625
0.32	0.346	0.161	0.355	0.198	0.365	0.315	0.376	0.669
0.33	0.357	0.171	0.366	0.211	0.377	0.336	0.389	0.713
0.34	0.367	0.182	0.377	0.224	0.389	0.356	0.401	0.757
0.35	0.378	0.192	0.388	0.237	0.400	0.377	0.414	0.801
0.36	0.388	0.203	0.400	0.250	0.412	0.397	0.426	0.845
0.37	0.399	0.213	0.411	0.263	0.424	0.418	0.439	0.888
0.38	0.409	0.224	0.422	0.276	0.436	0.439	0.451	0.932
0.39	0.420	0.234	0.433	0.289	0.447	0.459	0.464	0.976
0.40	0.430	0.245	0.444	0.302	0.459	0.480	0.476	1.020
0.41	0.441	0.255	0.455	0.315	0.471	0.501	0.489	1.060
0.42	0.451	0.266	0.466	0.328	0.483	0.521	0.501	1.110
0.43	0.462	0.277	0.477	0.341	0.495	0.542	0.514	1.150
0.44	0.473	0.287	0.488	0.354	0.506	0.562	0.526	1.200
0.45	0.483	0.298	0.500	0.367	0.518	0.583	0.539	1.240
0.46	0.494	0.308	0.511	0.380	0.530	0.604	0.551	1.280
0.47	0.504	0.319	0.522	0.393	0.542	0.624	0.564	1.330
0.48	0.515	0.329	0.533	0.406	0.553	0.645	0.576	1.370
0.49	0.525	0.340	0.544	0.419	0.565	0.666	0.589	1.410
0.50	0.536	0.350	0.555	0.432	0.577	0.686	0.601	1.460

Note: With values of μ_s less than the smaller value in this table see Tab. 5.22 Table 4_Concrete classes C55_Steel class S500.

Tab 5.26

TABLE 6_Concrete classes C12-C50_Steel class S500_x/d = 0.350

Bending and axial force of rectangular sections doubly reinforced with prescribed neutral axis depth

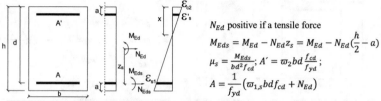

N_{Ed} positive if a tensile force

$$M_{Eds} = M_{Ed} - N_{Ed}z_s = M_{Ed} - N_{Ed}\left(\frac{h}{2} - a\right)$$

$$\mu_s = \frac{M_{Eds}}{bd^2 f_{cd}}; \; A' = \varpi_2 bd \frac{f_{cd}}{f_{yd}};$$

$$A = \frac{1}{f_{yd}}\left(\varpi_{1,s}bd f_{cd} + N_{Ed}\right)$$

	Prescribed neutral axis position, x/d = 0.350							
	$a/d = 0.05$		$a/d = 0.10$		$a/d = 0.15$		$a/d = 0.20$	
μ_s	$\varpi_{1,s}$	ϖ_2	$\varpi_{1,s}$	ϖ_2	μ_{Sds}	$\varpi_{1,s}$	ϖ_2	$\varpi_{1,s}$
0.25	0.292	0.008	0.292	0.009	0.293	0.010	0.293	0.014
0.26	0.302	0.019	0.303	0.020	0.304	0.023	0.306	0.032
0.27	0.313	0.029	0.314	0.031	0.316	0.036	0.318	0.051
0.28	0.323	0.040	0.325	0.042	0.328	0.049	0.331	0.069
0.29	0.334	0.050	0.337	0.053	0.340	0.061	0.343	0.087
0.30	0.344	0.061	0.348	0.064	0.351	0.074	0.356	0.105
0.31	0.355	0.072	0.359	0.076	0.363	0.087	0.368	0.123
0.32	0.365	0.082	0.370	0.087	0.375	0.100	0.381	0.141
0.33	0.376	0.093	0.381	0.098	0.387	0.112	0.393	0.159
0.34	0.386	0.103	0.392	0.109	0.399	0.125	0.406	0.177
0.35	0.397	0.114	0.403	0.120	0.410	0.138	0.418	0.196
0.36	0.407	0.124	0.414	0.131	0.422	0.151	0.431	0.214
0.37	0.418	0.135	0.425	0.142	0.434	0.164	0.443	0.232
0.38	0.429	0.145	0.437	0.153	0.446	0.176	0.456	0.250
0.39	0.439	0.156	0.448	0.164	0.457	0.189	0.468	0.268
0.40	0.450	0.166	0.459	0.175	0.469	0.202	0.481	0.286
0.41	0.460	0.177	0.470	0.187	0.481	0.215	0.493	0.304
0.42	0.471	0.187	0.481	0.198	0.493	0.228	0.506	0.322
0.43	0.481	0.198	0.492	0.209	0.504	0.240	0.518	0.340
0.44	0.492	0.208	0.503	0.220	0.516	0.253	0.531	0.359
0.45	0.502	0.219	0.514	0.231	0.528	0.266	0.543	0.377
0.46	0.513	0.229	0.525	0.242	0.540	0.279	0.556	0.395
0.47	0.523	0.240	0.537	0.253	0.551	0.291	0.568	0.413
0.48	0.534	0.250	0.548	0.264	0.563	0.304	0.581	0.431
0.49	0.544	0.261	0.559	0.275	0.575	0.317	0.593	0.449
0.50	0.555	0.271	0.570	0.287	0.587	0.330	0.606	0.467

Note: With values of μ_s less than the smaller value in this table see Tab. 5.23 Table 5_Concrete classes C12-C50_Steel class S500_x/d = 0.250.

Tab 5.27

TABLE 6_Concrete classes C12-C50_Steel class S400_x/d = 0.350

Bending and axial force of rectangular sections doubly reinforced with prescribed neutral axis depth

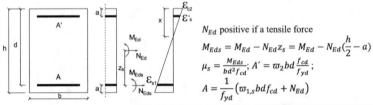

N_{Ed} positive if a tensile force

$$M_{Eds} = M_{Ed} - N_{Ed}z_s = M_{Ed} - N_{Ed}\left(\frac{h}{2} - a\right)$$

$$\mu_s = \frac{M_{Eds}}{bd^2 f_{cd}}; A' = \varpi_2 bd \frac{f_{cd}}{f_{yd}};$$

$$A = \frac{1}{f_{yd}}\left(\varpi_{1,s}bd f_{cd} + N_{Ed}\right)$$

	Prescribed neutral axis position, x/d = 0.350							
	$a/d = 0.05$		$a/d = 0.10$		$a/d = 0.15$		$a/d = 0.20$	
μ_s	$\varpi_{1,s}$	ϖ_2	$\varpi_{1,s}$	ϖ_2	μ_{Sds}	$\varpi_{1,s}$	ϖ_2	$\varpi_{1,s}$
0.25	0.292	0.008	0.292	0.009	0.293	0.009	0.293	0.011
0.26	0.302	0.019	0.303	0.020	0.304	0.021	0.306	0.026
0.27	0.313	0.029	0.314	0.031	0.316	0.033	0.318	0.040
0.28	0.323	0.040	0.325	0.042	0.328	0.045	0.331	0.055
0.29	0.334	0.050	0.337	0.053	0.340	0.056	0.343	0.069
0.30	0.344	0.061	0.348	0.064	0.351	0.068	0.356	0.084
0.31	0.355	0.072	0.359	0.076	0.363	0.080	0.368	0.098
0.32	0.365	0.082	0.370	0.087	0.375	0.092	0.381	0.113
0.33	0.376	0.093	0.381	0.098	0.387	0.103	0.393	0.127
0.34	0.386	0.103	0.392	0.109	0.399	0.115	0.406	0.142
0.35	0.397	0.114	0.403	0.120	0.410	0.127	0.418	0.156
0.36	0.407	0.124	0.414	0.131	0.422	0.139	0.431	0.171
0.37	0.418	0.135	0.425	0.142	0.434	0.150	0.443	0.185
0.38	0.429	0.145	0.437	0.153	0.446	0.162	0.456	0.200
0.39	0.439	0.156	0.448	0.164	0.457	0.174	0.468	0.214
0.40	0.450	0.166	0.459	0.175	0.469	0.186	0.481	0.229
0.41	0.460	0.177	0.470	0.187	0.481	0.198	0.493	0.243
0.42	0.471	0.187	0.481	0.198	0.493	0.209	0.506	0.258
0.43	0.481	0.198	0.492	0.209	0.504	0.221	0.518	0.272
0.44	0.492	0.208	0.503	0.220	0.516	0.233	0.531	0.287
0.45	0.502	0.219	0.514	0.231	0.528	0.245	0.543	0.301
0.46	0.513	0.229	0.525	0.242	0.540	0.256	0.556	0.316
0.47	0.523	0.240	0.537	0.253	0.551	0.268	0.568	0.330
0.48	0.534	0.250	0.548	0.264	0.563	0.280	0.581	0.345
0.49	0.544	0.261	0.559	0.275	0.575	0.292	0.593	0.359
0.50	0.555	0.271	0.570	0.287	0.587	0.303	0.606	0.374

Note: With values of μ_s less than the smaller value in this table see Tab. 5.24 Table 5_Concrete classes C12-C50_Steel class S400_x/d = 0.250.

Tab 5.28

TABLE 6_Concrete classes C55_Steel class S500_x/d = 0.350

Bending and axial force of rectangular sections doubly reinforced with prescribed neutral axis depth

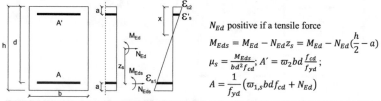

N_{Ed} positive if a tensile force

$$M_{Eds} = M_{Ed} - N_{Ed}z_s = M_{Ed} - N_{Ed}\left(\frac{h}{2} - a\right)$$

$$\mu_s = \frac{M_{Eds}}{bd^2 f_{cd}}; \; A' = \varpi_2 bd \frac{f_{cd}}{f_{yd}};$$

$$A = \frac{1}{f_{yd}}\left(\varpi_{1,s}bdf_{cd} + N_{Ed}\right)$$

	Prescribed neutral axis position, x/d = 0.350							
	$a/d = 0.05$		$a/d = 0.10$		$a/d = 0.15$		$a/d = 0.20$	
μ_s	$\varpi_{1,s}$	ϖ_2	$\varpi_{1,s}$	ϖ_2	μ_{sds}	$\varpi_{1,s}$	ϖ_2	$\varpi_{1,s}$
0.23	0.266	0.006	0.266	0.007	0.267	0.009	0.267	0.012
0.24	0.276	0.017	0.277	0.018	0.278	0.023	0.280	0.033
0.25	0.287	0.027	0.289	0.029	0.290	0.038	0.292	0.053
0.26	0.298	0.038	0.300	0.040	0.302	0.052	0.305	0.074
0.27	0.308	0.048	0.311	0.051	0.314	0.066	0.317	0.094
0.28	0.319	0.059	0.322	0.062	0.325	0.081	0.330	0.114
0.29	0.329	0.069	0.333	0.073	0.337	0.095	0.342	0.135
0.30	0.340	0.080	0.344	0.084	0.349	0.110	0.355	0.155
0.31	0.350	0.091	0.355	0.096	0.361	0.124	0.367	0.176
0.32	0.361	0.101	0.366	0.107	0.373	0.139	0.380	0.196
0.33	0.371	0.112	0.377	0.118	0.384	0.153	0.392	0.217
0.34	0.382	0.122	0.389	0.129	0.396	0.167	0.405	0.237
0.35	0.392	0.133	0.400	0.140	0.408	0.182	0.417	0.258
0.36	0.403	0.143	0.411	0.151	0.420	0.196	0.430	0.278
0.37	0.413	0.154	0.422	0.162	0.431	0.211	0.442	0.299
0.38	0.424	0.164	0.433	0.173	0.443	0.225	0.455	0.319
0.39	0.434	0.175	0.444	0.184	0.455	0.240	0.467	0.339
0.40	0.445	0.185	0.455	0.195	0.467	0.254	0.480	0.360
0.41	0.455	0.196	0.466	0.207	0.478	0.268	0.492	0.380
0.42	0.466	0.206	0.477	0.218	0.490	0.283	0.505	0.401
0.43	0.476	0.217	0.489	0.229	0.502	0.297	0.517	0.421
0.44	0.487	0.227	0.500	0.240	0.514	0.312	0.530	0.442
0.45	0.498	0.238	0.511	0.251	0.525	0.326	0.542	0.462
0.46	0.508	0.248	0.522	0.262	0.537	0.341	0.555	0.483
0.47	0.519	0.259	0.533	0.273	0.549	0.355	0.567	0.503
0.48	0.529	0.269	0.544	0.284	0.561	0.370	0.580	0.523
0.49	0.540	0.280	0.555	0.295	0.573	0.384	0.592	0.544
0.50	0.550	0.290	0.566	0.307	0.584	0.398	0.605	0.564

Note: With values of μ_s less than the smaller value in this table see Tab. 5.25 Table 5_Concrete classes C55_Steel class S500_x/d = 0.250.

Tab 5.29

TABLE 7_Concrete classes C12-C50_Steel class S500_x/d = 0.450

Bending and axial force of rectangular sections doubly reinforced with prescribed neutral axis depth

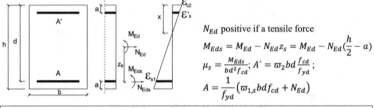

N_{Ed} positive if a tensile force

$$M_{Eds} = M_{Ed} - N_{Ed}z_s = M_{Ed} - N_{Ed}(\frac{h}{2} - a)$$

$$\mu_s = \frac{M_{Eds}}{bd^2 f_{cd}}; \quad A' = \varpi_2 bd \frac{f_{cd}}{f_{yd}};$$

$$A = \frac{1}{f_{yd}}(\varpi_{1,s}bd f_{cd} + N_{Ed})$$

	Prescribed neutral axis position, x/d = 0.450							
	$a/d = 0.05$		$a/d = 0.10$		$a/d = 0.15$		$a/d = 0.20$	
μ_s	$\varpi_{1,s}$	ϖ_2	$\varpi_{1,s}$	ϖ_2	μ_{Sds}	$\varpi_{1,s}$	ϖ_2	$\varpi_{1,s}$
0.30	0.368	0.004	0.369	0.004	0.369	0.005	0.369	0.005
0.31	0.379	0.015	0.380	0.015	0.381	0.016	0.382	0.019
0.32	0.389	0.025	0.391	0.027	0.392	0.028	0.394	0.033
0.33	0.400	0.036	0.402	0.038	0.404	0.040	0.407	0.047
0.34	0.411	0.046	0.413	0.049	0.416	0.052	0.419	0.061
0.35	0.421	0.057	0.424	0.060	0.428	0.063	0.432	0.075
0.36	0.432	0.067	0.435	0.071	0.439	0.075	0.444	0.089
0.37	0.442	0.078	0.446	0.082	0.451	0.087	0.457	0.103
0.38	0.453	0.088	0.458	0.093	0.463	0.099	0.469	0.117
0.39	0.463	0.099	0.469	0.104	0.475	0.110	0.482	0.131
0.40	0.474	0.109	0.480	0.115	0.487	0.122	0.494	0.145
0.41	0.484	0.120	0.491	0.127	0.498	0.134	0.507	0.159
0.42	0.495	0.130	0.502	0.138	0.510	0.146	0.519	0.173
0.43	0.505	0.141	0.513	0.149	0.522	0.158	0.532	0.187
0.44	0.516	0.151	0.524	0.160	0.534	0.169	0.544	0.201
0.45	0.526	0.162	0.535	0.171	0.545	0.181	0.557	0.215
0.46	0.537	0.173	0.546	0.182	0.557	0.193	0.569	0.229
0.47	0.547	0.183	0.558	0.193	0.569	0.205	0.582	0.243
0.48	0.558	0.194	0.569	0.204	0.581	0.216	0.594	0.257
0.49	0.568	0.204	0.580	0.215	0.592	0.228	0.607	0.271
0.50	0.579	0.215	0.591	0.227	0.604	0.240	0.619	0.285

Note: With values of μ_s less than the smaller value in this table see Tab. 5.26 Table 6_Concrete classes C12-C50_Steel class S500_x/d = 0.350.

Tab 5.30

TABLE 7_Concrete classes C12-C50_Steel class S400_x/d = 0.450

Bending and axial force of rectangular sections doubly reinforced with prescribed neutral axis depth

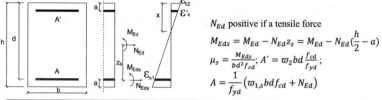

N_{Ed} positive if a tensile force

$$M_{Eds} = M_{Ed} - N_{Ed}z_s = M_{Ed} - N_{Ed}\left(\frac{h}{2} - a\right)$$

$$\mu_s = \frac{M_{Eds}}{bd^2 f_{cd}}; \quad A' = \varpi_2 bd \frac{f_{cd}}{f_{yd}};$$

$$A = \frac{1}{f_{yd}}\left(\varpi_{1,s} bd f_{cd} + N_{Ed}\right)$$

	Prescribed neutral axis position, x/d = 0.450							
	$a/d = 0.05$		$a/d = 0.10$		$a/d = 0.15$		$a/d = 0.20$	
μ_s	$\varpi_{1,s}$	ϖ_2	$\varpi_{1,s}$	ϖ_2	μ_{Sds}	$\varpi_{1,s}$	ϖ_2	$\varpi_{1,s}$
0.30	0.368	0.004	0.369	0.004	0.369	0.005	0.369	0.005
0.31	0.379	0.015	0.380	0.015	0.381	0.016	0.382	0.017
0.32	0.389	0.025	0.391	0.027	0.392	0.028	0.394	0.030
0.33	0.400	0.036	0.402	0.038	0.404	0.040	0.407	0.042
0.34	0.411	0.046	0.413	0.049	0.416	0.052	0.419	0.055
0.35	0.421	0.057	0.424	0.060	0.428	0.063	0.432	0.067
0.36	0.432	0.067	0.435	0.071	0.439	0.075	0.444	0.080
0.37	0.442	0.078	0.446	0.082	0.451	0.087	0.457	0.092
0.38	0.453	0.088	0.458	0.093	0.463	0.099	0.469	0.105
0.39	0.463	0.099	0.469	0.104	0.475	0.110	0.482	0.117
0.40	0.474	0.109	0.480	0.115	0.487	0.122	0.494	0.130
0.41	0.484	0.120	0.491	0.127	0.498	0.134	0.507	0.142
0.42	0.495	0.130	0.502	0.138	0.510	0.146	0.519	0.155
0.43	0.505	0.141	0.513	0.149	0.522	0.158	0.532	0.167
0.44	0.516	0.151	0.524	0.160	0.534	0.169	0.544	0.180
0.45	0.526	0.162	0.535	0.171	0.545	0.181	0.557	0.192
0.46	0.537	0.173	0.546	0.182	0.557	0.193	0.569	0.205
0.47	0.547	0.183	0.558	0.193	0.569	0.205	0.582	0.217
0.48	0.558	0.194	0.569	0.204	0.581	0.216	0.594	0.230
0.49	0.568	0.204	0.580	0.215	0.592	0.228	0.607	0.242
0.50	0.579	0.215	0.591	0.227	0.604	0.240	0.619	0.255

Note: With values of μ_s less than the smaller value in this table see Tab. 5.27 Table 6_Concrete classes C12-C50_Steel class S400_x/d = 0.350.

Tab 5.31

TABLE 7_Concrete classes C55_Steel class S500_x/d = 0.450

Bending and axial force of rectangular sections doubly reinforced with prescribed neutral axis depth

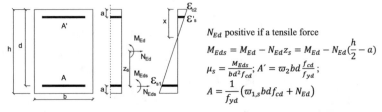

N_{Ed} positive if a tensile force

$$M_{Eds} = M_{Ed} - N_{Ed}z_s = M_{Ed} - N_{Ed}\left(\frac{h}{2} - a\right)$$

$$\mu_s = \frac{M_{Eds}}{bd^2 f_{cd}}; \quad A' = \varpi_2 bd\frac{f_{cd}}{f_{yd}};$$

$$A = \frac{1}{f_{yd}}\left(\varpi_{1,s}bdf_{cd} + N_{Ed}\right)$$

	Prescribed neutral axis position, x/d = 0.450							
	$a/d = 0.05$		$a/d = 0.10$		$a/d = 0.15$		$a/d = 0.20$	
μ_s	$\varpi_{1,s}$	ϖ_2	$\varpi_{1,s}$	ϖ_2	μ_{Sds}	$\varpi_{1,s}$	ϖ_2	$\varpi_{1,s}$
0.28	0.339	0.005	0.339	0.006	0.340	0.006	0.340	0.008
0.29	0.350	0.016	0.351	0.017	0.352	0.019	0.353	0.024
0.30	0.360	0.026	0.362	0.028	0.363	0.031	0.365	0.040
0.31	0.371	0.037	0.373	0.039	0.375	0.043	0.378	0.055
0.32	0.381	0.047	0.384	0.050	0.387	0.056	0.390	0.071
0.33	0.392	0.058	0.395	0.061	0.399	0.068	0.403	0.087
0.34	0.402	0.068	0.406	0.072	0.410	0.081	0.415	0.103
0.35	0.413	0.079	0.417	0.083	0.422	0.093	0.428	0.118
0.36	0.423	0.090	0.428	0.095	0.434	0.105	0.440	0.134
0.37	0.434	0.100	0.439	0.106	0.446	0.118	0.453	0.150
0.38	0.444	0.111	0.451	0.117	0.457	0.130	0.465	0.166
0.39	0.455	0.121	0.462	0.128	0.469	0.142	0.478	0.181
0.40	0.465	0.132	0.473	0.139	0.481	0.155	0.490	0.197
0.41	0.476	0.142	0.484	0.150	0.493	0.167	0.503	0.213
0.42	0.487	0.153	0.495	0.161	0.504	0.179	0.515	0.229
0.43	0.497	0.163	0.506	0.172	0.516	0.192	0.528	0.245
0.44	0.508	0.174	0.517	0.183	0.528	0.204	0.540	0.260
0.45	0.518	0.184	0.528	0.194	0.540	0.217	0.553	0.276
0.46	0.529	0.195	0.539	0.206	0.552	0.229	0.565	0.292
0.47	0.539	0.205	0.551	0.217	0.563	0.241	0.578	0.308
0.48	0.550	0.216	0.562	0.228	0.575	0.254	0.590	0.323
0.49	0.560	0.226	0.573	0.239	0.587	0.266	0.603	0.339
0.50	0.571	0.237	0.584	0.250	0.599	0.278	0.615	0.355

Note: With values of μ_s less than the smaller value in this table see Tab. 5.28 Table 6_Concrete classes C55_Steel class S500_x/d = 0.350.

Tab 5.32

TABLE 8_Concrete classes C12-C50_Steel class S500_x/d = 0.617

Bending and axial force of rectangular sections doubly reinforced with prescribed neutral axis depth

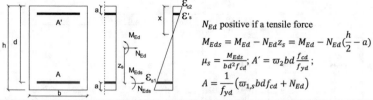

N_{Ed} positive if a tensile force

$$M_{Eds} = M_{Ed} - N_{Ed}z_s = M_{Ed} - N_{Ed}\left(\frac{h}{2} - a\right)$$

$$\mu_s = \frac{M_{Eds}}{bd^2 f_{cd}}; \quad A' = \varpi_2 bd \frac{f_{cd}}{f_{yd}};$$

$$A = \frac{1}{f_{yd}}\left(\varpi_{1,s} bd f_{cd} + N_{Ed}\right)$$

	Prescribed neutral axis position, x/d = 0.617							
	a/d = 0.05		a/d = 0.10		a/d = 0.15		a/d = 0.20	
μ_s	$\varpi_{1,s}$	ϖ_2	$\varpi_{1,s}$	ϖ_2	μ_{Sds}	$\varpi_{1,s}$	ϖ_2	$\varpi_{1,s}$
0.38	0.509	0.009	0.509	0.010	0.510	0.010	0.511	0.011
0.39	0.519	0.020	0.521	0.021	0.522	0.022	0.523	0.023
0.40	0.530	0.030	0.532	0.032	0.534	0.034	0.536	0.036
0.41	0.541	0.041	0.543	0.043	0.545	0.046	0.548	0.048
0.42	0.551	0.051	0.554	0.054	0.557	0.057	0.561	0.061
0.43	0.562	0.062	0.565	0.065	0.569	0.069	0.573	0.073
0.44	0.572	0.072	0.576	0.076	0.581	0.081	0.586	0.086
0.45	0.583	0.083	0.587	0.088	0.592	0.093	0.598	0.098
0.46	0.593	0.093	0.598	0.099	0.604	0.104	0.611	0.111
0.47	0.604	0.104	0.610	0.110	0.616	0.116	0.623	0.123
0.48	0.614	0.114	0.621	0.121	0.628	0.128	0.636	0.136
0.49	0.625	0.125	0.632	0.132	0.640	0.140	0.648	0.148
0.50	0.635	0.135	0.643	0.143	0.651	0.151	0.661	0.161

Note: With values of μ_s less than the smaller value in this table see Tab. 5.29 Table 7_Concrete classes C12-C50_Steel class S500_x/d = 0.450.

Tab 5.33

TABLE 8_Concrete classes C12-C50_Steel class S400_x/d = 0.617

Bending and axial force of rectangular sections doubly reinforced with neutral axis depth

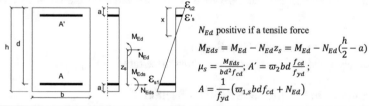

N_{Ed} positive if a tensile force

$$M_{Eds} = M_{Ed} - N_{Ed}z_s = M_{Ed} - N_{Ed}\left(\frac{h}{2} - a\right)$$

$$\mu_s = \frac{M_{Eds}}{bd^2 f_{cd}}; \quad A' = \varpi_2 bd \frac{f_{cd}}{f_{yd}};$$

$$A = \frac{1}{f_{yd}}\left(\varpi_{1,s} bd f_{cd} + N_{Ed}\right)$$

	Prescribed neutral axis position, x/d = 0.617							
	a/d = 0.05		a/d = 0.10		a/d = 0.15		a/d = 0.20	
μ_s	$\varpi_{1,s}$	ϖ_2	$\varpi_{1,s}$	ϖ_2	μ_{Sds}	$\varpi_{1,s}$	ϖ_2	$\varpi_{1,s}$
0.38	0.509	0.009	0.509	0.010	0.510	0.010	0.510	0.011
0.39	0.519	0.020	0.520	0.021	0.521	0.022	0.523	0.023
0.40	0.530	0.030	0.531	0.032	0.533	0.034	0.535	0.036
0.41	0.540	0.041	0.542	0.043	0.545	0.046	0.548	0.048
0.42	0.551	0.051	0.554	0.054	0.557	0.057	0.560	0.061
0.43	0.561	0.062	0.565	0.065	0.569	0.069	0.573	0.073
0.44	0.572	0.072	0.576	0.076	0.580	0.081	0.585	0.086
0.45	0.582	0.083	0.587	0.088	0.592	0.093	0.598	0.098
0.46	0.593	0.093	0.598	0.099	0.604	0.104	0.610	0.111
0.47	0.603	0.104	0.609	0.110	0.616	0.116	0.623	0.123
0.48	0.614	0.114	0.620	0.121	0.627	0.128	0.635	0.136
0.49	0.624	0.125	0.631	0.132	0.639	0.140	0.648	0.148
0.50	0.635	0.135	0.642	0.143	0.651	0.151	0.660	0.161

Note: With values of μ_s less than the smaller value in this table see Tab. 5.30 Table 7_Concrete classes C12-C50_Steel class S400_x/d = 0.450.

Tab 5.34

TABLE 8_Concrete class C55_Steel class S500_x/d = 0.617
Bending and axial force of rectangular sections doubly reinforced with neutral axis depth

μ_s	\multicolumn{8}{c}{Prescribed neutral axis position, x/d = 0.617}
	\multicolumn{2}{c}{$a/d = 0.05$}	\multicolumn{2}{c}{$a/d = 0.10$}	\multicolumn{2}{c}{$a/d = 0.15$}	\multicolumn{2}{c}{$a/d = 0.20$}
	$\varpi_{1,s}$	ϖ_2	$\varpi_{1,s}$	ϖ_2	μ_{sds}	$\varpi_{1,s}$	ϖ_2	$\varpi_{1,s}$
0.35	0.521	0.003	0.521	0.003	0.521	0.003	0.521	0.004
0.36	0.533	0.014	0.533	0.014	0.534	0.015	0.535	0.017
0.37	0.544	0.024	0.546	0.026	0.548	0.027	0.550	0.030
0.38	0.556	0.035	0.558	0.037	0.561	0.039	0.564	0.043
0.39	0.568	0.045	0.571	0.048	0.574	0.051	0.578	0.056
0.40	0.580	0.056	0.584	0.059	0.587	0.062	0.592	0.069
0.41	0.592	0.066	0.596	0.070	0.601	0.074	0.606	0.082
0.42	0.604	0.077	0.609	0.081	0.614	0.086	0.620	0.095
0.43	0.616	0.087	0.621	0.092	0.627	0.098	0.634	0.108
0.44	0.628	0.098	0.634	0.103	0.641	0.109	0.648	0.121
0.45	0.640	0.108	0.646	0.114	0.654	0.121	0.662	0.133
0.46	0.651	0.119	0.659	0.125	0.667	0.133	0.677	0.146
0.47	0.663	0.129	0.671	0.137	0.681	0.145	0.691	0.159
0.48	0.675	0.140	0.684	0.148	0.694	0.156	0.705	0.172
0.49	0.687	0.150	0.697	0.159	0.707	0.168	0.719	0.185
0.50	0.699	0.161	0.709	0.170	0.720	0.180	0.733	0.198

Note: With values of μ_s less than the smaller value in this table see Tab. 5.31 Table 7_Concrete classes C55_Steel class S500_x/d = 0.450.

5.5   Bending and Axial Force of Simply Reinforced T–Sections

Objetive of the Section
This section contains design tables to be used in the design of reinforced T–sections under bending moment and axial load. There are values of the ratio b/bw=4, 8 and 16.

How to Use the Tables of This Section
Examples 4 and 5 of Chapter 4 show how these tables are applied.

Precautions to Use the Tables of This Section
Note that both the bending moment and the axial load given at the centroid of the section should be evaluated at the reinforcement centroid (equivalent static system) to enter in the tables.

5.5.1 Bending and Axial Force of Simply Reinforced T–Sections—TABLES 9 to 11

Tab 5.35

TABLE 9_Concrete classes C12-C50_Steel classes S400;S500;S600_b/b$_w$=4
Bending and axial force of simply reinforced T-sections

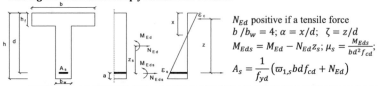

N_{Ed} positive if a tensile force

$b/b_w = 4; \ \alpha = x/d; \ \zeta = z/d$

$M_{Eds} = M_{Ed} - N_{Ed}z_s; \ \mu_s = \dfrac{M_{Eds}}{bd^2 f_{cd}};$

$A_s = \dfrac{1}{f_{yd}}\left(\varpi_{1,s}bd f_{cd} + N_{Ed}\right)$

	$h_f/d = 0.08$			$h_f/d = 0.10$			$h_f/d = 0.12$			$h_f/d = 0.14$		
μ_s	$\varpi_{1,s}$	α	ζ	$\varpi_{1,s}$	α	ζ	$\varpi_{1,s}$	α	ζ	$\varpi_{1,s}$	α	ζ
0.010	0.010	0.030	0.993	0.010	0.030	0.993	0.010	0.030	0.993	0.010	0.030	0.993
0.015	0.015	0.037	0.993	0.015	0.037	0.993	0.015	0.037	0.993	0.015	0.037	0.993
0.020	0.020	0.044	0.984	0.020	0.044	0.984	0.020	0.044	0.984	0.020	0.044	0.984
0.025	0.025	0.050	0.982	0.025	0.050	0.982	0.025	0.050	0.982	0.025	0.050	0.982
0.030	0.031	0.055	0.982	0.031	0.055	0.982	0.031	0.055	0.982	0.031	0.055	0.982
0.035	0.036	0.061	0.979	0.036	0.061	0.979	0.036	0.061	0.979	0.036	0.061	0.979
0.040	0.041	0.066	0.976	0.041	0.066	0.976	0.041	0.066	0.976	0.041	0.066	0.976
0.045	0.046	0.071	0.975	0.046	0.071	0.975	0.046	0.071	0.975	0.046	0.071	0.975
0.050	0.052	0.076	0.971	0.052	0.076	0.971	0.052	0.076	0.971	0.052	0.076	0.971
0.055	0.057	0.081	0.970	0.057	0.081	0.969	0.057	0.081	0.969	0.057	0.081	0.969
0.060	0.062	0.087	0.967	0.062	0.087	0.966	0.062	0.087	0.966	0.062	0.087	0.966
0.065	0.067	0.093	0.966	0.067	0.092	0.966	0.067	0.092	0.966	0.067	0.092	0.966
0.070	0.073	0.101	0.963	0.073	0.097	0.961	0.073	0.097	0.961	0.073	0.097	0.961
0.075	0.078	0.110	0.960	0.078	0.102	0.960	0.078	0.102	0.959	0.078	0.102	0.959
0.080	0.084	0.122	0.958	0.084	0.108	0.957	0.084	0.107	0.957	0.084	0.107	0.957
0.085	0.089	0.145	0.954	0.089	0.115	0.955	0.089	0.113	0.954	0.089	0.113	0.954
0.090	0.095	0.172	0.949	0.094	0.123	0.953	0.095	0.118	0.952	0.095	0.118	0.952
0.095	0.101	0.201	0.943	0.100	0.137	0.950	0.100	0.124	0.949	0.100	0.124	0.950
0.100	0.107	0.232	0.936	0.106	0.157	0.947	0.106	0.132	0.947	0.106	0.131	0.946
0.105	0.113	0.263	0.928	0.111	0.181	0.942	0.111	0.143	0.944	0.111	0.137	0.944
0.110	0.120	0.295	0.920	0.117	0.210	0.937	0.117	0.156	0.941	0.117	0.145	0.940
0.115	0.126	0.328	0.910	0.124	0.240	0.931	0.123	0.173	0.938	0.123	0.153	0.938
0.120	0.133	0.363	0.900	0.130	0.271	0.924	0.128	0.195	0.935	0.128	0.164	0.935
0.125	0.141	0.399	0.888	0.136	0.304	0.916	0.134	0.221	0.930	0.134	0.176	0.932
0.130	0.148	0.437	0.876	0.143	0.337	0.908	0.141	0.250	0.925	0.140	0.192	0.929
0.135	0.156	0.477	0.863	0.150	0.372	0.898	0.147	0.282	0.918	0.146	0.212	0.926
0.140	0.165	0.519	0.849	0.158	0.409	0.888	0.154	0.315	0.911	0.152	0.236	0.922
0.145	0.174	0.564	0.833	0.165	0.447	0.876	0.161	0.349	0.903	0.158	0.263	0.917
0.150				0.174	0.488	0.864	0.168	0.384	0.894	0.165	0.294	0.912
0.155				0.182	0.531	0.850	0.175	0.421	0.884	0.171	0.328	0.905
0.160							0.183	0.460	0.874	0.178	0.362	0.897
0.165							0.191	0.502	0.862	0.186	0.399	0.889
0.170							0.200	0.545	0.849	0.193	0.436	0.880
0.175										0.201	0.476	0.869
0.180										0.210	0.518	0.858
0.185										0.219	0.563	0.845
0.190												
0.195												
0.200												
0.205												
0.210												
0.215												
0.220												
0.225												
0.230												
0.235												
0.240												
0.245												
0.250												

TABLE 9_Concrete classes C12-C50_Steel classes S400;S500;S600_b/b$_w$=4 (cont.)
Bending and axial force of simply reinforced T-sections

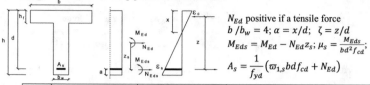

N_{Ed} positive if a tensile force
$b/b_w = 4;\ \alpha = x/d;\ \zeta = z/d$
$M_{Eds} = M_{Ed} - N_{Ed}z_s;\ \mu_s = \dfrac{M_{Eds}}{bd^2 f_{cd}};$
$A_s = \dfrac{1}{f_{yd}}\left(\varpi_{1,s}bd f_{cd} + N_{Ed}\right)$

μ_s	$h_f/d=0.16$			$h_f/d=0.18$			$h_f/d=0.20$			$h_f/d=0.25$		
	$\varpi_{1,s}$	α	ζ	$\varpi_{1,s}$	α	ζ	$\varpi_{1,s}$	α	ζ	$\varpi_{1,s}$	α	ζ
0.010	0.010	0.030	0.993	0.010	0.030	0.993	0.010	0.030	0.993	0.010	0.030	0.993
0.015	0.015	0.037	0.993	0.015	0.037	0.993	0.015	0.037	0.993	0.015	0.037	0.993
0.020	0.020	0.044	0.984	0.020	0.044	0.984	0.020	0.044	0.984	0.020	0.044	0.984
0.025	0.025	0.050	0.982	0.025	0.050	0.982	0.025	0.050	0.982	0.025	0.050	0.982
0.030	0.031	0.055	0.982	0.031	0.055	0.982	0.031	0.055	0.982	0.031	0.055	0.982
0.035	0.036	0.061	0.979	0.036	0.061	0.979	0.036	0.061	0.979	0.036	0.061	0.979
0.040	0.041	0.066	0.976	0.041	0.066	0.976	0.041	0.066	0.976	0.041	0.066	0.976
0.045	0.046	0.071	0.975	0.046	0.071	0.975	0.046	0.071	0.975	0.046	0.071	0.975
0.050	0.052	0.076	0.971	0.052	0.076	0.971	0.052	0.076	0.971	0.052	0.076	0.971
0.055	0.057	0.081	0.969	0.057	0.081	0.969	0.057	0.081	0.969	0.057	0.081	0.969
0.060	0.062	0.087	0.966	0.062	0.087	0.966	0.062	0.087	0.966	0.062	0.087	0.966
0.065	0.067	0.092	0.966	0.067	0.092	0.966	0.067	0.092	0.966	0.067	0.092	0.966
0.070	0.073	0.097	0.961	0.073	0.097	0.961	0.073	0.097	0.961	0.073	0.097	0.961
0.075	0.078	0.102	0.959	0.078	0.102	0.959	0.078	0.102	0.959	0.078	0.102	0.959
0.080	0.084	0.107	0.957	0.084	0.107	0.957	0.084	0.107	0.957	0.084	0.107	0.957
0.085	0.089	0.113	0.954	0.089	0.113	0.954	0.089	0.113	0.954	0.089	0.113	0.954
0.090	0.095	0.118	0.952	0.095	0.118	0.952	0.095	0.118	0.952	0.095	0.118	0.952
0.095	0.100	0.124	0.950	0.100	0.124	0.950	0.100	0.124	0.950	0.100	0.124	0.950
0.100	0.106	0.131	0.946	0.106	0.131	0.946	0.106	0.131	0.946	0.106	0.131	0.946
0.105	0.111	0.137	0.944	0.111	0.137	0.944	0.111	0.137	0.944	0.111	0.137	0.944
0.110	0.117	0.144	0.940	0.117	0.144	0.940	0.117	0.144	0.940	0.117	0.144	0.940
0.115	0.123	0.151	0.938	0.123	0.151	0.938	0.123	0.151	0.938	0.123	0.151	0.938
0.120	0.128	0.159	0.934	0.128	0.159	0.934	0.128	0.159	0.934	0.128	0.159	0.934
0.125	0.134	0.166	0.932	0.134	0.166	0.931	0.134	0.166	0.931	0.134	0.166	0.931
0.130	0.140	0.175	0.929	0.140	0.173	0.929	0.140	0.173	0.929	0.140	0.173	0.929
0.135	0.146	0.185	0.926	0.146	0.180	0.925	0.146	0.180	0.925	0.146	0.180	0.925
0.140	0.152	0.198	0.923	0.152	0.188	0.922	0.152	0.188	0.922	0.152	0.188	0.922
0.145	0.158	0.213	0.920	0.158	0.197	0.920	0.158	0.195	0.919	0.158	0.195	0.919
0.150	0.164	0.231	0.917	0.164	0.208	0.917	0.164	0.202	0.916	0.164	0.202	0.916
0.155	0.170	0.254	0.914	0.170	0.220	0.914	0.170	0.210	0.913	0.170	0.210	0.913
0.160	0.176	0.280	0.909	0.176	0.235	0.911	0.176	0.220	0.911	0.176	0.217	0.910
0.165	0.183	0.310	0.904	0.182	0.253	0.908	0.182	0.231	0.908	0.182	0.225	0.907
0.170	0.189	0.343	0.898	0.188	0.274	0.904	0.188	0.243	0.905	0.188	0.232	0.904
0.175	0.197	0.378	0.890	0.194	0.299	0.900	0.194	0.258	0.902	0.194	0.240	0.901
0.180	0.204	0.415	0.882	0.201	0.329	0.895	0.200	0.276	0.899	0.201	0.248	0.897
0.185	0.212	0.454	0.873	0.208	0.361	0.889	0.207	0.297	0.895	0.207	0.256	0.894
0.190	0.220	0.495	0.863	0.215	0.397	0.882	0.213	0.322	0.891	0.213	0.265	0.891
0.195	0.229	0.538	0.852	0.223	0.435	0.875	0.220	0.350	0.886	0.220	0.275	0.888
0.200				0.231	0.474	0.866	0.227	0.383	0.880	0.226	0.286	0.885
0.205				0.240	0.517	0.856	0.235	0.419	0.874	0.232	0.299	0.882
0.210				0.249	0.561	0.845	0.242	0.457	0.866	0.239	0.314	0.879
0.215							0.251	0.498	0.857	0.246	0.331	0.875
0.220							0.260	0.542	0.847	0.252	0.351	0.872
0.225										0.259	0.374	0.868
0.230										0.266	0.401	0.864
0.235										0.274	0.432	0.858
0.240										0.282	0.467	0.852
0.245										0.290	0.507	0.845
0.250										0.299	0.550	0.837

Tab 5.36

TABLE 9_Concrete classes C55_Steel classes S400;S500;S600_b/b$_w$=4

Bending and axial force of simply reinforced T sections

N_{Ed} positive if a tensile force
$b/b_w = 4$; $\alpha = x/d$; $\zeta = z/d$
$M_{Eds} = M_{Ed} - N_{Ed}z_s$; $\mu_s = \dfrac{M_{Eds}}{bd^2 f_{cd}}$;
$A_s = \dfrac{1}{f_{yd}}\left(\varpi_{1,s}bd f_{cd} + N_{Ed}\right)$

	$h_f/d = 0.08$			$h_f/d = 0.10$			$h_f/d = 0.12$			$h_f/d = 0.14$		
μ_s	$\varpi_{1,s}$	α	ζ	$\varpi_{1,s}$	α	ζ	$\varpi_{1,s}$	α	ζ	$\varpi_{1,s}$	α	ζ
0.010	0.010	0.033	0.993	0.010	0.033	0.993	0.010	0.033	0.993	0.010	0.033	0.993
0.015	0.015	0.041	0.987	0.015	0.041	0.987	0.015	0.041	0.987	0.015	0.041	0.987
0.020	0.020	0.048	0.984	0.020	0.048	0.984	0.020	0.048	0.984	0.020	0.048	0.984
0.025	0.025	0.054	0.983	0.025	0.054	0.983	0.025	0.054	0.983	0.025	0.054	0.983
0.030	0.031	0.060	0.981	0.031	0.060	0.981	0.031	0.060	0.981	0.031	0.060	0.981
0.035	0.036	0.065	0.979	0.036	0.065	0.979	0.036	0.065	0.979	0.036	0.065	0.979
0.040	0.041	0.071	0.975	0.041	0.071	0.975	0.041	0.071	0.975	0.041	0.071	0.975
0.045	0.046	0.076	0.975	0.046	0.076	0.975	0.046	0.076	0.975	0.046	0.076	0.975
0.050	0.052	0.081	0.970	0.052	0.081	0.970	0.052	0.081	0.970	0.052	0.081	0.970
0.055	0.057	0.086	0.969	0.057	0.086	0.968	0.057	0.086	0.968	0.057	0.086	0.968
0.060	0.062	0.092	0.968	0.062	0.091	0.967	0.062	0.091	0.967	0.062	0.091	0.967
0.065	0.067	0.099	0.965	0.067	0.096	0.964	0.067	0.096	0.964	0.067	0.096	0.964
0.070	0.073	0.107	0.962	0.073	0.102	0.961	0.073	0.102	0.961	0.073	0.102	0.961
0.075	0.078	0.120	0.960	0.078	0.107	0.959	0.078	0.107	0.959	0.078	0.107	0.959
0.080	0.084	0.139	0.957	0.084	0.115	0.957	0.084	0.113	0.957	0.084	0.113	0.957
0.085	0.089	0.163	0.953	0.089	0.126	0.954	0.089	0.120	0.954	0.089	0.120	0.954
0.090	0.095	0.190	0.948	0.095	0.140	0.952	0.095	0.128	0.951	0.095	0.128	0.951
0.095	0.101	0.221	0.942	0.100	0.158	0.949	0.100	0.138	0.948	0.100	0.135	0.948
0.100	0.107	0.253	0.935	0.106	0.179	0.945	0.106	0.149	0.946	0.106	0.143	0.945
0.105	0.113	0.288	0.927	0.112	0.204	0.941	0.111	0.163	0.943	0.111	0.151	0.942
0.110	0.120	0.323	0.917	0.118	0.232	0.936	0.117	0.179	0.940	0.117	0.161	0.939
0.115	0.127	0.360	0.908	0.124	0.264	0.930	0.123	0.199	0.937	0.123	0.173	0.937
0.120	0.134	0.398	0.897	0.130	0.297	0.923	0.129	0.222	0.933	0.129	0.186	0.934
0.125	0.141	0.438	0.885	0.137	0.333	0.914	0.135	0.248	0.928	0.134	0.202	0.931
0.130	0.149	0.480	0.872	0.144	0.370	0.905	0.141	0.278	0.923	0.140	0.221	0.928
0.135	0.157	0.525	0.858	0.151	0.409	0.895	0.147	0.310	0.917	0.146	0.243	0.924
0.140				0.158	0.449	0.884	0.154	0.345	0.909	0.152	0.267	0.920
0.145				0.166	0.492	0.872	0.161	0.383	0.901	0.158	0.296	0.915
0.150				0.175	0.537	0.859	0.168	0.422	0.891	0.165	0.327	0.909
0.155							0.176	0.463	0.881	0.172	0.361	0.903
0.160							0.184	0.507	0.870	0.179	0.398	0.895
0.165										0.186	0.438	0.886
0.170										0.194	0.480	0.876
0.175										0.202	0.524	0.865
0.180												
0.185												
0.190												
0.195												
0.200												
0.205												
0.210												
0.215												
0.220												
0.225												
0.230												
0.235												
0.240												
0.245												
0.250												

TABLE 9_Concrete classes C55_Steel classes S400;S500;S600_b/b_w=4 (cont.)
Bending and axial force of simply reinforced T sections

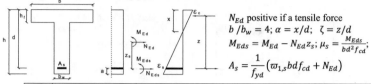

N_{Ed} positive if a tensile force
$b/b_w = 4;\ \alpha = x/d;\ \zeta = z/d$
$M_{Eds} = M_{Ed} - N_{Ed}z_s;\ \mu_s = \dfrac{M_{Eds}}{bd^2 f_{cd}};$
$A_s = \dfrac{1}{f_{yd}}\left(\varpi_{1,s}bd f_{cd} + N_{Ed}\right)$

μ_s	$h_f/d = 0.16$			$h_f/d = 0.18$			$h_f/d = 0.20$			$h_f/d = 0.25$		
	$\varpi_{1,s}$	α	ζ	$\varpi_{1,s}$	α	ζ	$\varpi_{1,s}$	α	ζ	$\varpi_{1,s}$	α	ζ
0.010	0.010	0.033	0.993	0.010	0.033	0.993	0.010	0.033	0.993	0.010	0.033	0.993
0.015	0.015	0.041	0.987	0.015	0.041	0.987	0.015	0.041	0.987	0.015	0.041	0.987
0.020	0.020	0.048	0.984	0.020	0.048	0.984	0.020	0.048	0.984	0.020	0.048	0.984
0.025	0.025	0.054	0.983	0.025	0.054	0.983	0.025	0.054	0.983	0.025	0.054	0.983
0.030	0.031	0.060	0.981	0.031	0.060	0.981	0.031	0.060	0.981	0.031	0.060	0.981
0.035	0.036	0.065	0.979	0.036	0.065	0.979	0.036	0.065	0.979	0.036	0.065	0.979
0.040	0.041	0.071	0.975	0.041	0.071	0.975	0.041	0.071	0.975	0.041	0.071	0.975
0.045	0.046	0.076	0.975	0.046	0.076	0.975	0.046	0.076	0.975	0.046	0.076	0.975
0.050	0.052	0.081	0.970	0.052	0.081	0.970	0.052	0.081	0.970	0.052	0.081	0.970
0.055	0.057	0.086	0.968	0.057	0.086	0.968	0.057	0.086	0.968	0.057	0.086	0.968
0.060	0.062	0.091	0.967	0.062	0.091	0.967	0.062	0.091	0.967	0.062	0.091	0.967
0.065	0.067	0.096	0.964	0.067	0.096	0.964	0.067	0.096	0.964	0.067	0.096	0.964
0.070	0.073	0.102	0.961	0.073	0.102	0.961	0.073	0.102	0.961	0.073	0.102	0.961
0.075	0.078	0.107	0.959	0.078	0.107	0.959	0.078	0.107	0.959	0.078	0.107	0.959
0.080	0.084	0.113	0.957	0.084	0.113	0.957	0.084	0.113	0.957	0.084	0.113	0.957
0.085	0.089	0.120	0.954	0.089	0.120	0.954	0.089	0.120	0.954	0.089	0.120	0.954
0.090	0.095	0.128	0.951	0.095	0.128	0.951	0.095	0.128	0.951	0.095	0.128	0.951
0.095	0.100	0.135	0.948	0.100	0.135	0.948	0.100	0.135	0.948	0.100	0.135	0.948
0.100	0.106	0.143	0.945	0.106	0.143	0.945	0.106	0.143	0.945	0.106	0.143	0.945
0.105	0.111	0.150	0.942	0.111	0.150	0.942	0.111	0.150	0.942	0.111	0.150	0.942
0.110	0.117	0.158	0.939	0.117	0.158	0.939	0.117	0.158	0.939	0.117	0.158	0.939
0.115	0.123	0.166	0.936	0.123	0.166	0.936	0.123	0.166	0.936	0.123	0.166	0.936
0.120	0.129	0.175	0.933	0.129	0.173	0.932	0.129	0.173	0.932	0.129	0.173	0.932
0.125	0.134	0.185	0.930	0.135	0.181	0.929	0.135	0.181	0.929	0.135	0.181	0.929
0.130	0.140	0.197	0.928	0.140	0.190	0.927	0.140	0.189	0.926	0.140	0.189	0.926
0.135	0.146	0.211	0.925	0.146	0.199	0.924	0.146	0.197	0.923	0.146	0.197	0.923
0.140	0.152	0.226	0.922	0.152	0.210	0.921	0.152	0.205	0.920	0.152	0.205	0.920
0.145	0.158	0.245	0.919	0.158	0.222	0.918	0.158	0.214	0.917	0.158	0.213	0.917
0.150	0.164	0.266	0.915	0.164	0.236	0.915	0.164	0.224	0.914	0.164	0.221	0.913
0.155	0.170	0.290	0.911	0.170	0.252	0.912	0.170	0.236	0.912	0.170	0.229	0.910
0.160	0.177	0.317	0.907	0.176	0.270	0.909	0.176	0.248	0.909	0.176	0.238	0.907
0.165	0.183	0.347	0.901	0.182	0.291	0.906	0.182	0.262	0.906	0.183	0.246	0.904
0.170	0.190	0.381	0.895	0.188	0.314	0.902	0.188	0.279	0.903	0.189	0.254	0.901
0.175	0.197	0.418	0.888	0.195	0.341	0.897	0.195	0.297	0.899	0.195	0.264	0.897
0.180	0.205	0.457	0.879	0.202	0.371	0.892	0.201	0.318	0.896	0.201	0.274	0.894
0.185	0.213	0.500	0.870	0.209	0.404	0.886	0.207	0.341	0.892	0.208	0.285	0.892
0.190				0.216	0.441	0.879	0.214	0.368	0.888	0.214	0.297	0.889
0.195				0.224	0.480	0.871	0.221	0.397	0.883	0.220	0.310	0.886
0.200				0.232	0.523	0.862	0.228	0.430	0.877	0.227	0.325	0.883
0.205							0.236	0.467	0.870	0.233	0.342	0.879
0.210							0.243	0.507	0.863	0.240	0.361	0.876
0.215										0.246	0.382	0.873
0.220										0.253	0.405	0.869
0.225										0.260	0.431	0.864
0.230										0.268	0.460	0.860
0.235										0.275	0.493	0.854
0.240										0.283	0.529	0.848
0.245												
0.250												

Tab 5.37

TABLE 10_Concrete classes C12-C50_Steel classes S400;S500;S600_b/b$_w$=8

Bending and axial force of simply reinforced T sections

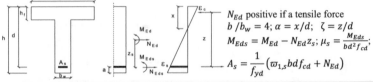

N_{Ed} positive if a tensile force
$b/b_w = 4; \alpha = x/d; \zeta = z/d$
$M_{Eds} = M_{Ed} - N_{Ed}z_s; \mu_s = \dfrac{M_{Eds}}{bd^2 f_{cd}};$

$A_s = \dfrac{1}{f_{yd}}\left(\varpi_{1,s}bd f_{cd} + N_{Ed}\right)$

	$h_f/d=0.08$			$h_f/d=0.10$			$h_f/d=0.12$			$h_f/d=0.14$		
μ_s	$\varpi_{1,s}$	α	ζ	$\varpi_{1,s}$	α	ζ	$\varpi_{1,s}$	α	ζ	$\varpi_{1,s}$	α	ζ
0.010	0.010	0.030	0.993	0.010	0.030	0.993	0.010	0.030	0.993	0.010	0.030	0.993
0.015	0.015	0.037	0.993	0.015	0.037	0.993	0.015	0.037	0.993	0.015	0.037	0.993
0.020	0.020	0.044	0.984	0.020	0.044	0.984	0.020	0.044	0.984	0.020	0.044	0.984
0.025	0.025	0.050	0.982	0.025	0.050	0.982	0.025	0.050	0.982	0.025	0.050	0.982
0.030	0.031	0.055	0.982	0.031	0.055	0.982	0.031	0.055	0.982	0.031	0.055	0.982
0.035	0.036	0.061	0.979	0.036	0.061	0.979	0.036	0.061	0.979	0.036	0.061	0.979
0.040	0.041	0.066	0.976	0.041	0.066	0.976	0.041	0.066	0.976	0.041	0.066	0.976
0.045	0.046	0.071	0.975	0.046	0.071	0.975	0.046	0.071	0.975	0.046	0.071	0.975
0.050	0.052	0.076	0.971	0.052	0.076	0.971	0.052	0.076	0.971	0.052	0.076	0.971
0.055	0.057	0.081	0.970	0.057	0.081	0.969	0.057	0.081	0.969	0.057	0.081	0.969
0.060	0.062	0.087	0.968	0.062	0.087	0.966	0.062	0.087	0.966	0.062	0.087	0.966
0.065	0.067	0.094	0.965	0.067	0.092	0.966	0.067	0.092	0.966	0.067	0.092	0.966
0.070	0.073	0.102	0.963	0.073	0.097	0.961	0.073	0.097	0.961	0.073	0.097	0.961
0.075	0.078	0.113	0.961	0.078	0.102	0.960	0.078	0.102	0.959	0.078	0.102	0.959
0.080	0.083	0.138	0.958	0.084	0.108	0.957	0.084	0.107	0.957	0.084	0.107	0.957
0.085	0.089	0.190	0.953	0.089	0.115	0.955	0.089	0.113	0.954	0.089	0.113	0.954
0.090	0.095	0.251	0.944	0.094	0.125	0.953	0.095	0.118	0.952	0.095	0.118	0.952
0.095	0.102	0.315	0.932	0.100	0.146	0.950	0.100	0.124	0.950	0.100	0.124	0.950
0.100	0.109	0.385	0.918	0.106	0.182	0.947	0.106	0.133	0.947	0.106	0.131	0.946
0.105	0.117	0.461	0.900	0.112	0.239	0.940	0.111	0.144	0.944	0.111	0.137	0.944
0.110	0.125	0.546	0.878	0.118	0.303	0.931	0.117	0.161	0.942	0.117	0.145	0.941
0.115				0.125	0.372	0.919	0.123	0.188	0.938	0.123	0.154	0.938
0.120				0.133	0.447	0.904	0.128	0.233	0.934	0.128	0.165	0.935
0.125				0.141	0.530	0.886	0.135	0.295	0.927	0.134	0.181	0.933
0.130							0.142	0.363	0.917	0.140	0.203	0.930
0.135							0.149	0.437	0.904	0.146	0.238	0.926
0.140							0.158	0.520	0.889	0.152	0.293	0.921
0.145										0.159	0.360	0.912
0.150										0.166	0.434	0.901
0.155										0.175	0.516	0.887
0.160												
0.165												
0.170												
0.175												
0.180												
0.185												
0.190												
0.195												
0.200												
0.205												
0.210												
0.215												
0.220												
0.225												
0.230												
0.235												
0.240												
0.245												
0.250												

TABLE 10_Concrete classes C12-C50_Steel classes S400;S500;S600_b/bw=8 (cont.)
Bending and axial force of simply reinforced T sections

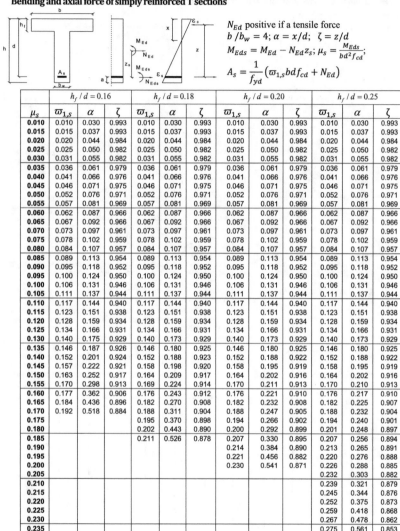

N_{Ed} positive if a tensile force
$b/b_w = 4;\ \alpha = x/d;\ \zeta = z/d$
$M_{Eds} = M_{Ed} - N_{Ed}z_s;\ \mu_s = \dfrac{M_{Eds}}{bd^2 f_{cd}};$
$A_s = \dfrac{1}{f_{yd}}\left(\varpi_{1,s}bd f_{cd} + N_{Ed}\right)$

	$h_f/d = 0.16$			$h_f/d = 0.18$			$h_f/d = 0.20$			$h_f/d = 0.25$		
μ_s	$\varpi_{1,s}$	α	ζ	$\varpi_{1,s}$	α	ζ	$\varpi_{1,s}$	α	ζ	$\varpi_{1,s}$	α	ζ
0.010	0.010	0.030	0.993	0.010	0.030	0.993	0.010	0.030	0.993	0.010	0.030	0.993
0.015	0.015	0.037	0.993	0.015	0.037	0.993	0.015	0.037	0.993	0.015	0.037	0.993
0.020	0.020	0.044	0.984	0.020	0.044	0.984	0.020	0.044	0.984	0.020	0.044	0.984
0.025	0.025	0.050	0.982	0.025	0.050	0.982	0.025	0.050	0.982	0.025	0.050	0.982
0.030	0.031	0.055	0.982	0.031	0.055	0.982	0.031	0.055	0.982	0.031	0.055	0.982
0.035	0.036	0.061	0.979	0.036	0.061	0.979	0.036	0.061	0.979	0.036	0.061	0.979
0.040	0.041	0.066	0.976	0.041	0.066	0.976	0.041	0.066	0.976	0.041	0.066	0.976
0.045	0.046	0.071	0.975	0.046	0.071	0.975	0.046	0.071	0.975	0.046	0.071	0.975
0.050	0.052	0.076	0.971	0.052	0.076	0.971	0.052	0.076	0.971	0.052	0.076	0.971
0.055	0.057	0.081	0.969	0.057	0.081	0.969	0.057	0.081	0.969	0.057	0.081	0.969
0.060	0.062	0.087	0.966	0.062	0.087	0.966	0.062	0.087	0.966	0.062	0.087	0.966
0.065	0.067	0.092	0.966	0.067	0.092	0.966	0.067	0.092	0.966	0.067	0.092	0.966
0.070	0.073	0.097	0.961	0.073	0.097	0.961	0.073	0.097	0.961	0.073	0.097	0.961
0.075	0.078	0.102	0.959	0.078	0.102	0.959	0.078	0.102	0.959	0.078	0.102	0.959
0.080	0.084	0.107	0.957	0.084	0.107	0.957	0.084	0.107	0.957	0.084	0.107	0.957
0.085	0.089	0.113	0.954	0.089	0.113	0.954	0.089	0.113	0.954	0.089	0.113	0.954
0.090	0.095	0.118	0.952	0.095	0.118	0.952	0.095	0.118	0.952	0.095	0.118	0.952
0.095	0.100	0.124	0.950	0.100	0.124	0.950	0.100	0.124	0.950	0.100	0.124	0.950
0.100	0.106	0.131	0.946	0.106	0.131	0.946	0.106	0.131	0.946	0.106	0.131	0.946
0.105	0.111	0.137	0.944	0.111	0.137	0.944	0.111	0.137	0.944	0.111	0.137	0.944
0.110	0.117	0.144	0.940	0.117	0.144	0.940	0.117	0.144	0.940	0.117	0.144	0.940
0.115	0.123	0.151	0.938	0.123	0.151	0.938	0.123	0.151	0.938	0.123	0.151	0.938
0.120	0.128	0.159	0.934	0.128	0.159	0.934	0.128	0.159	0.934	0.128	0.159	0.934
0.125	0.134	0.166	0.931	0.134	0.166	0.931	0.134	0.166	0.931	0.134	0.166	0.931
0.130	0.140	0.175	0.929	0.140	0.173	0.929	0.140	0.173	0.929	0.140	0.173	0.929
0.135	0.146	0.187	0.926	0.146	0.180	0.925	0.146	0.180	0.925	0.146	0.180	0.925
0.140	0.152	0.201	0.924	0.152	0.188	0.923	0.152	0.188	0.922	0.152	0.188	0.922
0.145	0.157	0.222	0.921	0.158	0.198	0.920	0.158	0.195	0.919	0.158	0.195	0.919
0.150	0.163	0.252	0.917	0.164	0.209	0.917	0.164	0.202	0.916	0.164	0.202	0.916
0.155	0.170	0.298	0.913	0.169	0.224	0.914	0.170	0.211	0.913	0.170	0.210	0.913
0.160	0.177	0.362	0.906	0.176	0.243	0.912	0.176	0.221	0.910	0.176	0.217	0.910
0.165	0.184	0.436	0.896	0.182	0.270	0.908	0.182	0.232	0.908	0.182	0.225	0.907
0.170	0.192	0.518	0.884	0.188	0.311	0.904	0.188	0.247	0.905	0.188	0.232	0.904
0.175				0.195	0.370	0.898	0.194	0.266	0.902	0.194	0.240	0.901
0.180				0.202	0.443	0.890	0.200	0.292	0.899	0.201	0.248	0.897
0.185				0.211	0.526	0.878	0.207	0.330	0.895	0.207	0.256	0.894
0.190							0.214	0.384	0.890	0.213	0.265	0.891
0.195							0.221	0.456	0.882	0.220	0.276	0.888
0.200							0.230	0.541	0.871	0.226	0.288	0.885
0.205										0.232	0.303	0.882
0.210										0.239	0.321	0.879
0.215										0.245	0.344	0.876
0.220										0.252	0.375	0.873
0.225										0.259	0.418	0.868
0.230										0.267	0.478	0.862
0.235										0.275	0.561	0.853
0.240												
0.245												
0.250												

Tab 5.38

TABLE 10_Concrete classes C55_Steel classes S400;S500;S600_b/b$_w$=8

Bending and axial force of simply reinforced T sections

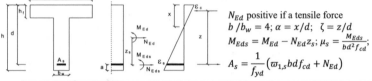

N_{Ed} positive if a tensile force
$b/b_w = 4; \alpha = x/d; \ \zeta = z/d$
$M_{Eds} = M_{Ed} - N_{Ed}z_s; \ \mu_s = \dfrac{M_{Eds}}{bd^2f_{cd}};$
$A_s = \dfrac{1}{f_{yd}}(\varpi_{1,s}bdf_{cd} + N_{Ed})$

	$h_f/d=0.08$			$h_f/d=0.10$			$h_f/d=0.12$			$h_f/d=0.14$		
μ_s	$\varpi_{1,s}$	α	ζ	$\varpi_{1,s}$	α	ζ	$\varpi_{1,s}$	α	ζ	$\varpi_{1,s}$	α	ζ
0.010	0.010	0.033	0.993	0.010	0.033	0.993	0.010	0.033	0.993	0.010	0.033	0.993
0.015	0.015	0.041	0.987	0.015	0.041	0.987	0.015	0.041	0.987	0.015	0.041	0.987
0.020	0.020	0.048	0.984	0.020	0.048	0.984	0.020	0.048	0.984	0.020	0.048	0.984
0.025	0.025	0.054	0.983	0.025	0.054	0.983	0.025	0.054	0.983	0.025	0.054	0.983
0.030	0.031	0.060	0.977	0.031	0.060	0.977	0.031	0.060	0.977	0.031	0.060	0.977
0.035	0.036	0.065	0.975	0.036	0.065	0.975	0.036	0.065	0.975	0.036	0.065	0.975
0.040	0.041	0.071	0.975	0.041	0.071	0.975	0.041	0.071	0.975	0.041	0.071	0.975
0.045	0.046	0.076	0.972	0.046	0.076	0.972	0.046	0.076	0.972	0.046	0.076	0.972
0.050	0.052	0.081	0.970	0.052	0.081	0.970	0.052	0.081	0.970	0.052	0.081	0.970
0.055	0.057	0.087	0.968	0.057	0.086	0.968	0.057	0.086	0.968	0.057	0.086	0.968
0.060	0.062	0.093	0.966	0.062	0.091	0.965	0.062	0.091	0.965	0.062	0.091	0.965
0.065	0.067	0.100	0.964	0.067	0.096	0.964	0.067	0.096	0.964	0.067	0.096	0.964
0.070	0.073	0.108	0.962	0.073	0.102	0.961	0.073	0.102	0.961	0.073	0.102	0.961
0.075	0.078	0.128	0.960	0.078	0.107	0.958	0.078	0.107	0.959	0.078	0.107	0.959
0.080	0.084	0.160	0.957	0.084	0.116	0.956	0.084	0.113	0.955	0.084	0.113	0.955
0.085	0.089	0.211	0.951	0.089	0.129	0.954	0.089	0.120	0.952	0.089	0.120	0.952
0.090	0.096	0.275	0.942	0.095	0.146	0.952	0.095	0.128	0.951	0.095	0.128	0.950
0.095	0.102	0.346	0.930	0.100	0.172	0.949	0.100	0.139	0.948	0.100	0.135	0.948
0.100	0.109	0.424	0.915	0.106	0.210	0.945	0.106	0.151	0.945	0.106	0.143	0.944
0.105				0.112	0.265	0.939	0.111	0.168	0.943	0.112	0.152	0.941
0.110				0.118	0.333	0.929	0.117	0.191	0.940	0.117	0.162	0.939
0.115				0.125	0.409	0.917	0.123	0.222	0.937	0.123	0.175	0.936
0.120				0.133	0.492	0.901	0.129	0.267	0.932	0.129	0.191	0.934
0.125							0.135	0.327	0.925	0.134	0.212	0.931
0.130							0.142	0.400	0.915	0.140	0.240	0.928
0.135							0.150	0.482	0.902	0.146	0.279	0.924
0.140										0.152	0.331	0.919
0.145										0.159	0.398	0.910
0.150										0.167	0.478	0.899
0.155												
0.160												
0.165												
0.170												
0.175												
0.180												
0.185												
0.190												
0.195												
0.200												
0.205												
0.210												
0.215												
0.220												
0.225												
0.230												
0.235												
0.240												
0.245												
0.250												

TABLE 10_Concrete classes C55_Steel classes S400;S500;S600_b/bw=8 (cont.)

Bending and axial force of simply reinforced T sections

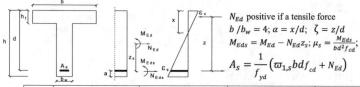

N_{Ed} positive if a tensile force

$b/b_w = 4$; $\alpha = x/d$; $\zeta = z/d$

$M_{Eds} = M_{Ed} - N_{Ed}z_s$; $\mu_s = \dfrac{M_{Eds}}{bd^2f_{cd}}$;

$$A_s = \frac{1}{f_{yd}}\left(\varpi_{1,s}bdf_{cd} + N_{Ed}\right)$$

μ_s	$h_f/d=0.16$			$h_f/d=0.18$			$h_f/d=0.20$			$h_f/d=0.25$		
	$\varpi_{1,s}$	α	ζ	$\varpi_{1,s}$	α	ζ	$\varpi_{1,s}$	α	ζ	$\varpi_{1,s}$	α	ζ
0.010	0.010	0.033	0.993	0.010	0.033	0.993	0.010	0.033	0.993	0.010	0.033	0.993
0.015	0.015	0.041	0.987	0.015	0.041	0.987	0.015	0.041	0.987	0.015	0.041	0.987
0.020	0.020	0.048	0.984	0.020	0.048	0.984	0.020	0.048	0.984	0.020	0.048	0.984
0.025	0.025	0.054	0.983	0.025	0.054	0.983	0.025	0.054	0.983	0.025	0.054	0.983
0.030	0.031	0.060	0.977	0.031	0.060	0.977	0.031	0.060	0.977	0.031	0.060	0.977
0.035	0.036	0.065	0.975	0.036	0.065	0.975	0.036	0.065	0.975	0.036	0.065	0.975
0.040	0.041	0.071	0.975	0.041	0.071	0.975	0.041	0.071	0.975	0.041	0.071	0.975
0.045	0.046	0.076	0.972	0.046	0.076	0.972	0.046	0.076	0.972	0.046	0.076	0.972
0.050	0.052	0.081	0.970	0.052	0.081	0.970	0.052	0.081	0.970	0.052	0.081	0.970
0.055	0.057	0.086	0.968	0.057	0.086	0.968	0.057	0.086	0.968	0.057	0.086	0.968
0.060	0.062	0.091	0.965	0.062	0.091	0.965	0.062	0.091	0.965	0.062	0.091	0.965
0.065	0.067	0.096	0.964	0.067	0.096	0.964	0.067	0.096	0.964	0.067	0.096	0.964
0.070	0.073	0.102	0.961	0.073	0.102	0.961	0.073	0.102	0.961	0.073	0.102	0.961
0.075	0.078	0.107	0.959	0.078	0.107	0.959	0.078	0.107	0.959	0.078	0.107	0.959
0.080	0.084	0.113	0.955	0.084	0.113	0.955	0.084	0.113	0.955	0.084	0.113	0.955
0.085	0.089	0.120	0.952	0.089	0.120	0.952	0.089	0.120	0.952	0.089	0.120	0.952
0.090	0.095	0.128	0.950	0.095	0.128	0.950	0.095	0.128	0.950	0.095	0.128	0.950
0.095	0.100	0.135	0.948	0.100	0.135	0.948	0.100	0.135	0.948	0.100	0.135	0.948
0.100	0.106	0.143	0.945	0.106	0.143	0.945	0.106	0.143	0.945	0.106	0.143	0.945
0.105	0.112	0.150	0.941	0.112	0.150	0.941	0.112	0.150	0.941	0.112	0.150	0.941
0.110	0.117	0.158	0.938	0.117	0.158	0.938	0.117	0.158	0.938	0.117	0.158	0.938
0.115	0.123	0.166	0.935	0.123	0.166	0.935	0.123	0.166	0.935	0.123	0.166	0.935
0.120	0.129	0.175	0.932	0.129	0.173	0.932	0.129	0.173	0.932	0.129	0.173	0.932
0.125	0.134	0.186	0.930	0.135	0.181	0.929	0.135	0.181	0.929	0.135	0.181	0.929
0.130	0.140	0.199	0.927	0.140	0.190	0.926	0.140	0.189	0.926	0.140	0.189	0.926
0.135	0.146	0.216	0.925	0.146	0.200	0.923	0.146	0.197	0.923	0.146	0.197	0.923
0.140	0.152	0.236	0.922	0.152	0.211	0.921	0.152	0.205	0.920	0.152	0.205	0.920
0.145	0.158	0.263	0.919	0.158	0.225	0.918	0.158	0.215	0.917	0.158	0.213	0.917
0.150	0.164	0.297	0.915	0.164	0.242	0.915	0.164	0.225	0.914	0.164	0.221	0.913
0.155	0.170	0.344	0.911	0.170	0.262	0.913	0.170	0.237	0.911	0.170	0.230	0.910
0.160	0.177	0.405	0.904	0.176	0.287	0.910	0.176	0.252	0.909	0.176	0.238	0.907
0.165	0.185	0.481	0.894	0.182	0.320	0.906	0.182	0.269	0.906	0.183	0.246	0.904
0.170				0.188	0.363	0.902	0.188	0.289	0.903	0.189	0.255	0.900
0.175				0.195	0.420	0.896	0.194	0.315	0.900	0.195	0.264	0.897
0.180				0.203	0.492	0.887	0.201	0.347	0.897	0.201	0.274	0.894
0.185							0.207	0.388	0.892	0.208	0.286	0.892
0.190							0.214	0.442	0.887	0.214	0.299	0.889
0.195							0.222	0.511	0.879	0.220	0.315	0.886
0.200										0.227	0.332	0.883
0.205										0.233	0.353	0.880
0.210										0.239	0.378	0.877
0.215										0.246	0.408	0.874
0.220										0.253	0.446	0.870
0.225										0.260	0.493	0.865
0.230												
0.235												
0.240												
0.245												
0.250												

Tab 5.39

TABLE 11_Concrete classes C12-C50_Steel classes S400;S500;S600_b/b$_w$=16

Bending and axial force of simply reinforced T sections

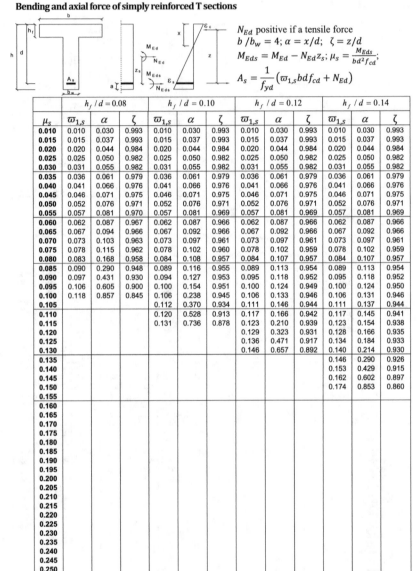

N_{Ed} positive if a tensile force
$b/b_w = 4$; $\alpha = x/d$; $\zeta = z/d$
$M_{Eds} = M_{Ed} - N_{Ed}z_s$; $\mu_s = \dfrac{M_{Eds}}{bd^2 f_{cd}}$;
$A_s = \dfrac{1}{f_{yd}}(\varpi_{1,s}bd f_{cd} + N_{Ed})$

	$h_f/d = 0.08$			$h_f/d = 0.10$			$h_f/d = 0.12$			$h_f/d = 0.14$		
μ_s	$\varpi_{1,s}$	α	ζ	$\varpi_{1,s}$	α	ζ	$\varpi_{1,s}$	α	ζ	$\varpi_{1,s}$	α	ζ
0.010	0.010	0.030	0.993	0.010	0.030	0.993	0.010	0.030	0.993	0.010	0.030	0.993
0.015	0.015	0.037	0.993	0.015	0.037	0.993	0.015	0.037	0.993	0.015	0.037	0.993
0.020	0.020	0.044	0.984	0.020	0.044	0.984	0.020	0.044	0.984	0.020	0.044	0.984
0.025	0.025	0.050	0.982	0.025	0.050	0.982	0.025	0.050	0.982	0.025	0.050	0.982
0.030	0.031	0.055	0.982	0.031	0.055	0.982	0.031	0.055	0.982	0.031	0.055	0.982
0.035	0.036	0.061	0.979	0.036	0.061	0.979	0.036	0.061	0.979	0.036	0.061	0.979
0.040	0.041	0.066	0.976	0.041	0.066	0.976	0.041	0.066	0.976	0.041	0.066	0.976
0.045	0.046	0.071	0.975	0.046	0.071	0.975	0.046	0.071	0.975	0.046	0.071	0.975
0.050	0.052	0.076	0.971	0.052	0.076	0.971	0.052	0.076	0.971	0.052	0.076	0.971
0.055	0.057	0.081	0.970	0.057	0.081	0.969	0.057	0.081	0.969	0.057	0.081	0.969
0.060	0.062	0.087	0.967	0.062	0.087	0.966	0.062	0.087	0.966	0.062	0.087	0.966
0.065	0.067	0.094	0.966	0.067	0.092	0.966	0.067	0.092	0.966	0.067	0.092	0.966
0.070	0.073	0.103	0.963	0.073	0.097	0.961	0.073	0.097	0.961	0.073	0.097	0.961
0.075	0.078	0.115	0.962	0.078	0.102	0.960	0.078	0.102	0.959	0.078	0.102	0.959
0.080	0.083	0.168	0.958	0.084	0.108	0.957	0.084	0.107	0.957	0.084	0.107	0.957
0.085	0.090	0.290	0.948	0.089	0.116	0.955	0.089	0.113	0.954	0.089	0.113	0.954
0.090	0.097	0.431	0.930	0.094	0.127	0.953	0.095	0.118	0.952	0.095	0.118	0.952
0.095	0.106	0.605	0.900	0.100	0.154	0.951	0.100	0.124	0.949	0.100	0.124	0.950
0.100	0.118	0.857	0.845	0.106	0.238	0.945	0.106	0.133	0.946	0.106	0.131	0.946
0.105				0.112	0.370	0.934	0.111	0.146	0.944	0.111	0.137	0.944
0.110				0.120	0.528	0.913	0.117	0.166	0.942	0.117	0.145	0.941
0.115				0.131	0.736	0.878	0.123	0.210	0.939	0.123	0.154	0.938
0.120							0.129	0.323	0.931	0.128	0.166	0.935
0.125							0.136	0.471	0.917	0.134	0.184	0.933
0.130							0.146	0.657	0.892	0.140	0.214	0.930
0.135										0.146	0.290	0.926
0.140										0.153	0.429	0.915
0.145										0.162	0.602	0.897
0.150										0.174	0.853	0.860
0.155												
0.160												
0.165												
0.170												
0.175												
0.180												
0.185												
0.190												
0.195												
0.200												
0.205												
0.210												
0.215												
0.220												
0.225												
0.230												
0.235												
0.240												
0.245												
0.250												

TABLE 11_Concrete classes C12-C50_Steel classes S400;S500;S600_b/b$_w$=16 (cont.)
Bending and axial force of simply reinforced T sections

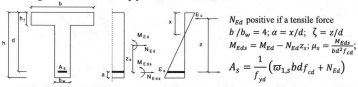

N_{Ed} positive if a tensile force

$b/b_w = 4$; $\alpha = x/d$; $\zeta = z/d$

$M_{Eds} = M_{Ed} - N_{Ed}z_s$; $\mu_s = \dfrac{M_{Eds}}{bd^2 f_{cd}}$;

$A_s = \dfrac{1}{f_{yd}}\left(\varpi_{1,s}bdf_{cd} + N_{Ed}\right)$

	$h_f/d = 0.16$			$h_f/d = 0.18$			$h_f/d = 0.20$			$h_f/d = 0.25$		
μ_s	$\varpi_{1,s}$	α	ζ	$\varpi_{1,s}$	α	ζ	$\varpi_{1,s}$	α	ζ	$\varpi_{1,s}$	α	ζ
0.010	0.010	0.030	0.993	0.010	0.030	0.993	0.010	0.030	0.993	0.010	0.030	0.993
0.015	0.015	0.037	0.993	0.015	0.037	0.993	0.015	0.037	0.993	0.015	0.037	0.993
0.020	0.020	0.044	0.984	0.020	0.044	0.984	0.020	0.044	0.984	0.020	0.044	0.984
0.025	0.025	0.050	0.982	0.025	0.050	0.982	0.025	0.050	0.982	0.025	0.050	0.982
0.030	0.031	0.055	0.982	0.031	0.055	0.982	0.031	0.055	0.982	0.031	0.055	0.982
0.035	0.036	0.061	0.979	0.036	0.061	0.979	0.036	0.061	0.979	0.036	0.061	0.979
0.040	0.041	0.066	0.976	0.041	0.066	0.976	0.041	0.066	0.976	0.041	0.066	0.976
0.045	0.046	0.071	0.975	0.046	0.071	0.975	0.046	0.071	0.975	0.046	0.071	0.975
0.050	0.052	0.076	0.971	0.052	0.076	0.971	0.052	0.076	0.971	0.052	0.076	0.971
0.055	0.057	0.081	0.969	0.057	0.081	0.969	0.057	0.081	0.969	0.057	0.081	0.969
0.060	0.062	0.087	0.966	0.062	0.087	0.966	0.062	0.087	0.966	0.062	0.087	0.966
0.065	0.067	0.092	0.966	0.067	0.092	0.966	0.067	0.092	0.966	0.067	0.092	0.966
0.070	0.073	0.097	0.961	0.073	0.097	0.961	0.073	0.097	0.961	0.073	0.097	0.961
0.075	0.078	0.102	0.959	0.078	0.102	0.959	0.078	0.102	0.959	0.078	0.102	0.959
0.080	0.084	0.107	0.957	0.084	0.107	0.957	0.084	0.107	0.957	0.084	0.107	0.957
0.085	0.089	0.113	0.954	0.089	0.113	0.954	0.089	0.113	0.954	0.089	0.113	0.954
0.090	0.095	0.118	0.952	0.095	0.118	0.952	0.095	0.118	0.952	0.095	0.118	0.952
0.095	0.100	0.124	0.950	0.100	0.124	0.950	0.100	0.124	0.950	0.100	0.124	0.950
0.100	0.106	0.131	0.946	0.106	0.131	0.946	0.106	0.131	0.946	0.106	0.131	0.946
0.105	0.111	0.137	0.944	0.111	0.137	0.944	0.111	0.137	0.944	0.111	0.137	0.944
0.110	0.117	0.144	0.940	0.117	0.144	0.940	0.117	0.144	0.940	0.117	0.144	0.940
0.115	0.123	0.151	0.938	0.123	0.151	0.938	0.123	0.151	0.938	0.123	0.151	0.938
0.120	0.128	0.159	0.934	0.128	0.159	0.934	0.128	0.159	0.934	0.128	0.159	0.934
0.125	0.134	0.166	0.932	0.134	0.166	0.931	0.134	0.166	0.931	0.134	0.166	0.931
0.130	0.140	0.176	0.929	0.140	0.173	0.929	0.140	0.173	0.929	0.140	0.173	0.929
0.135	0.146	0.188	0.926	0.146	0.180	0.925	0.146	0.180	0.925	0.146	0.180	0.925
0.140	0.152	0.204	0.924	0.152	0.188	0.922	0.152	0.188	0.922	0.152	0.188	0.922
0.145	0.157	0.230	0.921	0.158	0.198	0.920	0.158	0.195	0.919	0.158	0.195	0.919
0.150	0.163	0.281	0.918	0.164	0.210	0.917	0.164	0.202	0.916	0.164	0.202	0.916
0.155	0.170	0.401	0.910	0.169	0.226	0.915	0.170	0.211	0.913	0.170	0.210	0.913
0.160	0.179	0.566	0.896	0.175	0.250	0.912	0.176	0.221	0.911	0.176	0.217	0.910
0.165	0.190	0.794	0.868	0.182	0.291	0.909	0.182	0.233	0.908	0.182	0.225	0.907
0.170				0.188	0.385	0.903	0.188	0.250	0.905	0.188	0.232	0.904
0.175				0.196	0.546	0.891	0.194	0.273	0.902	0.194	0.240	0.901
0.180				0.207	0.763	0.868	0.200	0.310	0.899	0.201	0.248	0.897
0.185							0.207	0.386	0.895	0.207	0.256	0.894
0.190							0.215	0.539	0.885	0.213	0.265	0.891
0.195							0.226	0.753	0.864	0.220	0.276	0.888
0.200										0.226	0.289	0.885
0.205										0.232	0.305	0.882
0.210										0.239	0.326	0.880
0.215										0.245	0.355	0.877
0.220										0.252	0.402	0.873
0.225										0.259	0.498	0.867
0.230										0.269	0.687	0.855
0.235												
0.240												
0.245												
0.250												

Tab 5.40

TABLE 11_Concrete classes C55_Steel classes S400;S500;S600_b/b$_w$=16

Bending and axial force of simply reinforced T sections

N_{Ed} positive if a tensile force

$b/b_w = 4;\ \alpha = x/d;\ \zeta = z/d$

$M_{Eds} = M_{Ed} - N_{Ed}z_s;\ \mu_s = \dfrac{M_{Eds}}{bd^2 f_{cd}};$

$A_s = \dfrac{1}{f_{yd}}\left(\varpi_{1,s}bdf_{cd} + N_{Ed}\right)$

μ_s	$h_f/d=0.08$			$h_f/d=0.10$			$h_f/d=0.12$			$h_f/d=0.14$		
	$\varpi_{1,s}$	α	ζ	$\varpi_{1,s}$	α	ζ	$\varpi_{1,s}$	α	ζ	$\varpi_{1,s}$	α	ζ
0.010	0.010	0.033	0.993	0.010	0.033	0.993	0.010	0.033	0.993	0.010	0.033	0.993
0.015	0.015	0.041	0.987	0.015	0.041	0.987	0.015	0.041	0.987	0.015	0.041	0.987
0.020	0.020	0.048	0.984	0.020	0.048	0.984	0.020	0.048	0.984	0.020	0.048	0.984
0.025	0.025	0.054	0.983	0.025	0.054	0.983	0.025	0.054	0.983	0.025	0.054	0.983
0.030	0.031	0.060	0.977	0.031	0.060	0.977	0.031	0.060	0.977	0.031	0.060	0.977
0.035	0.036	0.065	0.975	0.036	0.065	0.975	0.036	0.065	0.975	0.036	0.065	0.975
0.040	0.041	0.071	0.975	0.041	0.071	0.975	0.041	0.071	0.975	0.041	0.071	0.975
0.045	0.046	0.076	0.972	0.046	0.076	0.972	0.046	0.076	0.972	0.046	0.076	0.972
0.050	0.052	0.081	0.970	0.052	0.081	0.970	0.052	0.081	0.970	0.052	0.081	0.970
0.055	0.057	0.087	0.968	0.057	0.086	0.968	0.057	0.086	0.968	0.057	0.086	0.968
0.060	0.062	0.093	0.966	0.062	0.091	0.965	0.062	0.091	0.965	0.062	0.091	0.965
0.065	0.067	0.100	0.965	0.067	0.096	0.964	0.067	0.096	0.964	0.067	0.096	0.964
0.070	0.073	0.109	0.963	0.073	0.102	0.961	0.073	0.102	0.961	0.073	0.102	0.961
0.075	0.078	0.135	0.961	0.078	0.107	0.959	0.078	0.107	0.959	0.078	0.107	0.959
0.080	0.084	0.194	0.957	0.084	0.116	0.956	0.084	0.113	0.955	0.084	0.113	0.955
0.085	0.090	0.319	0.947	0.089	0.130	0.954	0.089	0.120	0.952	0.089	0.120	0.952
0.090	0.097	0.476	0.927	0.095	0.151	0.952	0.095	0.129	0.950	0.095	0.128	0.950
0.095	0.106	0.672	0.895	0.100	0.188	0.949	0.100	0.139	0.948	0.100	0.135	0.948
0.100				0.106	0.266	0.944	0.106	0.153	0.945	0.106	0.143	0.944
0.105				0.113	0.408	0.932	0.111	0.172	0.943	0.112	0.152	0.942
0.110				0.121	0.584	0.910	0.117	0.201	0.941	0.117	0.162	0.939
0.115							0.123	0.253	0.937	0.123	0.176	0.937
0.120							0.129	0.357	0.930	0.128	0.195	0.934
0.125							0.137	0.520	0.915	0.134	0.221	0.932
0.130							0.146	0.732	0.888	0.140	0.261	0.929
0.135										0.146	0.336	0.924
0.140										0.153	0.474	0.914
0.145										0.162	0.669	0.894
0.150												
0.155												
0.160												
0.165												
0.170												
0.175												
0.180												
0.185												
0.190												
0.195												
0.200												
0.205												
0.210												
0.215												
0.220												
0.225												
0.230												
0.235												
0.240												
0.245												
0.250												

TABLE 11_Concrete classes C55_Steel classes S400;S500;S600_b/b$_w$=16 (cont.)
Bending and axial force of simply reinforced T sections

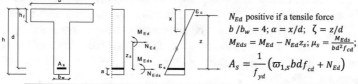

N_{Ed} positive if a tensile force

$b/b_w = 4;\ \alpha = x/d;\ \zeta = z/d$

$M_{Eds} = M_{Ed} - N_{Ed}z_s;\ \mu_s = \dfrac{M_{Eds}}{bd^2 f_{cd}};$

$A_s = \dfrac{1}{f_{yd}}\left(\varpi_{1,s}bd f_{cd} + N_{Ed}\right)$

μ_s	$h_f/d=0.16$			$h_f/d=0.18$			$h_f/d=0.20$			$h_f/d=0.25$		
	$\varpi_{1,s}$	α	ζ	$\varpi_{1,s}$	α	ζ	$\varpi_{1,s}$	α	ζ	$\varpi_{1,s}$	α	ζ
0.010	0.010	0.033	0.993	0.010	0.033	0.993	0.010	0.033	0.993	0.010	0.033	0.993
0.015	0.015	0.041	0.987	0.015	0.041	0.987	0.015	0.041	0.987	0.015	0.041	0.987
0.020	0.020	0.048	0.984	0.020	0.048	0.984	0.020	0.048	0.984	0.020	0.048	0.984
0.025	0.025	0.054	0.983	0.025	0.054	0.983	0.025	0.054	0.983	0.025	0.054	0.983
0.030	0.031	0.060	0.977	0.031	0.060	0.977	0.031	0.060	0.977	0.031	0.060	0.977
0.035	0.036	0.065	0.975	0.036	0.065	0.975	0.036	0.065	0.975	0.036	0.065	0.975
0.040	0.041	0.071	0.975	0.041	0.071	0.975	0.041	0.071	0.975	0.041	0.071	0.975
0.045	0.046	0.076	0.972	0.046	0.076	0.972	0.046	0.076	0.972	0.046	0.076	0.972
0.050	0.052	0.081	0.970	0.052	0.081	0.970	0.052	0.081	0.970	0.052	0.081	0.970
0.055	0.057	0.086	0.968	0.057	0.086	0.968	0.057	0.086	0.968	0.057	0.086	0.968
0.060	0.062	0.091	0.965	0.062	0.091	0.965	0.062	0.091	0.965	0.062	0.091	0.965
0.065	0.067	0.096	0.964	0.067	0.096	0.964	0.067	0.096	0.964	0.067	0.096	0.964
0.070	0.073	0.102	0.961	0.073	0.102	0.961	0.073	0.102	0.961	0.073	0.102	0.961
0.075	0.078	0.107	0.959	0.078	0.107	0.959	0.078	0.107	0.959	0.078	0.107	0.959
0.080	0.084	0.113	0.955	0.084	0.113	0.955	0.084	0.113	0.955	0.084	0.113	0.955
0.085	0.089	0.120	0.952	0.089	0.120	0.952	0.089	0.120	0.952	0.089	0.120	0.952
0.090	0.095	0.128	0.950	0.095	0.128	0.950	0.095	0.128	0.950	0.095	0.128	0.950
0.095	0.100	0.135	0.948	0.100	0.135	0.948	0.100	0.135	0.948	0.100	0.135	0.948
0.100	0.106	0.143	0.945	0.106	0.143	0.945	0.106	0.143	0.945	0.106	0.143	0.945
0.105	0.112	0.150	0.941	0.112	0.150	0.941	0.112	0.150	0.941	0.112	0.150	0.941
0.110	0.117	0.158	0.938	0.117	0.158	0.938	0.117	0.158	0.938	0.117	0.158	0.938
0.115	0.123	0.166	0.935	0.123	0.166	0.935	0.123	0.166	0.935	0.123	0.166	0.935
0.120	0.129	0.176	0.932	0.129	0.173	0.932	0.129	0.173	0.932	0.129	0.173	0.932
0.125	0.134	0.187	0.930	0.135	0.181	0.929	0.135	0.181	0.929	0.135	0.181	0.929
0.130	0.140	0.201	0.927	0.140	0.190	0.926	0.140	0.189	0.926	0.140	0.189	0.926
0.135	0.146	0.219	0.925	0.146	0.200	0.924	0.146	0.197	0.923	0.146	0.197	0.923
0.140	0.152	0.244	0.923	0.152	0.212	0.921	0.152	0.205	0.920	0.152	0.205	0.920
0.145	0.158	0.280	0.920	0.158	0.227	0.918	0.158	0.215	0.917	0.158	0.213	0.917
0.150	0.164	0.339	0.916	0.164	0.245	0.916	0.164	0.225	0.914	0.164	0.221	0.913
0.155	0.171	0.447	0.909	0.170	0.269	0.913	0.170	0.238	0.911	0.170	0.230	0.910
0.160	0.179	0.628	0.893	0.176	0.303	0.911	0.176	0.254	0.909	0.176	0.238	0.907
0.165				0.182	0.355	0.907	0.182	0.272	0.906	0.183	0.246	0.904
0.170				0.189	0.444	0.901	0.188	0.297	0.904	0.189	0.255	0.900
0.175				0.197	0.605	0.889	0.194	0.330	0.901	0.195	0.264	0.897
0.180							0.201	0.379	0.898	0.201	0.275	0.894
0.185							0.207	0.457	0.893	0.208	0.287	0.892
0.190							0.215	0.599	0.883	0.214	0.301	0.889
0.195										0.220	0.317	0.886
0.200										0.226	0.337	0.884
0.205										0.233	0.361	0.881
0.210										0.239	0.392	0.878
0.215										0.246	0.433	0.875
0.220										0.253	0.493	0.871
0.225										0.260	0.589	0.865
0.230										0.270	0.769	0.852
0.235												
0.240												
0.245												
0.250												

5.6   Summary

The tables presented herein have the purpose of designing the reinforcement of rect-angular or T–sections of beams or columns. Besides satisfying the failure conditions, the tables allow the quick search for the most economical solutions, as well as the verification of the ductility conditions. The advantages of the tables are the preci-sion of the results withal the informations about the ductility behaviour and the ease of calculation for usual geometrical ratios and material properties. Tables for high strength concrete are also included.

Chapter 6
Design Charts

6.1 Introduction

This Chapter compiles a set of charts to be used in the ultimate limit state design of reinforced concrete sections according to Eurocode 2 recommendations[2]. The design charts are applicable to the safety verification and design of reinforced concrete rectangular and circular cross-sections. Dedicated charts are provided for both uniaxial and biaxial bending moment with axial force. Concrete classes from C12 to C50 and the steel strength class S500 are considered.

Several reinforcement layouts are considered covering the current situations of columns in concrete building structures. For uniaxial bending in rectangular sections, symmetric ($A'/A = 1.0$) and non-symmetric ($A'/A = 0.5$) reinforcement in two opposite faces are included. Uniformely distributed reinforcement along the contour is considered for circular sections. For biaxial bending of rectangular sections, five cases of reinforcement distribution along the section contour are available, namely: (i) at section corners; (ii) uniformly distributed along the contour; (iii) 1/3 of the total reinforcement in each of two opposite sides and 1/6 in each of the others sides; (iv) 4/10 of the total reinforcement in each of two opposite sides and 1/10 in each of the others sides; and (v) 1/2 of the total reinforcement in each of two opposite sides.

The charts are presented in figures and designated as in the following examples:

- Figure 6.2 *CHART 1_C12-C50_A'/A = 1_a/h = 0.10* is the figure containing the Chart 1 computed for concrete classes C12-C50 and steel class S500, with symmetric reinforcement (A'/A = 1.0) and concrete cover ratio a/h = 0.10;
- Figure 6.17 *CHART 8_C12-C50 S500_opposite faces_a1/h = c1/b = 0.10*, meaning that this chart is applied to concrete classes from C12 to C50, steel class S500, reinforcing steel placed in two opposite faces and concrete cover ratio a/h = 0.10.

To streamline the choice to which chart to consider, the schedule of the charts content is summarized in the Table 6.1.

Table 6.1 Design charts

Material	Force	Section	Reinforcement Chart	Concrete cover a/h, a/(2r) or $a_1/h = c_1/b$	Figure
Concrete classes C12-C50 and steel class S500	Bending and Axial force	Rectangular	Symmetric: $A'/A = 1.0$ Chart 1	a/h=0.10	Fig. 6.2
				a/h=0.15	Fig. 6.3
			Non-symmetric: $A'/A = 0.5$ Chart 2	a/h=0.10	Fig. 6.4
				a/h=0.15	Fig. 6.5
		Circular	Distributed along the contour Chart 3	a/(2r)=0.10	Fig. 6.6
				a/(2r)=0.15	Fig. 6.7
	Biaxial bending and axial force	Rectangular	At section corners Chart 4	$a_1/h = c_1/b = 0.10$	Fig. 6.9
				$a_1/h = c_1/b = 0.15$	Fig. 6.10
			Distributed along the contour Chart 5	$a_1/h = c_1/b = 0.10$	Fig. 6.11
				$a_1/h = c_1/b = 0.15$	Fig. 6.12
			Ratios: 1/3; 1/6 Chart 6	$a_1/h = c_1/b = 0.10$	Fig. 6.13
				$a_1/h = c_1/b = 0.15$	Fig. 6.14
			Ratios: 0.4; 0.1 Chart 7	$a_1/h = c_1/b = 0.10$	Fig. 6.15
				$a_1/h = c_1/b = 0.15$	Fig. 6.16
			In two opposite faces Chart 8	$a_1/h = c_1/b = 0.10$	Fig. 6.17
				$a_1/h = c_1/b = 0.15$	Fig. 6.18

6.2 Bending and Axial Force of Rectangular and Circular Sections of Concrete Classes C12_C50 and Steel Class S500

Purpose of the Charts in this Section

The objective of this section is to design rectangular and circular cross-sections under bending and axial force to evaluate the renforcing steel area for the considered layout by using the design charts.

How to use the Charts in this Section

The example 3 of Chap. 4 shows the application of charts n° 1 and n° 2 to a rectangular section. The design of a circular section using Chart 3 is presented in example n° 9. The use of these charts is schematically illustrated in Fig. 6.1. Each line of the chart represents a vertical section of the interaction surface described in Fig. 2.4 and corresponds to a different mechanical reinforcement ratio, ϖ. There are two possible concrete covers (a/d = 0.10 and a/d = 0.15) allowing interpolation for intermediate values.

Cautions when using the Charts of this Section

The bending moment and axial force are considered at the centroid of the section and compression axial force is taken as positive. The charts are only valid for the indicated reinforcement layout.

Fig. 6.1 Method for determining the required mechanical reinforcement ratio ϖ in a cross-section subjected to a non-dimensional axial force $v = 0.75$ and non-dimensional bending moment $\mu = 0.20$

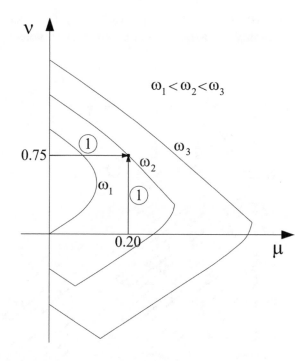

6.2.1 Rectangular Sections with Symmetric Reinforcement

See Fig. 6.2

See Fig. 6.3

6.2.2 Rectangular Sections with Non-Symmetric Reinforcement

See Fig. 6.4

See Fig. 6.5

6.2.3 Circular Sections with Distributed Reinforcement in the Contour

See Fig. 6.6

See Fig. 6.7

CHART 1_Concrete classes C12-C50_Steel class S500_A′/A=1.0_a/h=0.10
Bending and axial force of rectangular sections with symmetric reinforcement

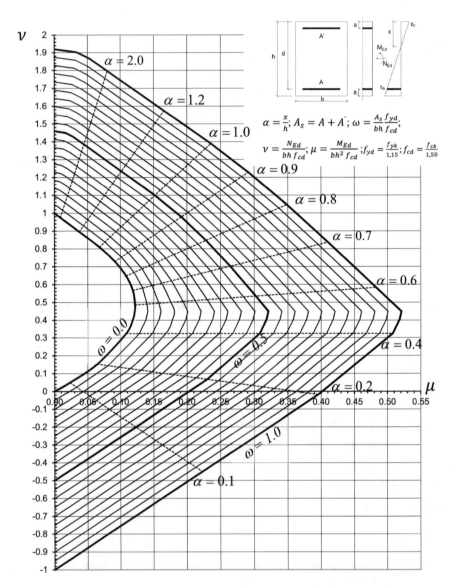

Fig. 6.2 CHART 1_C12-C50_S500_A′/A = 1.0_a/h = 0.10

CHART 1_Concrete classes C12-C50_Steel class S500_A'/A=1.0_a/h=0.15
Bending and axial force of rectangular sections with symmetric reinforcement

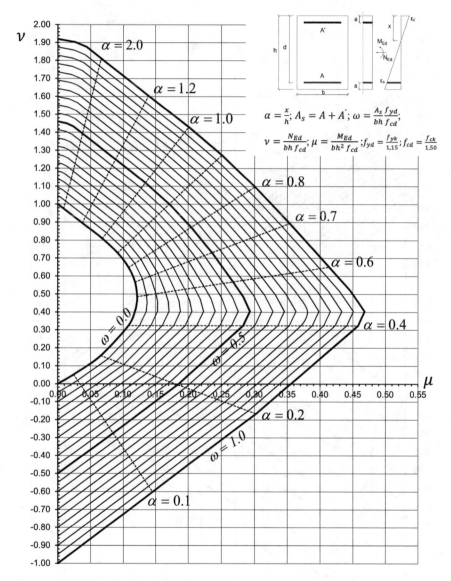

Fig. 6.3 CHART 1_C12-C50_S500_A'/A = 1.0_a/h = 0.15

CHART 2_Concrete classes C12-C50_Steel class S500_A'/A=0.5_a/h=0.10
Bending and axial force of rectangular sections with non-symmetric reinforcement

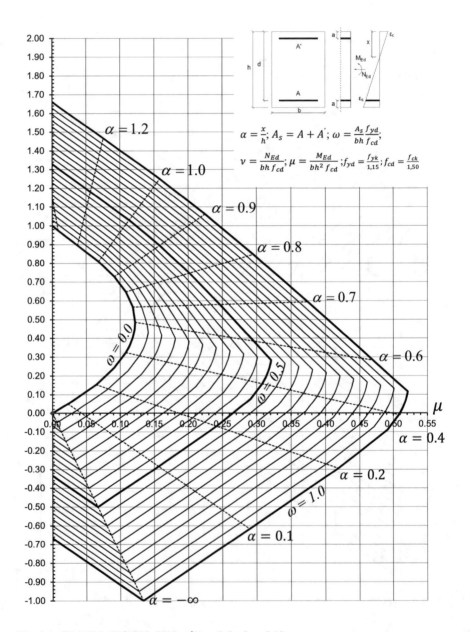

Fig. 6.4 CHART 2_C12-C50_S500_A'/A = 0.5_a/h = 0.10

CHART 2_Concrete classes C12-C50_Steel class S500_A′/A=0.5_a/h=0.15

Bending and axial force of rectangular sections with non-symmetric reinforcement

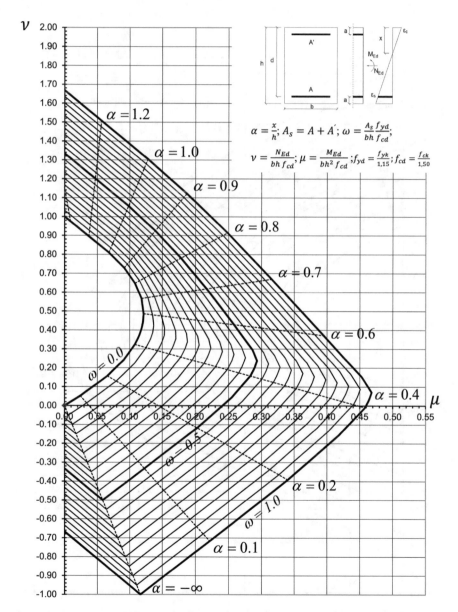

Fig. 6.5 CHART 2_C12-C50_S500_A′/A = 0.5_a/h = 0.15

CHART 3_Concrete classes C12-C50_Steel class S500_a/(2r)=0.10
Bending and axial force of circular sections with distributed reinforcement in the contour

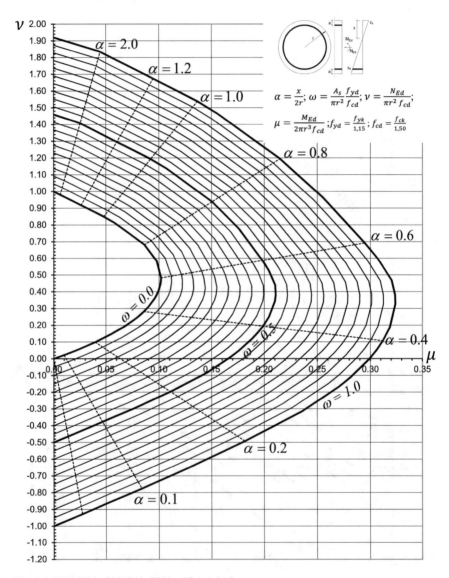

Fig. 6.6 CHART 3_C12-C50_S500_a/(2r) = 0.10

CHART 3_Concrete classes C12-C50_Steel class S500_a/(2r)=0.15

Bending and axial force of circular sections with distributed reinforcement in the contour

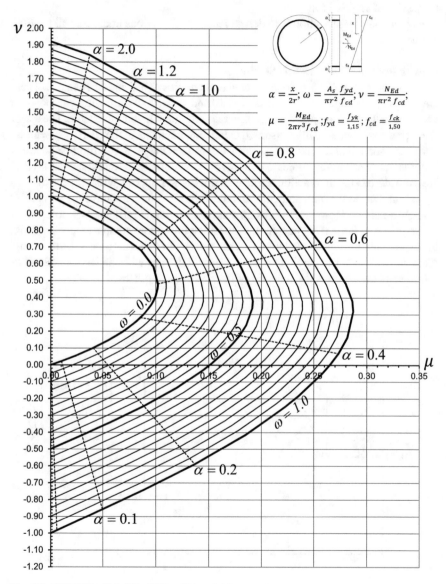

Fig. 6.7 CHART 3_C12-C50_S500_a/(2r) = 0.15

6.3   Biaxial Bending and Axial Force of Rectangular Sections of Concrete Classes C12-C50 and Steel Class S500

Objective of the Section

The objective of this section is to design rectangular sections under biaxial bending moment and axial force.

How to use the Charts in this Section

Applications of these charts are presented in Examples n° 6, 7 and 8 of Chap. 4. There are two possible concrete covers (a/d = 0.10 and a/d = 0.15), allowing interpolation for intermediate values, and five different reinforcement layouts available. The use of these charts is schematically illustrated in Fig. 6.8. Each line of the chart represents an horizontal section of the interaction surface described in Fig. 2.4 and corresponds to a different mechanical reinforcement ratio, ϖ. As symmetrical reinforcement layouts are considered, the interaction surface is doubly symmetric with respect to both $\mu_x = 0$ and $\mu_y = 0$ planes and the horizontal intersections are completely defined by a single quadrant. However, the design charts are divided in octants, providing eight intersections of the surface per chart. While allowing a more compact representation, the charts only provide the solution for the case in which the larger non-dimensional moment (μ_y) vector is parallel to the cross-section faces with the largest amount of reinforcement, which is the scenario of interest in most practical situations.

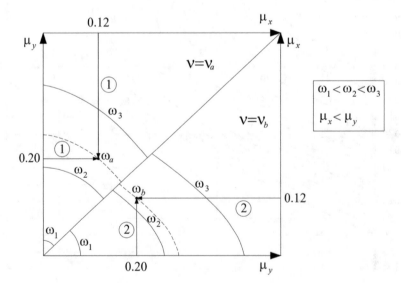

Fig. 6.8 Method for determining the required mechanical reinforcement ratio ϖ in a cross-section subjected to a non-dimensional axial force $\nu (\nu_a \leq \nu \leq \nu_b)$ and non-dimensional bending moments $\mu_y = 0.20$ and $\mu_x = 0.12$

Cautions when using the Charts of this Section

The bending moment, projected in the two principal axes −x and y, and the axial force are considered at the centroid of the section. The compression axial force is taken as positive. The charts are only valid for the indicated reinforcement layout. Note that the maximum non-dimensional bending moment vector, chosen as μ_y, is parallel to the faces with the larger amount of reinforcement.

6.3.1 Reinforcement at the Section Corners

See Fig. 6.9.

See Fig. 6.10.

6.3.2 Reinforcement Distributed Along the Section Contour

See Fig. 6.11.

See Fig. 6.12.

CHART 4_Concrete classes C12-C50_Steel class S500_1/4As at section corner_$a_1/h = c_1/b = 0.10$
Biaxial bending and axial force of rectangular sections with reinforcement at section corners

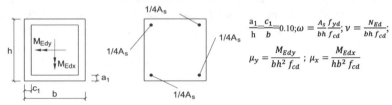

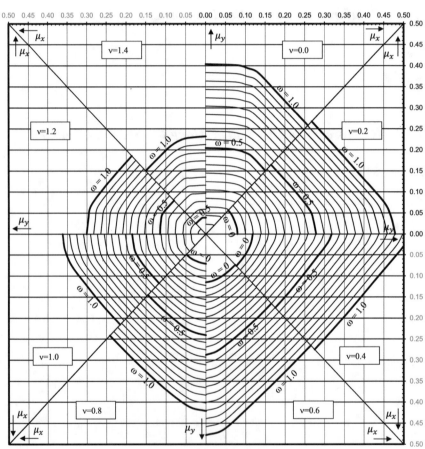

Fig. 6.9 CHART 4 C12-C50_S500_1/4A$_S$ at section corner_ $a_1/h = c_1/b = 0.10$

CHART 4_Concrete classes C12-C50_Steel class S500_1/4As at section corner_$a_1/h= c_1/b= 0.15$
Biaxial bending and axial force of rectangular sections with reinforcement at section corners

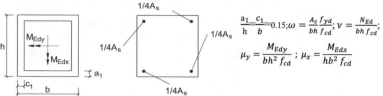

$$\frac{a_1}{h}\frac{c_1}{b}=0.15; \omega = \frac{A_s\,f_{yd}}{bh\,f_{cd}}; \nu = \frac{N_{Ed}}{bh\,f_{cd}};$$

$$\mu_y = \frac{M_{Edy}}{bh^2\,f_{cd}} \; ; \; \mu_x = \frac{M_{Edx}}{hb^2\,f_{cd}}$$

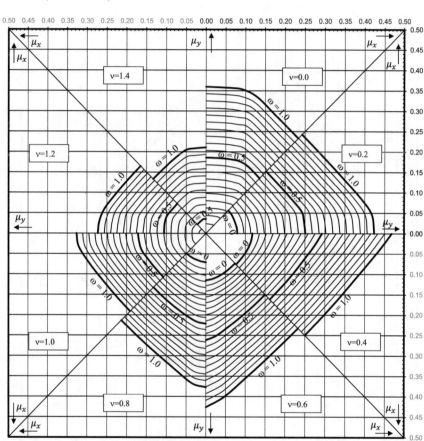

Fig. 6.10 CHART 4 C12-C50_S500_1/4A$_s$ at section corner_$a_1/h = c_1/b = 0.15$

CHART 5_Concrete classes C12-C50_Steel class S500_1/4As at side contour_$a_1/h = c_1/b = 0.10$
Biaxial bending and axial force of rectangular sections with reinforcement along the section contour

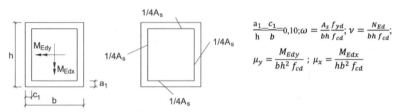

$$\frac{a_1}{h} = \frac{c_1}{b} = 0,10; \omega = \frac{A_s\, f_{yd}}{bh\, f_{cd}}; \nu = \frac{N_{Ed}}{bh\, f_{cd}};$$

$$\mu_y = \frac{M_{Edy}}{bh^2\, f_{cd}}\ ;\ \mu_x = \frac{M_{Edx}}{hb^2\, f_{cd}}$$

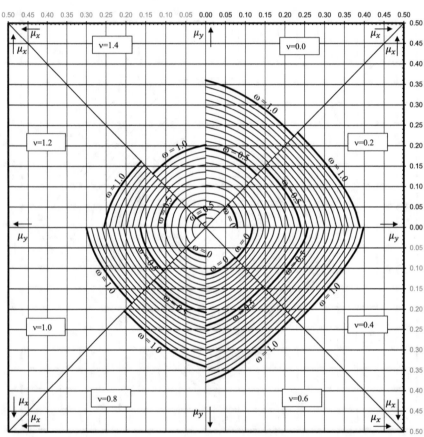

Fig. 6.11 CHART 5 C12-C50_S500_1/4A$_S$ at side contour_ $a_1/h = c_1/b = 0.10$

CHART 5_Concrete classes C12-C50_Steel class S500_1/4As at side contour_$a_1/h = c_1/b = 0.15$
Biaxial bending and axial force of rectangular sections with reinforcement along the section contour

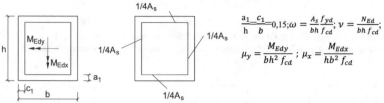

$$\frac{a_1}{h}=\frac{c_1}{b}=0.15; \omega = \frac{A_s\,f_{yd}}{bh\,f_{cd}}; \nu = \frac{N_{Ed}}{bh\,f_{cd}};$$

$$\mu_y = \frac{M_{Edy}}{bh^2\,f_{cd}}\;;\; \mu_x = \frac{M_{Edx}}{hb^2\,f_{cd}}$$

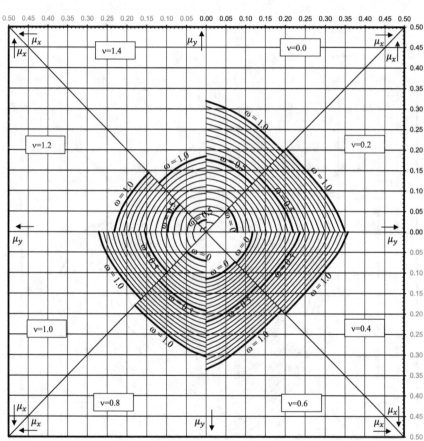

Fig. 6.12   CHART 5 C12-C50_S500_1/4A$_S$ at side contour_ $a_1/h = c_1/b = 0.15$

6.3.3 Reinforcement Ratios

CHART 6_Concrete classes C12-C50_Steel class S500_ As/3, As/6_a_1/h= c_1/b= 0.10
Biaxial bending and axial force of rectangular sections with contour reinforcement ratios 1/3, 1/6

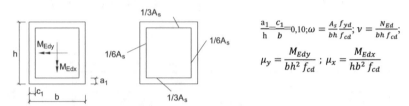

$$\frac{a_1}{h}=\frac{c_1}{b}=0,10; \omega = \frac{A_s}{bh}\frac{f_{yd}}{f_{cd}}; \nu = \frac{N_{Ed}}{bh\,f_{cd}};$$

$$\mu_y = \frac{M_{Edy}}{bh^2\,f_{cd}}\;;\;\mu_x = \frac{M_{Edx}}{hb^2\,f_{cd}}$$

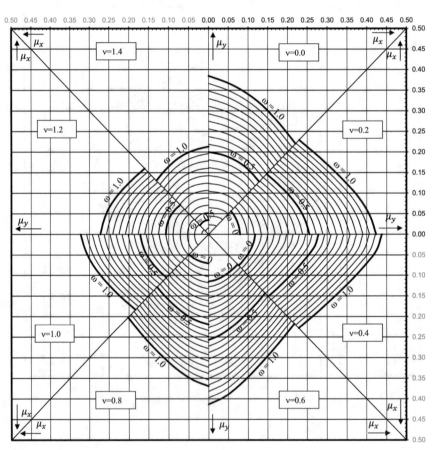

Fig. 6.13 CHART 6 C12-C50 S500_A_s/3, A_s/6_a_1/h $= c_1$/b $= 0.10$

CHART 6_Concrete classes C12-C50_Steel class S500_ As/3, As/6_a₁/h= c₁/b= 0.15
Biaxial bending and axial force of rectangular sections with contour reinforcement ratios 1/3, 1/6

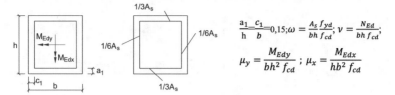

$$\frac{a_1}{h} \quad \frac{c_1}{b} = 0,15; \omega = \frac{A_s \, f_{yd}}{bh \, f_{cd}}; \nu = \frac{N_{Ed}}{bh \, f_{cd}};$$

$$\mu_y = \frac{M_{Edy}}{bh^2 f_{cd}} \; ; \; \mu_x = \frac{M_{Edx}}{hb^2 f_{cd}}$$

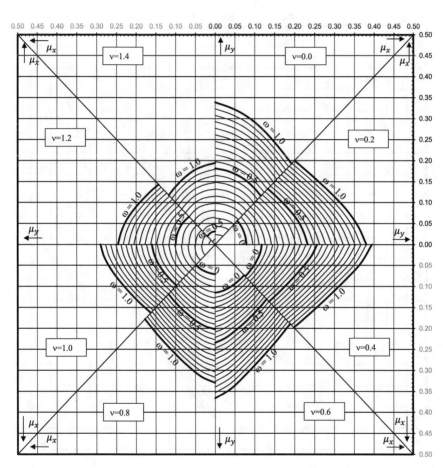

Fig. 6.14   CHART 6 C12-C50_S500_A_s/3, A_s/6_a_1/h $= c_1$/b $= 0.15$

CHART 7_Concrete classes C12-C50_Steel class S500_ 0.4As, 0.1As_a₁/h= c₁/b= 0.10
Biaxial bending and axial force of rectangular sections with contour reinforcement ratios 0.4, 0.1

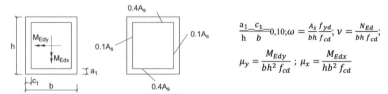

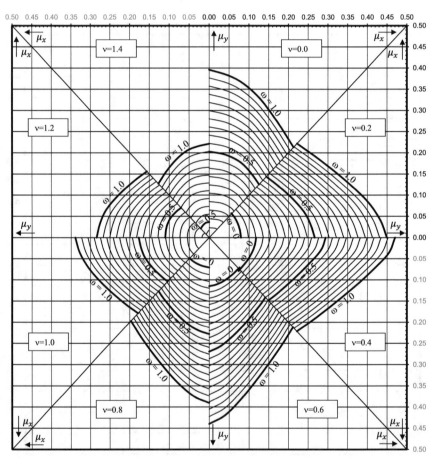

Fig. 6.15   CHART 7 C12-C50_S500_4A$_s$/10, A$_s$/10_a$_1$/h = c$_1$/b = 0.10

CHART 7_Concrete classes C12-C50_Steel class S500_ 0.4As, 0.1As_a_1/h= c_1/b= 0.15
Biaxial bending and axial force of rectangular sections with contour reinforcement ratios 0.4, 0.1

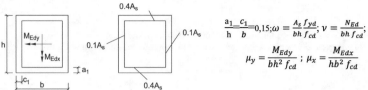

$$\frac{a_1}{h} = \frac{c_1}{b} = 0,15; \omega = \frac{A_s\, f_{yd}}{bh\, f_{cd}}; \nu = \frac{N_{Ed}}{bh\, f_{cd}};$$

$$\mu_y = \frac{M_{Edy}}{bh^2\, f_{cd}}; \mu_x = \frac{M_{Edx}}{hb^2\, f_{cd}}$$

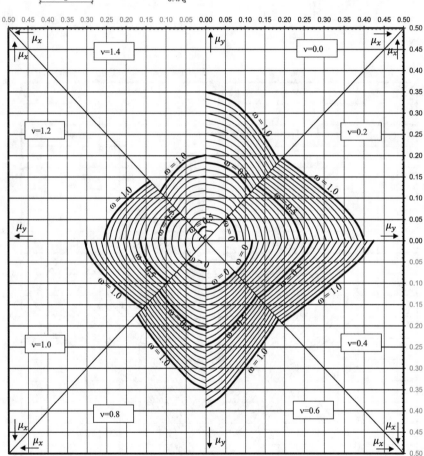

Fig. 6.16 CHART 7 C12-C50_S500_4A$_s$/10, A$_s$/10_a_1/h = c_1/b = 0.15

6.3.4   Reinforcement in Two Opposite Faces

CHART 8_Concrete classes C12-C50_Steel class S500_opposite faces_$a_1/h = c_1/b = 0.10$
Biaxial bending and axial force of rectangular sections with reinforcement in two opposite faces

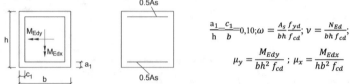

$$\frac{a_1}{h} = \frac{c_1}{b} = 0,10; \omega = \frac{A_s\,f_{yd}}{bh\,f_{cd}}; \nu = \frac{N_{Ed}}{bh\,f_{cd}};$$

$$\mu_y = \frac{M_{Edy}}{bh^2\,f_{cd}}\ ;\ \mu_x = \frac{M_{Edx}}{hb^2\,f_{cd}}$$

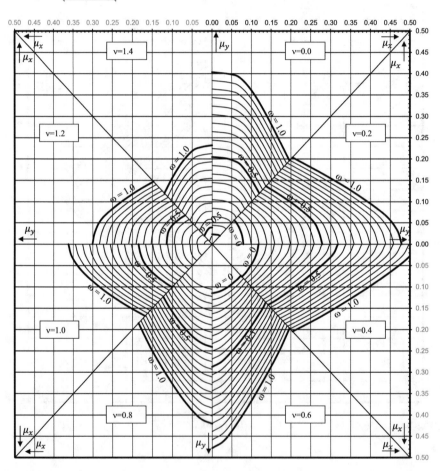

Fig. 6.17   CHART 8 C12-C50_S500_ opposite faces_$a_1/h = c_1/b = 0.10$

CHART 8_Concrete classes C12-C50_Steel class S500_opposite faces_$a_1/h = c_1/b = 0.15$
Biaxial bending and axial force of rectangular sections with reinforcement in two opposite faces

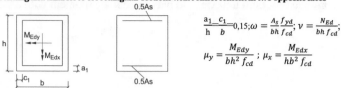

$$\frac{a_1}{h}\frac{c_1}{b} = 0{,}15; \omega = \frac{A_s}{bh}\frac{f_{yd}}{f_{cd}}; \nu = \frac{N_{Ed}}{bh\,f_{cd}};$$

$$\mu_y = \frac{M_{Edy}}{bh^2 f_{cd}} \; ; \; \mu_x = \frac{M_{Edx}}{hb^2 f_{cd}}$$

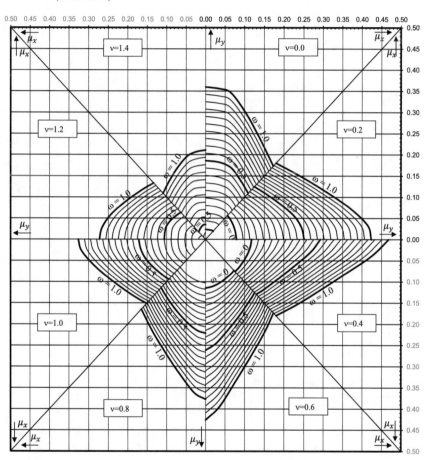

Fig. 6.18 CHART 8 C12-C50_S500_ opposite faces_$a_1/h = c_1/b = 0.15$

6.4 Summary

A set of charts to be used in the ultimate limit state design of reinforced concrete sections according to Eurocode 2 [2] are presented for rectangular and circular cross-sections with varying reinforcement layouts and concrete cover ratios. The charts are valid for the concrete strength classes C12 to C50 and steel class S500. Dedicated charts are provided for both uniaxial and biaxial bending moment with axial force.

Chapter 7
Diagrams for Verification of Service Limite States

7.1 Introduction

The diagrams presented herein are useful to verify the serviceability limit states of structural concrete elements, which require the determination of the maximum concrete compressive stresses and the of tensile stresses in the reinforcement steel. These stresses, and the corresponding depth of the neutral axis, are evaluated for cracked cross-sections subjected to bending moment and axial force. Linear elastic behaviour is assumed for steel and for compressed concrete. Each diagram provides results for varying tensile reinforcement ratios. The diagrams are available for simply or doubly reinforced rectangular sections, simply reinforced T-sections and circular sections with reinforcement uniformely distributed along the contour and for the concrete cover ratio a/d = 0.10.

The Diagrams 1 and 2 are applicable to the determination of the maximum stress in the concrete and in the steel reinforcement of simply and doubly reinforced rectangular sections, respectively, while the Diagrams 3 allow the evaluation of the reinforcing amount that corresponds to maximum compression stresses in concrete of $0.45 f_{ck}$ and $0.60 f_{ck}$ (limits adopted in EC2), in case of rectangular sections with symmetric reinforcement [5]. For the T-sections, two values of the flange-web ratio of b/bw = 4 and 8 (Diagrams 4 and 5, respectively) are available. The flange height/effective depth ratio varies from 0.10 to 0.25. Diagram 6 is applicable for circular sections with uniformly distributed reinforcement along the contour. To streamline the diagram to use, and the corresponding figure and page location, the scheme is summarized in Table 7.1.

Table 7.1 Design diagrams

Force	Materials	Section			Reinforcement DIAGRAM	Reinf. Ratio A'/A	Figure
Bending and Axial Force	Concrete class C12-C50 and Steel class S500	Rectangular			Simply DIAGRAM 1	0.0	Figure 7.2
					Doubly DIAGRAM 2	0.2	Figure 7.3
						0.4	Figure 7.4
						0.6	Figure 7.5
						0.8	Figure 7.6
						1.0	Figure 7.7
					Symmetric DIAGRAM 3	1.0	Figure 7.8
						1.0	Figure 7.9
		T-section	DIAGRAM 4 $b/b_w=4$	$h_f/d=0.10$	Simply DIAGRAM 4	0.0	Figure 7.10
				$h_f/d=0.12$			Figure 7.11
				$h_f/d=0.14$			Figure 7.12
				$h_f/d=0.16$			Figure 7.13
				$h_f/d=0.18$			Figure 7.14
				$h_f/d=0.20$			Figure 7.15
				$h_f/d=0.25$			Figure 7.16
			DIAGRAM 5 $b/b_w=8$	$h_f/d=0.10$	Simply DIAGRAM 5		Figure 7.17
				$h_f/d=0.12$			Figure 7.18
				$h_f/d=0.14$			Figure 7.19
				$h_f/d=0.16$			Figure 7.20
				$h_f/d=0.18$			Figure 7.21
				$h_f/d=0.20$			Figure 7.23
				$h_f/d=0.25$			Figure 7.24
		Circular			In contour DIAGRAM 6	-	Figure 7.25

7.2   Bending and Axial Force of Rectangular Cross Sections

Purpose of the Diagrams in this Section

The Diagrams 1 and 2 provided in this section allow determining the stresses in the concrete and in the steel that are required for the serviceability limit state safety verifications of rectangular sections. Diagram 3 applies a different procedure to verify the limits of compressive concrete stresses indicated in EC2 ($\sigma_{c,max} = 0.45 f_{ck}$ and $\sigma_{c,max} = 0.60 f_{ck}$), by using interaction diagrams defined according those limits.

How to use the Diagrams in this Section

The application of these diagrams is shown in the Examples 10 and 12 of Chap. 4. The use of these diagrams follows the procedure illustrated in Fig. 7.1 and can be summarized as follows: (1) for a given reinforcement ratio calculate $\alpha_e \rho$ and choose the convenient curve ($\alpha_e \rho = 0.04$ in the example); (2) find the intersection of the curve with the horizontal line defined by the ordinate $N_{Ed}d/M_{Ed}$ ($N_{Ed}d/M_{Ed} = -0.30$ in the example); (3) the vertical line allows finding the relative depth of the neutral axis ($\alpha = 0.28$ in the example); (4) following the vertical alignment find the intersection with curve of the second diagram; (5) the horizontal line allows finding the C_c

value in the vertical axis ($C_c = 8.0$ in the example); (6) following once again the vertical alignment towards the third diagram find the intersection with the curve corresponding to the $\alpha_e\rho$; (7) the horizontal line allows determining the product $C_s\rho$ ($C_s\rho = 0.85$ in the example). With C_c and C_s the concrete stresses and the steel stresses can be determined.

Cautions when using the Diagrams in this Section

The bending moment is evaluated at the level of the tensile reinforcing steel using the formula provided in the top right corner. The compressive axial force is considered to be negative.

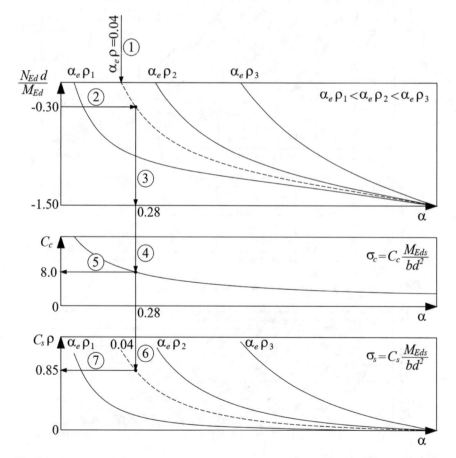

Fig. 7.1 Method for determining the concrete and steel stresses in a cracked section subjected to bending and axial force

7.2.1 Simply Reinforced

DIAGRAM 1_A'/A=0_a/d=0.10_ $0.01 \leq \alpha_e \rho \leq 0.50$
Simply reinforced rectangular sections under bending and axial force

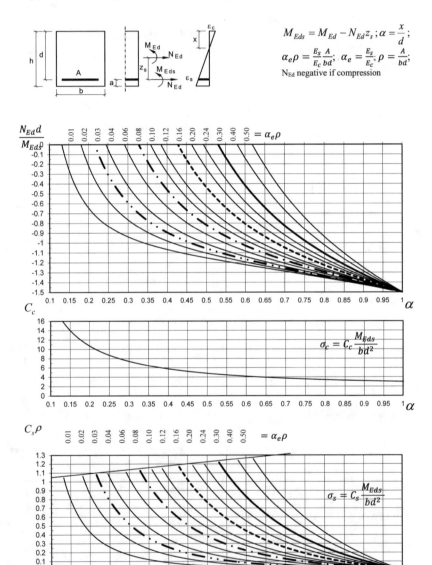

$$M_{Eds} = M_{Ed} - N_{Ed} z_s \, ; \alpha = \frac{x}{d} \, ;$$

$$\alpha_e \rho = \frac{E_s}{E_c} \frac{A}{bd} \, ; \quad \alpha_e = \frac{E_s}{E_c} \, ; \rho = \frac{A}{bd} \, ;$$

N_{Ed} negative if compression

$$\sigma_c = C_c \frac{M_{Eds}}{bd^2}$$

$$\sigma_s = C_s \frac{M_{Eds}}{bd^2}$$

Fig. 7.2 DIAGRAM 1_A'/A = 0

7.2.2 Doubly Reinforced

DIAGRAM 2_A′/A=0.2_a/d=0.10_$0.01 \leq \alpha_e \rho \leq 0.50$

Doubly reinforced rectangular sections under bending and axial force

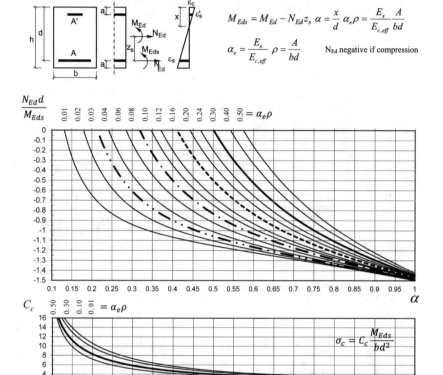

$$M_{Eds} = M_{Ed} - N_{Ed} z_s \quad \alpha = \frac{x}{d} \quad \alpha_e \rho = \frac{E_s}{E_{c,eff}} \frac{A}{bd}$$

$$\alpha_e = \frac{E_s}{E_{c,eff}} \quad \rho = \frac{A}{bd} \qquad N_{Ed} \text{ negative if compression}$$

$$\sigma_c = C_c \frac{M_{Eds}}{bd^2}$$

$$\sigma_s = C_s \frac{M_{Eds}}{bd^2}$$

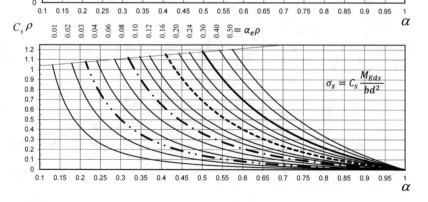

Fig. 7.3 DIAGRAM 2_$A/A' = 0.2$

DIAGRAM 2_A′/A=0.4_a/d=0.10_0.01≤$\alpha_e\rho$ ≤0.50
Doubly reinforced rectangular sections under bending and axial force

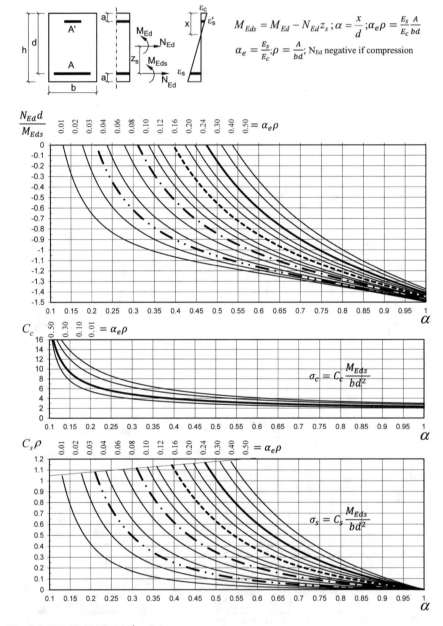

Fig. 7.4 DIAGRAM 2_A/A′ = 0.4

DIAGRAM 2_A′/A=0.6_a/d=0.10_0.01≤$\alpha_e\rho$ ≤0.50
Doubly reinforced rectangular sections under bending and axial force

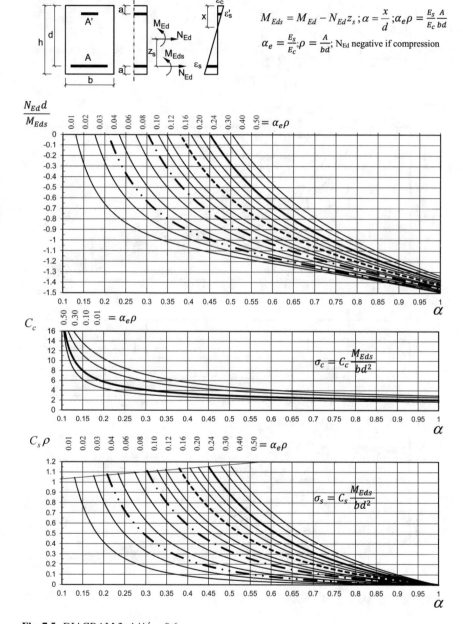

Fig. 7.5 DIAGRAM 2_A/A′ = 0.6

DIAGRAM 2_A'/A=0.8_a/d=0.10_0.01≤$\alpha_e\rho$ ≤0.50
Doubly reinforced rectangular sections under bending and axial force

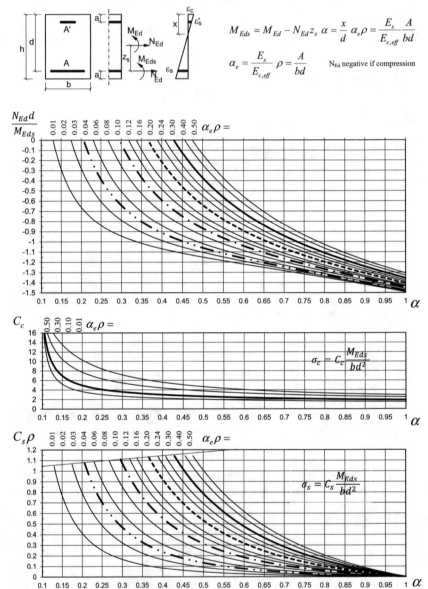

Fig. 7.6   DIAGRAM 2_A/A' = 0.8

DIAGRAM 2_A′/A=1.0_a/d=0.10_0.01≤$\alpha_e\rho$ ≤0.50
Doubly reinforced rectangular sections under bending and axial force

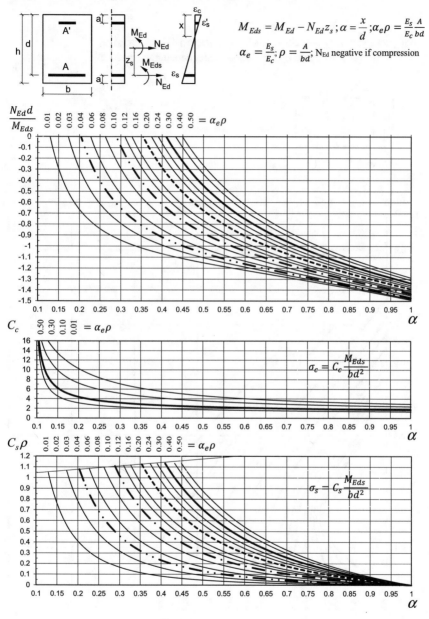

$$M_{Eds} = M_{Ed} - N_{Ed}z_s \; ; \alpha = \frac{x}{d} \; ; \alpha_e\rho = \frac{E_s}{E_c}\frac{A}{bd}$$

$$\alpha_e = \frac{E_s}{E_c}; \rho = \frac{A}{bd}; \; N_{Ed} \text{ negative if compression}$$

$$\sigma_c = C_c \frac{M_{Eds}}{bd^2}$$

$$\sigma_s = C_s \frac{M_{Eds}}{bd^2}$$

Fig. 7.7   DIAGRAM 2_A/A′ = 1.0

7.2.3 Symmetric Reinforcement

DIAGRAM 3_$\sigma_{c,max}$= 0.45f_{ck}_A/A$'$=1.0_a/d=0.10_0.00≤$\alpha_e\rho$ ≤0.50
Symmetric reinforced rectangular sections under bending and axial force

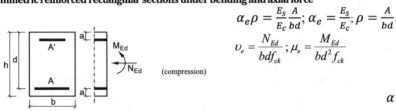

$$\alpha_e\rho = \frac{E_s}{E_c}\frac{A}{bd}; \; \alpha_e = \frac{E_s}{E_c}; \rho = \frac{A}{bd}$$

$$\upsilon_e = \frac{N_{Ed}}{bdf_{ck}} \; ; \mu_e = \frac{M_{Ed}}{bd^2 f_{ck}}$$

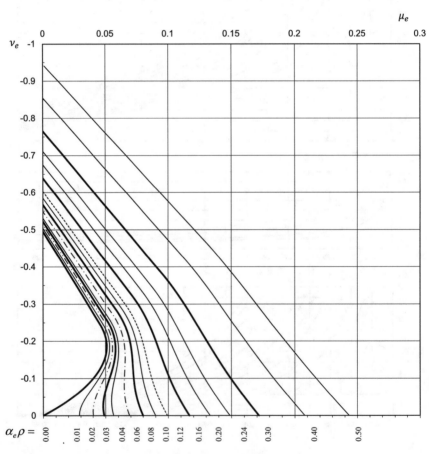

Fig. 7.8 DIAGRAM 3_$\sigma_{c.max}$ = 0.45 f_{ck}_A/A$'$ = 1.0

DIAGRAM 3_$\sigma_{c,max}$= 0.60f_{ck}_A/A´=1.0_a/d=0.10_0.00≤$\alpha_e\rho$ ≤0.50
Symmetric reinforced rectangular sections under bending and axial force

$$\alpha_e\rho = \frac{E_s}{E_c}\frac{A}{bd}; \alpha_e = \frac{E_s}{E_c}; \rho = \frac{A}{bd}$$

$$\upsilon_e = \frac{N_{Ed}}{bdf_{ck}}; \mu_e = \frac{M_{Ed}}{bd^2f_{ck}}$$

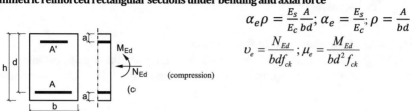

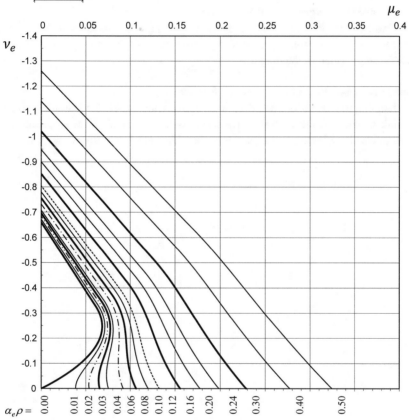

Fig. 7.9   DIAGRAM 3_$\sigma_{c.max} = 0.60\,f_{ck}$_A/A´ = 1.0

7.3 Bending and Axial Force of Simply Reinforced T-Sections

Puproose of the Diagrams in this Section
The Diagrams provided in this section allow determining the stresses in the concrete and in the steel that are required for the serviceability limit state safety verifications of T-sections.

How to use the Diagrams in this Section
The Example 11 of Chap. 4 concerns the use of these diagrams. Reference is also made to Fig. 7.1. Interpolation is possible for beams with b/b_w and h_f/d ratios different than those provided in the diagrams.

Cautions to use the Diagrams in this Section
The bending moment is evaluated at the level of the tensile reinforcing steel using the formula provided in the top right corner. The compressive axial force is considered to be negative. The diagram with the appropriate b/b_w and h_f/d ratios show be selected.

7.3.1 Flange-Web Ratio $b/b_w = 4$

DIAGRAM 4_ $b/b_w=4$_$h_f/d=0.10$_$0.01\leq\alpha_e\rho\leq0.50$

Simply reinforced T- sections under bending and axial force

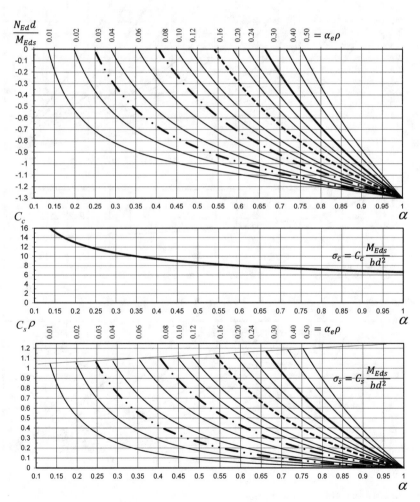

$$M_{Eds} = M_{Ed} - N_{Ed}z_s \; ; \alpha = \frac{x}{d} \; ; \alpha_e\rho = \frac{E_s}{E_c}\frac{A}{bd}$$

$$\alpha_e = \frac{E_s}{E_c}; \rho = \frac{A}{bd}; \; N_{Ed} \text{ negative if compression}$$

$$\sigma_c = C_c \frac{M_{Eds}}{bd^2}$$

$$\sigma_s = C_s \frac{M_{Eds}}{bd^2}$$

Fig. 7.10 DIAGRAM 4_$b/b_w = 4$_ $h_f/d = 0.10$

DIAGRAM 4_b/b_w=4_h_f/d=0.12_0.01≤$\alpha_e\rho$≤0.50

Simply reinforced T- sections under bending and axial force

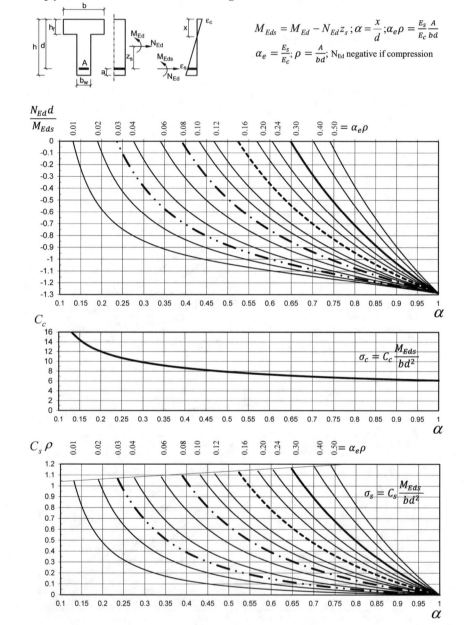

$$M_{Eds} = M_{Ed} - N_{Ed}z_s \; ; \alpha = \frac{x}{d} ; \alpha_e\rho = \frac{E_s}{E_c}\frac{A}{bd}$$

$$\alpha_e = \frac{E_s}{E_c} ; \rho = \frac{A}{bd}; \text{N}_{Ed} \text{ negative if compression}$$

$$\sigma_c = C_c \frac{M_{Eds}}{bd^2}$$

$$\sigma_s = C_s \frac{M_{Eds}}{bd^2}$$

Fig. 7.11 DIAGRAM 4_$b/b_w = 4$_ $h_f/d = 0.12$

DIAGRAM 4_b/b_w=4_h_f/d=0.14_$0.01 \leq \alpha_e \rho \leq 0.50$

Simply reinforced T- sections under bending and axial force

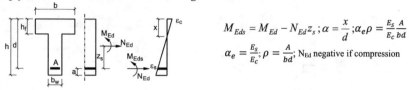

$$M_{Eds} = M_{Ed} - N_{Ed}z_s \; ; \; \alpha = \frac{x}{d} \; ; \alpha_e\rho = \frac{E_s}{E_c}\frac{A}{bd}$$

$$\alpha_e = \frac{E_s}{E_c}; \rho = \frac{A}{bd}; \; N_{Ed} \text{ negative if compression}$$

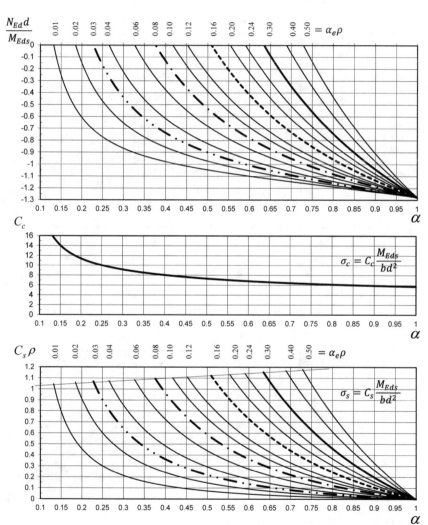

Fig. 7.12   DIAGRAM 4_$b/b_w = 4$_ $h_f/d = 0.14$

DIAGRAM 4_b/b_w=4_h_f/d=0.16_0.01≤$\alpha_e\rho$ ≤0.50

Simply reinforced T- sections under bending and axial force

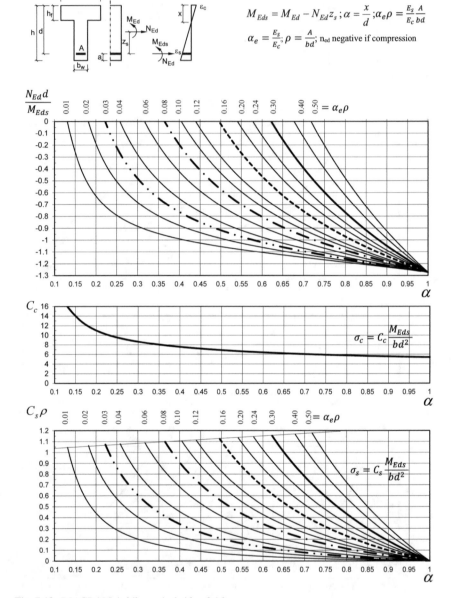

Fig. 7.13 DIAGRAM 4_$b/b_w = 4$_ $h_f/d = 0.16$

DIAGRAM 4_b/b_w=4_h_f/d=0.18_0.01≤$α_eρ$ ≤0.50

Simply reinforced T- sections under bending and axial force

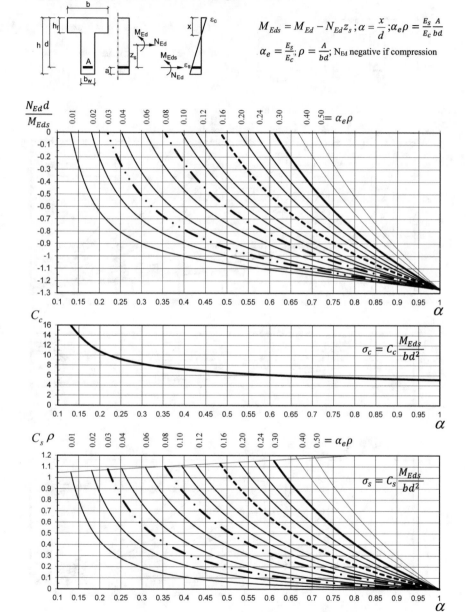

$$M_{Eds} = M_{Ed} - N_{Ed}z_s \; ; \alpha = \frac{x}{d} ; \alpha_e\rho = \frac{E_s}{E_c}\frac{A}{bd}$$

$$\alpha_e = \frac{E_s}{E_c}; \rho = \frac{A}{bd}; N_{Ed} \text{ negative if compression}$$

$$\sigma_c = C_c \frac{M_{Eds}}{bd^2}$$

$$\sigma_s = C_s \frac{M_{Eds}}{bd^2}$$

Fig. 7.14 DIAGRAM 4_$b/b_w = 4$_ $h_f/d = 0.18$

DIAGRAM 4_b/b_w=4_h_f/d=0.20_0.01≤$\alpha_e\rho$ ≤0.50

Simply reinforced T- sections under bending and axial force

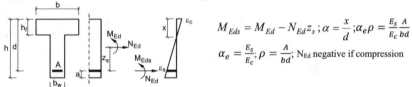

$$M_{Eds} = M_{Ed} - N_{Ed}z_s \; ; \alpha = \frac{x}{d} \; ; \alpha_e\rho = \frac{E_s}{E_c}\frac{A}{bd}$$

$$\alpha_e = \frac{E_s}{E_c}; \rho = \frac{A}{bd}; \; N_{Ed} \text{ negative if compression}$$

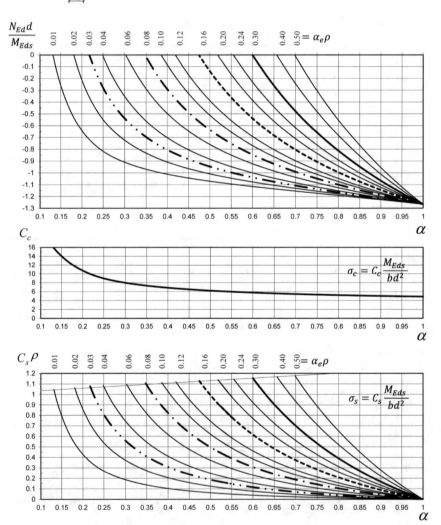

Fig. 7.15 DIAGRAM 4_$b/b_w = 4$_ $h_f/d = 0.20$

DIAGRAM 4_b/b_w=4_h_f/d=0.25_0.01≤$\alpha_e\rho$ ≤0.50

Simply reinforced T- sections under bending and axial force

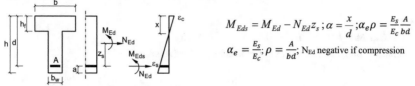

$$M_{Eds} = M_{Ed} - N_{Ed}z_s \; ; \alpha = \frac{x}{d} \; ; \alpha_e\rho = \frac{E_s}{E_c}\frac{A}{bd}$$

$$\alpha_e = \frac{E_s}{E_c}; \rho = \frac{A}{bd}; \; N_{Ed} \text{ negative if compression}$$

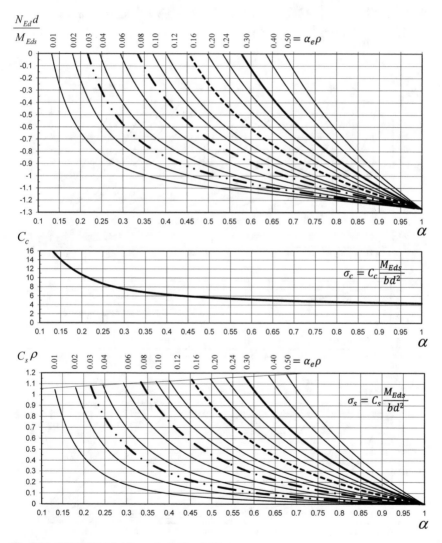

Fig. 7.16 DIAGRAM 4_$b/b_w = 4$_ $h_f/d = 0.25$

7.3.2 Flange-Web Ratio $b/b_w = 8$

DIAGRAM 5_b/b_w=8_h_f/d=0.10_0.01≤$\alpha_e\rho$ ≤0.50

Simply reinforced T- sections under bending and axial force

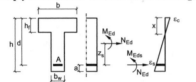

$$M_{Eds} = M_{Ed} - N_{Ed}z_s\,;\alpha = \frac{x}{d}\,;\alpha_e\rho = \frac{E_s}{E_c}\frac{A}{bd}$$

$$\alpha_e = \frac{E_s}{E_c}; \rho = \frac{A}{bd};\ N_{Ed}\ \text{negative if compression}$$

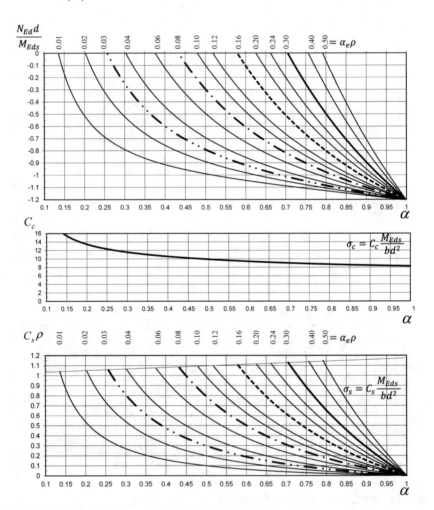

Fig. 7.17 DIAGRAM 5_$b/b_w = 8$_$h_f/d = 0.10$

DIAGRAM 5_b/b_w=8_h_f/d=0.12_0.01$\leq\alpha_e\rho \leq$0.50

Simply reinforced T- sections under bending and axial force

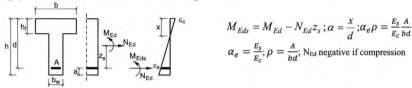

$$M_{Eds} = M_{Ed} - N_{Ed}z_s \; ; \alpha = \frac{x}{d} \; ; \alpha_e\rho = \frac{E_s}{E_c}\frac{A}{bd}$$

$$\alpha_e = \frac{E_s}{E_c}; \rho = \frac{A}{bd}; \; N_{Ed} \text{ negative if compression}$$

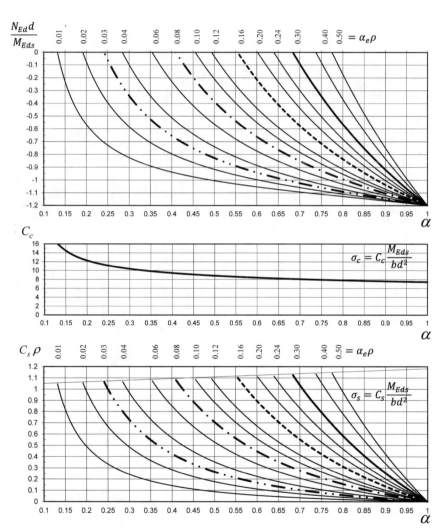

Fig. 7.18 DIAGRAM 5_$b/b_w = 8$_ $h_f/d = 0.12$

DIAGRAM 5_b/b_w=8_h_f/d=0.14_$0.01 \leq \alpha_e\rho \leq 0.50$

Simply reinforced T- sections under bending and axial force

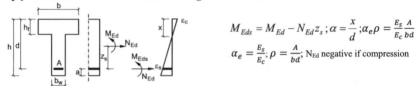

$$M_{Eds} = M_{Ed} - N_{Ed}z_s \, ; \alpha = \frac{x}{d} \, ; \alpha_e\rho = \frac{E_s}{E_c}\frac{A}{bd}$$

$$\alpha_e = \frac{E_s}{E_c} ; \rho = \frac{A}{bd}; \text{ } N_{Ed} \text{ negative if compression}$$

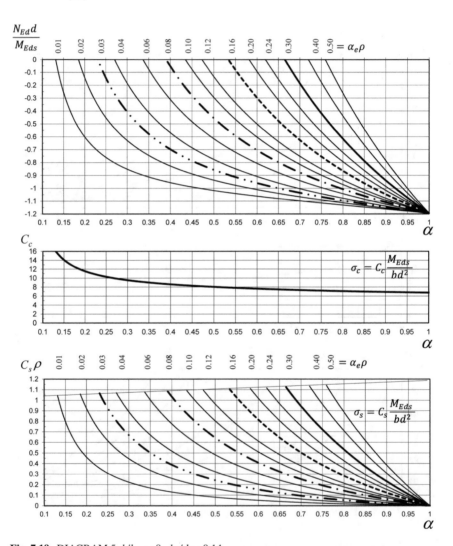

Fig. 7.19 DIAGRAM 5_$b/b_w = 8$_ $h_f/d = 0.14$

DIAGRAM 5_b/b_w=8_h_f/d=0.16_0.01≤$\alpha_e\rho$ ≤0.50

Simply reinforced T- sections under bending and axial force

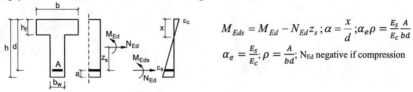

$$M_{Eds} = M_{Ed} - N_{Ed}z_s \ ; \alpha = \frac{x}{d} \ ; \alpha_e\rho = \frac{E_s}{E_c}\frac{A}{bd}$$

$$\alpha_e = \frac{E_s}{E_c}; \rho = \frac{A}{bd}; \text{N}_{Ed} \text{ negative if compression}$$

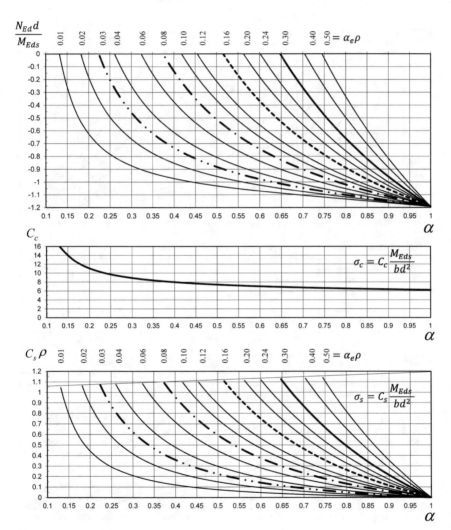

Fig. 7.20 DIAGRAM 5_$b/b_w = 8$_ $h_f/d = 0.16$

DIAGRAM 5_b/b_w=8_h_f/d=0.18_$0.01 \le \alpha_e \rho \le 0.50$

Simply reinforced T- sections under bending and axial force

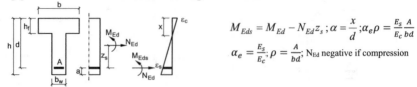

$$M_{Eds} = M_{Ed} - N_{Ed}z_s \,;\, \alpha = \frac{x}{d}\,;\, \alpha_e \rho = \frac{E_s}{E_c}\frac{A}{bd}$$

$$\alpha_e = \frac{E_s}{E_c}\,;\, \rho = \frac{A}{bd}\,;\, N_{Ed}\text{ negative if compression}$$

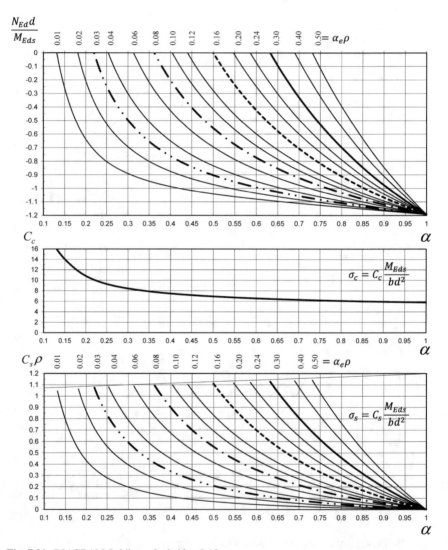

Fig. 7.21 DIAGRAM 5_$b/b_w = 8\_ h_f/d = 0.18$

DIAGRAM 5_b/b_w=8_h_f/d=0.20_0.01≤$\alpha_e\rho$ ≤0.50

Simply reinforced T- sections under bending and axial force

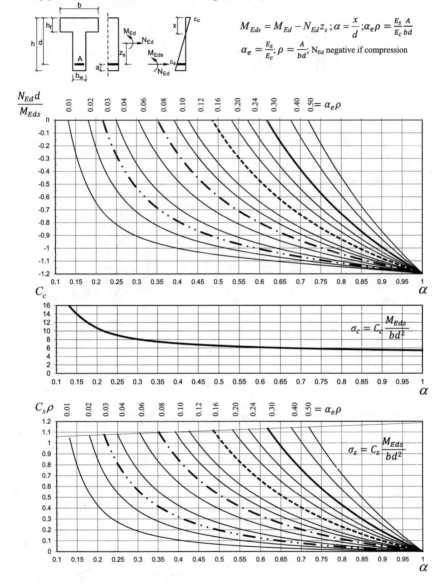

Fig. 7.22 DIAGRAM 5_$b/b_w = 8$_ $h_f/d = 0.20$

DIAGRAM 5_b/b_w=8_h_f/d=0.25_$0.01 \leq \alpha_e\rho \leq 0.50$

Simply reinforced T- sections under bending and axial force

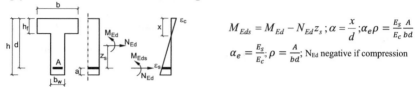

$$M_{Eds} = M_{Ed} - N_{Ed}z_s \, ; \alpha = \frac{x}{d}; \alpha_e\rho = \frac{E_s}{E_c}\frac{A}{bd}$$

$$\alpha_e = \frac{E_s}{E_c}; \rho = \frac{A}{bd}; \text{N}_{Ed} \text{ negative if compression}$$

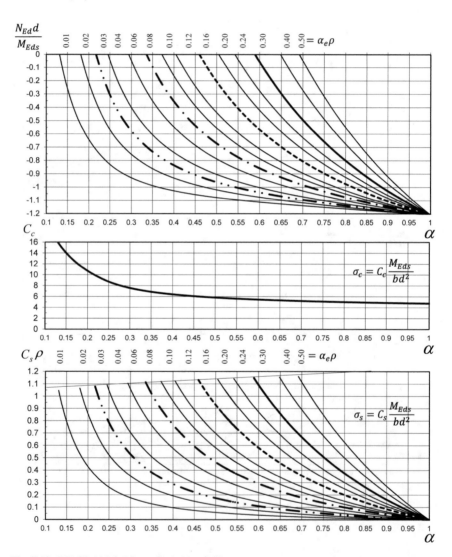

$$\sigma_c = C_c\frac{M_{Eds}}{bd^2}$$

$$\sigma_s = C_s\frac{M_{Eds}}{bd^2}$$

Fig. 7.23 DIAGRAM 5_$b/b_w = 8$_ $h_f/d = 0.20$

7.4   Bending and Axial Force of Circular Sections with Distributed Reinforcement in the Contour

Pupose of the Diagram in this Section

The Diagrams provided in this section allows determining the stresses in the concrete and in the steel that are required for the serviceability limit state safety verifications of circular cross-sections.

How to use the Diagram in this Section

Example 13 of Chap. 4 concerns the use of this diagram.

Cautions when using the Diagram in the Section

The bending moment and axial load should be evaluated at the centroid of the section.

DIAGRAM 6_ a/(2r)=0.10_0.01≤$\alpha_e \rho$ ≤0.50
Circular sections under bending and axial force

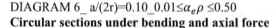

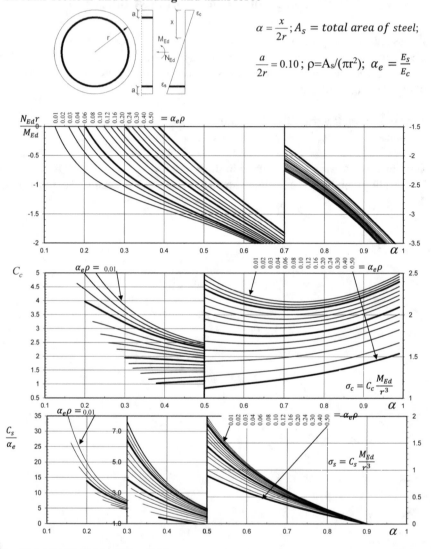

Fig. 7.24 DIAGRAM 6_a/(2r) = 0.10

7.5 Summary

This Chapter presents a set of diagrams allowing the easy evaluation of the maximum compressive stress in the concrete, the tensile stress in the steel and the neutral axis depth of reinforced concrete rectangular, T- and circular sections, under bending and axial force. The obtained concrete and steel stresses are useful to verify the serviceability limit states according EC2, namely the limitation of stresses, control of cracking and deflections.

References

1. EC1 EN 1991–1–1: Eurocode 1—Actions on Structures—Part 1–1: General Actions: Densities, Self-weight and Imposed Loads (2001)
2. EN 1992–1–1: Eurocode 2—Design of Concrete Structures—Part 1–1: General Rules and Rules for Buildings (2004)
3. Barros, M.H.F.M., Barros, A.F.M., Ferreira, C.C.: Closed form solution of optimal design of rectangular reinforced concrete sections. Eng. Comput. **21**(7), 761–776 (2004)
4. Barros, M.H.F.M., Martins, R.A.F.: Nonlinear analysis of service stresses in reinforced concrete sections using closed form solutions. In: 5th International Conference on Mechanics and Materials in Design, M2D'2006, 24–26 July Porto-Portugal (2006)
5. Barros, M.H.F.M., Ferreira, C.C., Figueiras J.A.: Serviceability design of reinforced concrete members—closed form solution. Eng. Struct. **49**, 600–605 (2013)

Printed in the United States
by Baker & Taylor Publisher Services